Pavlov's Legacy

How and What Animals Learn

Pavlov claimed that his experiments with dogs would transform the study of psychology and the treatment of mental illness. His work inspired researchers to study how animals learn to traverse mazes, avoid shocks or press levers to obtain food and also to compare the learning and cognitive abilities of different species, ranging from apes and dolphins to rats and pigeons. This book describes five decades of research into animal learning and comparative psychology, examining Pavlov's influence on this research and discoveries made by scientists who accepted many of his claims, while others looked for evidence to reject them. Drawing together diverse strands of research and providing historical and biographical information to bring the details to life, this is an ideal resource for graduate students and researchers in behavioral neuroscience, as well as for anyone in adjacent fields with an interest in learning theory.

Robert A. Boakes is Emeritus Professor of Psychology at the University of Sydney. He has held positions at universities in the UK and USA, focusing his research efforts on the study of learning, mainly in animals. He has published numerous articles on animal learning and coauthored a Cambridge handbook on animal training, *Carrots and Sticks* (2007), following his highly regarded previous Cambridge book, *From Darwin to Behaviourism* (1984).

T0335586

Pavlov's Legacy

How and What Animals Learn

ROBERT A. BOAKES
University of Sydney

Shaftesbury Road, Cambridge CB2 8EA, United Kingdom

One Liberty Plaza, 20th Floor, New York, NY 10006, USA

477 Williamstown Road, Port Melbourne, VIC 3207, Australia

314–321, 3rd Floor, Plot 3, Splendor Forum, Jasola District Centre, New Delhi – 110025, India

103 Penang Road, #05–06/07, Visioncrest Commercial, Singapore 238467

Cambridge University Press is part of Cambridge University Press & Assessment, a department of the University of Cambridge.

We share the University's mission to contribute to society through the pursuit of education, learning and research at the highest international levels of excellence.

www.cambridge.org
Information on this title: www.cambridge.org/9781316512074

DOI: 10.1017/9781009057530

© Robert A. Boakes 2023

This publication is in copyright. Subject to statutory exception and to the provisions of relevant collective licensing agreements, no reproduction of any part may take place without the written permission of Cambridge University Press & Assessment.

First published 2023

A catalogue record for this publication is available from the British Library

A Cataloging-in-Publication data record for this book is available from the Library of Congress

ISBN 978-1-316-51207-4 Hardback

Cambridge University Press & Assessment has no responsibility for the persistence or accuracy of URLs for external or third-party internet websites referred to in this publication and does not guarantee that any content on such websites is, or will remain, accurate or appropriate.

Contents

Preface

In the late 1890s in St. Petersburg, Ivan Pavlov began to study what he came to call "*conditional* reflexes," a term that became converted into "*conditioned* reflexes" in translations from Russian into English. The importance of Pavlov's research was appreciated by several American psychologists, to the extent that John Watson proposed that it should provide the theoretical framework for the behaviorist movement in psychology that Watson was the first to promote. However, at the time knowledge of Pavlov's studies among those unable to understand Russian was limited to translations of the presentations given by Pavlov at a handful of international conferences. It was not until 1927, with the first translation into English of a set of lectures that Pavlov had given a few years earlier in St. Petersburg, that the non-Russian speaking world could appreciate the enormous amount of experimental work over the previous three decades that Pavlov and his many collaborators and students had completed. This body of research could now be fully appreciated in the English-speaking world.

The present book provides an account of how Pavlov's research influenced either directly or indirectly studies of how and what animals learn over the 50 years or so years after 1927. Some theories and research programs followed Watson in that they were based on the belief that the study of conditioning – even if not Pavlov's own version – was fundamental to understanding how animals learn and what they learn. Others completely rejected this belief and to a large extent sought in their experiments for evidence that would demonstrate the limitations of conditioning theory.

The nine chapters in this book are organized around topics. They can be read independently of each other. However, in most cases, it will help the reader to have at least skimmed through one or two of the earlier chapters. In addition to describing key projects, accounts of the lives of many of the more influential researchers are included.

Chapter 1 describes Pavlov's life and major research achievements. It also follows research on one of the topics he was the first to investigate. Pavlov believed that his experiments on dogs contributed to understanding various aspects of human psychology. One such topic was *experimental neurosis*. The chapter describes how his approach influenced other researchers and eventually led to what has become a standard tool used by many clinical psychologists. Despite such interests, Pavlov always described himself as a physiologist and maintained that the main point of his experiments was to obtain greater understanding of the brain. Yet, as described in this chapter, his concept of how the brain works ignored mainstream developments in neurophysiology and was a major target for a critical admirer, the Polish scientist, Jerzy Konorski.

Chapter 2 describes the theories of Clark Hull, who had an enormous influence on American psychology from the time he first wrote about – and reinterpreted – Pavlov's work in the early 1930s to well into the 1960s. Hull's ambition was to develop a general theory of learning and motivation based on the development of habits, as studied mainly in rats; Hull's theory was intended to provide psychology with the equivalent of Newton's contribution to physics. His vision inspired some of the brightest and most productive researchers into animal learning in the 1950s and 1960s.

Chapter 3 centers on the work of Hull's arch critic, Edward Tolman. He was convinced that the Pavlov-inspired approach to the study of learning was far too narrow. Using a variety of mazes, Tolman and his students obtained evidence that their rats could anticipate events – rather just make conditioned responses – and could learn about the spatial properties of environments they were placed in. In the late 1960s, various researchers – most with little connection to Tolman – discovered the important role of the context in which an animal was conditioned. In the 1970s, experiments by neuroscientists on the function of the hippocampus led to a revival of interest in spatial learning and renewed appreciation of Tolman's suggestion that animals form representations – "maps" – of their environment.

Hull's and other theories of habit learning were based on the idea that "rewards" – or "reinforcements" following a response – "stamp in" connections between whatever stimulus or stimuli are present and the response. Such S-R-Reinforcement theories face a problem in explaining avoidance learning. As described in Chapter 4, the original experiments were first performed in the St. Petersburg laboratory of Pavlov's arch-rival, Vladimir Bekhterev. Researchers there arranged that, if when a signal was given, a dog failed to flex a leg, a shock would be delivered; if the leg was flexed in time, the shock was avoided. In general, these dogs learned quickly to flex the target leg as soon as the signal was given. Bekhterev considered this kind of learning to be a variant of Pavlov's conditioned reflexes. It took over 30 years before a widely accepted explanation was developed of how the absence of an event could promote learning. A key contribution was the "two-factor theory" developed by Hobart Mowrer. His studies of avoidance learning and those that followed, mainly by Richard Solomon and his students, laid the foundation for breakthroughs in the study of associative learning in the late 1960s.

At least until very late in his career, Pavlov believed that what was true of the dog's capacity for learning was true of any other vertebrate. However, he suggested that humans have the extra benefit of a "second signaling system," that is, language. Attempts to show that species differ in their learning and problem-solving abilities are described in Chapter 5. This chapter on comparative psychology starts with studies that attempted to teach a language to chimpanzees and examined their apparently remarkable ability to learn by imitation. The chapter then describes studies of complex learning in monkeys and dolphins. It ends with accounts of experiments in the 1970s that compared the learning abilities of corvids with those of pigeons.

While psychologists and neuroscientists have almost always studied learning in a laboratory setting, other scientists have mainly studied learning in more natural environments. One such form of learning was termed *imprinting* by the Austrian ethologist, Konrad Lorenz. Chapter 6 describes his claims about this phenomenon and how researchers that were more experimentally oriented than Lorenz tested these claims and rejected many of them. The interaction between ethologists and learning theorists led to studies of how species differed in regard to *constraints* on what they could learn, a topic that is covered in the second half of Chapter 6.

The beginning the twentieth century saw a steady increase in the number of experimental studies aimed at the question of how animals perceive their world. These included ones that, for example, tested whether a dog, cat, or rat could learn to discriminate between two stimuli that differed only in color. Chapter 7 describes developments in the study of discrimination learning that became increasingly theoretical. For example, one important issue had to do with the possible role of attention-like processes in such learning. This was a topic that was central to the work in the 1950s and 1960s of Stuart Sutherland and Nick Mackintosh in the UK.

Chapter 8 provides an account of the life and work of B.F. Skinner, a behaviorist who became one of the most famous and influential psychologists of his generation. It describes his development of the operant chamber – widely known as a *Skinner box* – and the increasingly sophisticated equipment used to control events with the chamber. These enabled increasingly complex experiments to be run, not only by "Skinnerians" who agreed with Skinner's radical views on science and on the preeminent role of operant conditioning in human life but also by those without any sympathy for his views. The chapter also traces the expansion of the movement, the Experimental Analysis of Behavior founded by Skinner and his close friend and associate, Charles Ferster, both within and beyond the United States.

Chapter 9, the final chapter, starts by describing the handful of key findings from the late 1960s that were obtained by Leon Kamin, Robert Rescorla and Allan Wagner from experiments on discrimination learning and fear conditioning. These led to a revolution in the way learning by animals was studied. Instead of concentrating of how their behavior changed, it took such changes as an index of what associations the animals had formed. A number of theories of associative learning that were developed in the 1970s remained influential for the next 50 years. In the 1980s, an important and influential distinction was drawn between *actions*, which are sensitive to the value of their consequences, and automatic habits.

In 1984, I published a book with the title, *From Darwin to Behaviorism*. As the title suggests, it examined the influence of late nineteenth-century theories of evolution and the new ideas about animal behavior and learning that these theories prompted on research into how animals learn. The latter led to the emergence of behaviorism as a major movement within American psychology. The present book is in some sense a sequel that could have been given the clumsy title, *From Pavlov to Associative Learning Theory*.

Writing the present book has been an entirely different experience from writing my book of 40 or so years ago. One obvious difference is provided by the Internet. Previously I needed to obtain a travel grant to visit various libraries in North America and to obtain permission, for example, from Cambridge University Library to examine

Darwin's notebooks. Now only a few clicks on my laptop are needed to access almost any source I need.

The amount of easily available material is so abundant that I have had to be very selective in ways that reflect my own history of studying psychology. I was an undergraduate in the UK and then a Ph.D. student in the USA. I then taught psychology and carried out research in the UK for many years, for two years in the USA and, finally, for over 30 years in Australia. Someone with a different history would have written a very different book.

Another and more important way in which the present book differs from its predecessor is that in writing the latter, I could adopt the pose of an objective assessor, since I had had no direct or even indirect contact with any of the people I wrote about. This is not true of the present book. I have met many of the people whose lives and work are described in this book. Several were, or still are, good friends. Consequently, I have not tried to obscure the fact that this is in some ways a personal book. It contains potted biographies of many of the people who made major contributions. The rule I have maintained is to provide these only for researchers who have died.

I am an experimental psychologist with an interest in the history of this branch of science that goes back to when I was a graduate student. But I am not an historian. When I was a boy, my friends and I would go on long cycle rides and afterwards record where we had been and what happened – usually, very little! – in a book we called *The Chronicles*. The present book could be seen as a "chronicle" of theories and experiments concerned with understanding how animals learn. I believe these to be of considerable interest but fear they could well be entirely forgotten except by a handful of specialists. Thus, this book can be seen to celebrate a 50-year-long golden age of research on animal learning. It is an age that is unlikely to be repeated. Setting up and maintaining an animal research facility is expensive. Since the 1980s, it has become steadily more difficult to obtain funding for purely behavioral research. Instead, to be successful, the main focus of a project needs to be on underlying brain mechanisms and to use "cutting edge" neuroscientific techniques.

Acknowledgments

The reader I often had in mind when writing these chapters was a student or early career researcher who already had an interest in learning or behavioral neuroscience. However, I also thought about whether some topic would be of more general interest and whether my treatment to a large extent be comprehensible to someone with no background at all in this area. Such a reader is my wife, Margaret Kirkwood, who read every draft chapter and let me know what sections were obscure or impossibly technical, and so needed re-writing, as well as pointing out typos and other minor errors. Without her amazing support for more than 25 years, this book would never have been written.

I also am pleased to give heartfelt thanks to the many colleagues, friends and previously unknown individuals who have also read and commented on drafts ranging from short sections to several whole chapters, have helped me locate photos or have let me use photos they own. They include – in no particular order – Vin LoLordo, Fred Westbrook, Tony Dickinson, Geoff Hall, Euan Macphail, David Booth, Billy Baum, Fred Toates, John Staddon, Charlie Catania, Terry Davidson, Ludy Benjamin, Don Dewsbury, Patricia Courvillon, Herb Jenkins, Bill Whitlow, Jerry Rudy, Sarah Shettleworth, Peter Holland, David Dickins, Donald Heth, John Pearce, Edgar Vogel, Jacqueline Glynn, Dick Stevenson, Marie-Claire Kamin, Gavin McNally, Shirley Steele, Lynn Nadel, Tony Riley and Herb Terrace.

I have normally relied on primary sources such as experimental reports, theoretical papers, obituaries and biographical sketches, all of which can be accessed via the Internet. The major exception has been in writing the first chapter, where I have relied heavily on an extraordinary, and the only complete, biography of Pavlov, the massive work of Daniel Todes (2014).

Finally, I would like to thank two groups of people. One consists of the many supportive colleagues at the University of Sydney, especially Evan, Justin and Ben. The other group are members of the Patonga non-fiction book club; Paul, Dain, Brad, Rob and Richard have taught me the different ways in which books can be read and the different features that can make for a "good book."

1 Ivan Pavlov, Conditioned Reflexes and Experimental Neuroses

Pavlov was in his 50s when he and his young collaborators began to study conditioned reflexes in dogs. Chapter 1 starts with an account of how their research began and describes some of the major findings that followed from the decades of intense work that continued right up until Pavlov's death at the age of 86. Many of the topics that he was the first to study later became the focus of research both inside and outside of Russia. Subsequent chapters in this book describe the development of research in English-speaking countries, predominantly the USA, on many of the aspects of learning by animals that were first examined in Pavlov's experiments on conditioning.

The latter part of the present chapter focuses on a topic for which Pavlov is less well known, *experimental neurosis*. Among those actively inspired by his ideas on neuroses were two US-based researchers, Horsley Gantt and Howard Liddell, who became the staunchest proponents of Pavlov's theories in the English-speaking world. An account of their work and that of those that followed in promoting the application of Pavlov's ideas to the study of neurosis is followed by an important critique of Pavlov's theories of how the brain works by the Polish scientist, Jerzy Konorski.

Problems with Digestion Research

The story of how Pavlov began to study conditioning is an unusual one. To start with, Pavlov's reputation for research on a very different topic was at its peak when he made the major shift from mainstream experimental physiology to study what was essentially a psychological problem.

Over the many years from when Pavlov worked as a lone scientist – usually working in someone else's laboratory – till when he headed a large team of research workers in his own well-equipped laboratory, he developed at least two important skills. One was surgical: Pavlov was one of the most accomplished physiologists of his era in terms of his ability to isolate surgically different parts of a dog's digestive system and insert fistulae – tubes – into various levels of this system. Importantly, he was able to carry out such operations in such a way as to achieve – at least on most occasions – his aim that the dog would survive and live in good health for many more years. During this time, such a dog could be the subject of a series of chronic experiments on how its digestive system worked.[1]

A very different skill was that of training, managing, and inspiring the 10–15 poorly-prepared medical students who arrived each year to work in his laboratory. What they wanted was to carry out enough research for a thesis that would earn them a doctoral degree. The content of these theses was limited and more comparable to the kind of report that a final year undergraduate in a Western university might submit than to a Ph.D. thesis. A student in Pavlov's laboratory would normally be allocated a single dog that may have already undergone surgery. Alternatively, a student would be incorporated within a team that might include Pavlov or a skilled research assistant to carry out some surgical procedure on a new dog. What Pavlov wanted from a student was data that would contribute to Pavlov's focused research strategy.[2]

From his earliest research on the heart, Pavlov had consistently embraced a theoretical position known as *nervism*. As applied to the digestive system, this was the belief that every stage in the digestion of food is coordinated by the central nervous system. Experimental support for this view consisted in demonstrating, for example, that an isolated segment of a dog's stomach – the *Pavlovian pouch* – would secrete gastric juices in response to food that entered the mouth, but fell out through a fistula in the esophagus – but only if the nerves connecting the brain to the pouch were intact. Such demonstrations were based in many cases on combining results from several student projects, and many were reported in Pavlov's first book, *Lectures on the Digestive System*, which was published in Russian in 1897. Subsequent translations into German, then French and English, gave Pavlov an international reputation and eventually led to him being awarded the Nobel Prize in 1904. This was the first to be awarded in physiology, and Pavlov was the first Russian to be honored in this way.[3]

Well before his Nobel Prize and soon after publication of his 1897 *Lectures*, doubts began to emerge concerning the claims that Pavlov had made in his book. One source was a discovery concerning the activity of the pancreas by one of his students, a result that was reluctantly confirmed by Pavlov. This study revealed that claims made about the pancreas contained in his 1897 book were incorrect. The need to retract previous claims in the light of subsequent research is common enough in any kind of scientific program. What was far more disturbing and led to one of the most violent outbursts of Pavlov's renowned fury was a critical analysis of the data reported in the *Lectures* of 1897 by a former student. Popel'skii was older and more independent-minded than most of the other students when he began to carry out experiments under Pavlov's direction. In Pavlov's *Lectures*, it was not readily apparent that the results reported in this book were mainly obtained from just two dogs. Popel'skii re-examined the theses on which Pavlov had based his claims about the pancreas and concluded that the reported data were selected to support the claims Pavlov wished to make and that another, more objective reading of these data would support conclusions opposite to those favored by Pavlov. Popel'skii even had the temerity to publish articles containing his criticisms of Pavlov in foreign language journals.[4]

The most important development to undermine Pavlov's claims regarding the digestive system came from a discovery made by two British physiologists that was prompted by a study from Pavlov's laboratory and that used procedures he had pioneered. With his long commitment to nervism, Pavlov had always rejected the idea that hormones played any role in the control of digestive processes. In 1902, Bayliss and Starling reported their discovery of secretin, a hormone that influences the action of the pancreas. Pavlov immediately set a student to attempt a replication of the critical experiment reported by Bayliss and Starling. As Pavlov was forced to acknowledge, the replication indicated that Bayliss and Starling's claim was correct.[5]

Psychic Reflexes

In experiments in which gastric juice was obtained by giving a dog meat powder or some other food, Pavlov and his students consistently observed that simply waving the food in front of the animal – 'teasing' – would start the juice to flow. By 1892, Pavlov had begun to refer to such effects as psychic reflexes. He explained them as being the product of mental processes such as 'choosing' or 'deciding.' In 1896, a similar phenomenon was found by Vulf'son, one of the first students assigned by Pavlov to study the salivary glands. Vulf'son first established that whether or not one of his four dogs produced mucus-rich or thin, watery saliva depended on whether plain meat or something noxious – including meat covered in mustard – was placed in its mouth. Most importantly, when teased with meat, the dog's psychic reflex produced mucus saliva but, when teased with something that the dog had learned was noxious, watery saliva was collected.

Pavlov became increasingly interested in psychic reflexes. Lacking any background in psychology, in 1900, he took the unusual step of taking on a student, Snarskii, who had received some training in another laboratory, that of Vladimir Bekhterev. A few years later, Bekhterev also began to study conditioning but used a very different approach to Pavlov's (see Chapter 4). For this and other reasons, Bekhterev became Pavlov's greatest rival.[6] While in Bekhterev's lab, Snarskii had also gained some expertise in psychology. After arriving in Pavlov's lab, Snarskii first extended Vulf'son's study by using a black-tinted solution of mild acid that, when injected into a dog's mouth, produced copious amounts of watery saliva. Once a dog had experienced this treatment several times, it began to salivate as soon as it was shown the bottle containing the acid.

A key finding followed. When Snarskii repeatedly showed the dog the bottle without injecting its contents into the dog's mouth, he obtained a decreasing amount of saliva. This could be seen as the first ever extinction experiment. It showed that this psychic reflex was *conditional* on maintaining a pairing between a dog seeing the bottle and then experiencing the acid within its mouth.

Snarskii was critical of Pavlov's use of the term 'psychic' and Pavlov's generally anthropomorphic approach to his dogs' personalities and presumed mental processes.

Snarskii preferred to describe his dogs as forming 'associations' between 'representations' of events, a process in which "the consciousness of the dog plays no important role." Bekhterev was a member of Snarskii's thesis committee, as well as Pavlov; at its meeting Bekhterev allegedly told Snarskii: "Your duty and mine is to teach physiologists psychology!"[7]

In 1901, Pavlov found another student to work on this topic. Tolochinov, llke Snarksii, had previously worked in Bekhterev's laboratory, but also had considerable clinical experience working with patients suffering from various mental disorders. He was already in his 40s when he started to work in Pavlov's laboratory. Thus, he was far older and, more importantly, like Snarskii, Tolochinov had much greater knowledge of research outside of Pavlov's domain than most other students. In particular, he knew about studies of human 'reflexes at a distance' that had demonstrated that a knee jerk or an eye-blink could occur in anticipation of the stimulus normally needed to elicit such responses.

Starting in February, 1902 Tolochinov systematically examined the extinction effect that Snarskii had reported. Furthermore, he discovered what many decades later was re-discovered and labelled *reinstatement*. After repeated 'teasing' by, for example, showing, but not giving, his dog some meat so that salivation had virtually ceased, letting the dog eat the meat on a single occasion would then restore the effectiveness of the sight of the meat to elicit saliva. Discussion of these results led Pavlov to coin the term *conditional reflex.* The first public use of the term was in a presentation by Tolochinov at a meeting in Helsinki in June 1902.[8]

In the meantime, most experiments undertaken within Pavlov's laboratory continued to focus on the physiology of the digestive system. However, the increasing importance that Pavlov gave to the conditional reflex is shown by two events. First, this was the topic he chose for his invited lecture to the meeting of the International Congress of Medicine that took place in Madrid in 1903. Second, in the same year, he pulled one of his most promising students, Babkin, from studying the pancreas and directed him to study conditional reflexes instead. The transition to the eventual situation whereby all the laboratory's resources were devoted to the study of conditioning was not complete until 1907. By that time Pavlov had completely adopted the 'objective' language that Snarskii had argued for and instituted for some years a system whereby students were fined for using the mentalist vocabulary that Pavlov himself had happily used only a few years earlier. Indeed, from 1906 onwards, Pavlov promoted the story that it was he, and not Snarskii, who had first wanted to exclude the everyday language of human mental processes from the quest to understand conditional reflexes.[9]

At some level, Pavlov must have recognized during this transition period that he did not have the skills to remain at the new cutting edge of research on the digestive system. On the other hand, he became more confident in the belief that the study of conditioning would provide a tool for examining "the seeming chaos of relations" with which the behavior of an animal comes to adapt to its world and for identifying general laws that govern changes in behavior. And even more important, it would lead to an understanding of how the brain worked (Figure 1.1).

Figure 1.1 Ivan Pavlov in 1890. Public domain.

Discovering the Properties of Conditioned Reflexes

Prior to 1890, Pavlov had only part-time, short-term academic appointments that failed to earn him enough to keep his family out of poverty. He also had limited access to lab facilities. His situation dramatically improved in 1890 so that until the outbreak of World War I in 1914, the resources at Pavlov's disposal were considerable; see Figure 1.2.

In the late 1880s, a wealthy aristocrat related to the Czar wanted to establish an institute for the study of infectious diseases – one concentrating on rabies – that would rival the world-renowned Pasteur Institute in Paris. Finding the considerable amount of money to fund what would become the largest research institute in Russia proved to be easier than finding top scientists, preferably experts in disease, to head its various laboratories. Partly by being on the right committee at the right time and having important contacts, Pavlov was appointed the Director of its Laboratory of Experimental Physiology when the Imperial Institute for Experimental Medicine opened in 1890. This provided him with as much space and with facilities as good as any physiological laboratory in the world at that time.[10] In addition, the income to the Laboratory was sufficient to provide Pavlov with a good salary for the first time in his life and, in most years, to pay the salaries of two full-time research assistants and those of two attendants who cared for the dogs and often assisted with experiments.

Figure 1.2 The special laboratory built for Pavlov in the Institute of Experimental Medicine in St. Petersburg.
Image credit: Topical Press Agency / Stringer / Hulton Archive / Getty Images.

Just as important a resource was mentioned earlier. His appointment from 1890 onwards at the Military Medical Academy led to a steady flow of medical students to work in his laboratory. Their aim was to obtain a doctoral degree that would advance their official position in Russian society and for the majority improve their chances for a favorable appointment within the Russian army. In a paper on conditioning that Pavlov wrote just before the outbreak of World War I, he acknowledged the contributions of over 100 "collaborators."[11]

Starting in 1898, an added boost to the budget came from the sale of gastric juice obtained from dogs whose sole purpose was commercial rather than scientific. Gastric juice from these dogs was supplied both for research purposes to other laboratories in Russia and elsewhere in Europe and to meet the considerable local demand for the juice as an aid to digestion. This enterprise was so successful that in 1904, it increased by over 65% the income to a laboratory that was already far more richly supported than any other Russian physiology laboratory.[12]

No one else in the world had anything like these resources for studying how animals learn. Even when, as described in Chapter 2, Clark Hull was set up in the Institute for Human Relations at Yale University, the laboratories in which his co-workers and students worked and the resources at their disposal in the 1930s hardly compared to Pavlov's laboratory prior to World War I; see Figure 1.2.

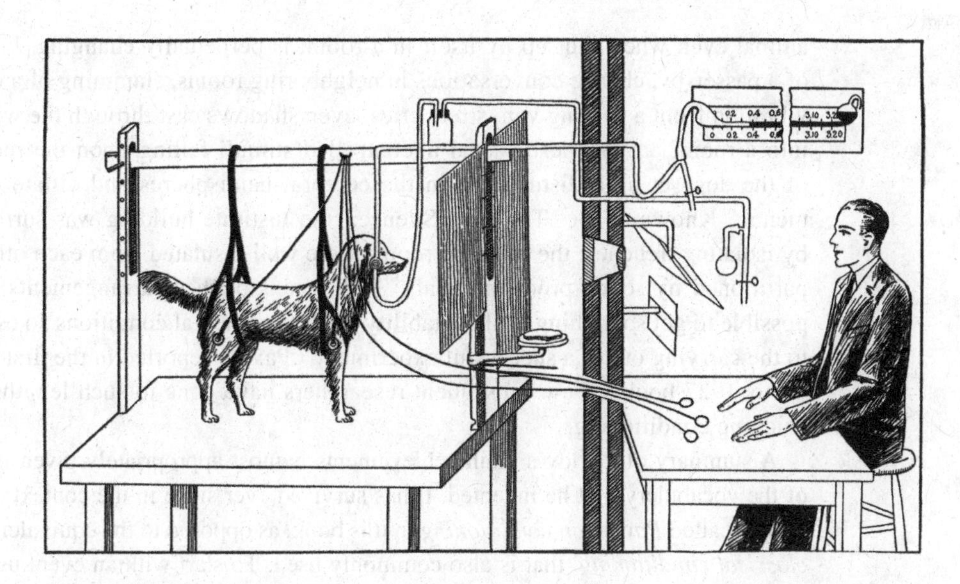

Figure 1.3 Sketch from 1928 of the standard arrangement used for salivary conditioning experiments in Pavlov's lab.
From Pavlov (1928). Reproduced with permission from Alamy.

Pavlov's research barely survived during the war and, following the Bolshevik revolution of 1917, his laboratory had to shut down for two years.[13] During 1918 and 1919, Pavlov and his family had barely enough food, let alone enough extra to maintain a colony of dogs; several had to be sacrificed.[14] Unexpectedly for someone who had been critical of the communist movement in Russia, Pavlov's fortunes improved even before the nation's political and economic situation had begun to stabilize. Lenin wanted to show that the new communist government supported science, and Pavlov was Russia's only Nobel Laureate. In 1921, Lenin signed a decree authorizing a committee to "create as soon as possible the most favorable conditions for safeguarding the scientific work of Academician Pavlov and his collaborators."[15] This resulted in Pavlov, now 72 years old, enjoying ample funding for the rest of his life.

The method used in most of Pavlov's experiments on conditioning was based on his previous studies of the digestive system. Surgery was first performed to insert a permanent fistula in a dog's cheek through which saliva could drain out through a tube. Then the dog was trained to stand on a bench where it was lightly restrained by a harness. Once a dog had completed such initial training, it served in experiment after experiment. Most of Pavlov's varied mongrels lived for many years; see Figure 1.3.

Considerable effort went into ensuring that a dog was unable to detect movements, even "blinking of the eye lids," or sounds made by the experimenter. Pavlov believed that it was extremely important to eliminate extraneous stimuli that might distract a dog and compete with the experimental stimuli. "In order to exclude this undue influence on the part of the experimenter as far as possible, he had to be stationed outside the room in which the dog was placed. … The environment of the animal even when shut up by itself in a room, is perpetually changing. Footfalls of a passer-by, chance conversations in neighboring rooms, slamming of a door or vibration from a passing van, street cries, even shadows cast through the windows into a room, any of these casual uncontrolled stimuli falling upon the receptors of the dog set up a disturbance in the cerebral hemispheres and vitiate experiments." Known as the 'Tower of Silence,' the Institute building was surrounded by isolating trenches; the research rooms were well insulated from each other and partitioned by sound-proof material.[16] "By means of these arrangements, it was possible to get something of that stability of environmental conditions so essential to the carrying out of a successful experiment," Pavlov reported in the first lecture of his 1927 book.[17] Few subsequent researchers have gone to such lengths when studying conditioning.

A summary of Pavlov's main achievements is most appropriately given in terms of the vocabulary that he invented. It has survived ever since in the context of what will be called *Pavlovian conditioning* in this book, as opposed to the equivalent label, *classical conditioning*, that is also commonly used. To start with an event used in a large number of Pavlov's experiments, presenting a dog with a small amount of meat served as the *Unconditioned Stimulus* (UCS or US), an action that elicited the preexisting reflexive response of producing saliva as the *Unconditioned Response* (UCR or UR). Pavlov used a variety of neutral stimuli in his lab; 'neutral' in the sense that they did not at first elicit any salivation. A favorite was a metronome whose ticking for a preset time was set at a particular frequency. When this sound was made just before a dog was given food, the ticking of the metronome was said to serve as a *Conditioned Stimulus* (CS). After many such pairings, the CS would typically come to elicit salivation as the *Conditioned Response* (CR).

Some of the first conditioning experiments performed in Pavlov's lab used the procedure that came to be known as *extinction.* Once a CR had been established to a CS by pairing the latter with the UCS, the CS was presented repeatedly in the absence of the UCS with the result that the CR occurred with decreasing frequency. This led Pavlov to view the occurrence of the CR as 'conditional' upon its continued pairing with the UCS and hence introduced the term (in Russian), 'conditional reflexes.' When his lectures were translated into English, 'conditional' became 'conditioned'; hence the term, *conditioning*; see Figure 1.4.

Pairing of two events can be arranged in a variety of ways. They can, for example, occur at exactly the same time, the *simultaneous* condition shown in Figure 1.5. Despite the historic claims by associationist philosophers that this was the optimal arrangement for the formation of associations between two events, Pavlov did not find this arrangement effective for establishing a conditioned reflex. Instead, he found that the most effective form of pairing was the *delayed* arrangement; here the onset of the CS precedes that of the UCS and they terminate together. Also extensively used in Pavlov's lab was the *trace* arrangement, whereby the CS is presented for a short time, followed by an empty interval before the UCS arrives. The term 'trace' reflects the idea that a memory trace of the CS becomes connected to the UCS. The final arrangement shown in Figure 1.5 is termed *backward* conditioning, in that the CS follows the

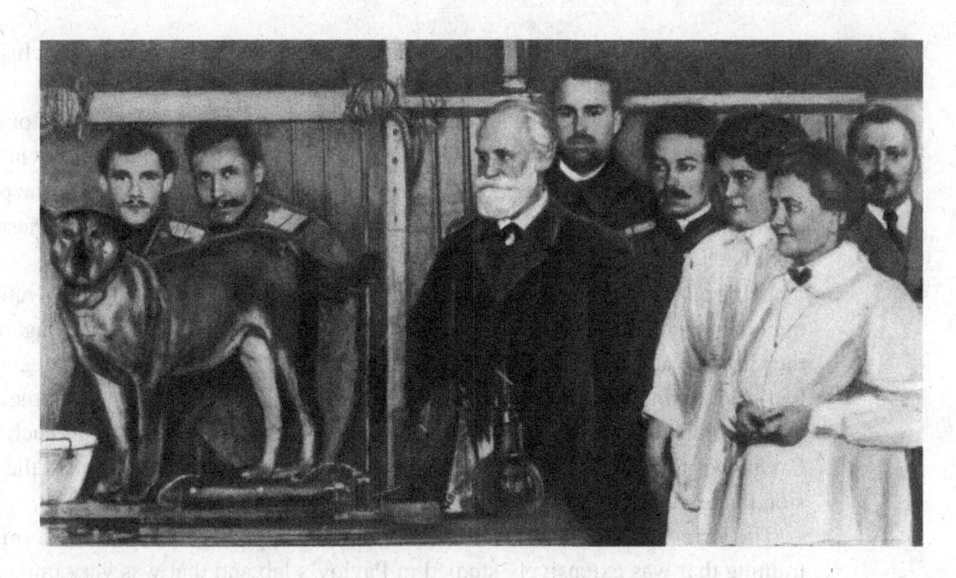

Figure 1.4 Pavlov, plus two students, three co-workers, two assistants and a dog. Reproduced with permission from the Granger Historical Picture Archive.

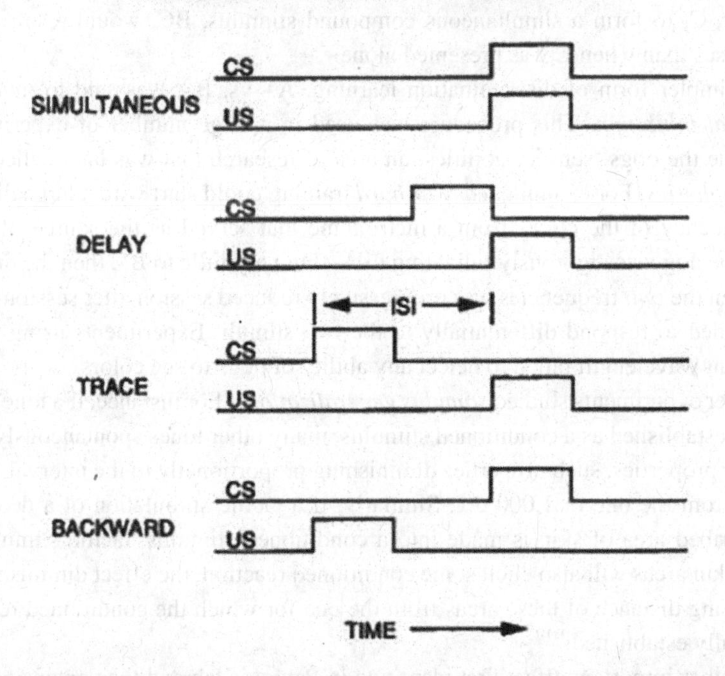

Figure 1.5 Different ways to present the conditioned stimulus (CS) and unconditioned stimulus (US) in time. This diagram from a book published nearly 60 years after Pavlov's death shows how the terminology that he introduced lives on.
From Schindler (1993). Reproduced with permission from Elsevier.

UCS. Decades later, many attempts were made to show that Pavlov's conclusion that no conditioning occurred under this arrangement was wrong.

Once again following the approach in his physiological studies, usually only a single dog, but sometimes two, were used in an experiment. When Pavlov was not totally convinced by a set of results, a new student was set the task of replicating the previous experiment. And almost all of the phenomena that Pavlov discovered by these means have been replicated ever since.

In the very early studies of extinction it was noted that, when a dog was returned to the lab after a delay of a few days, the presentation of a CS would once again evoke the CR; this effect was called *spontaneous recovery*. The CR could also recur after it had been extinguished if some unexpected stimulus occurred – for example, Pavlov walking into the room; this was called *disinhibition*. Commenting on such effects, Pavlov wrote: "By ruling out one interpretation after another we arrived at the conclusion that extinction must be regarded as a special form of inhibition."[18]

The term 'inhibition' was also used in a label applied to a form of discrimination training that was extensively studied in Pavlov's lab and that was very important for theoretical developments many decades later (see Chapter 9). One stimulus, A, was followed by food when it was presented on its own, A+, but not when a second stimulus, B, was present at the same time. A+ vs. AB– was termed *conditioned inhibition* training and B termed a *conditioned inhibitor*. To check that B had acquired inhibitory properties, a *summation test* was used; this asked whether adding B to a second excitor, C, to form a simultaneous compound stimulus, BC, would result in fewer responses than when C was presented alone.

A simpler form of discrimination learning, A+ vs. B–, was said to involve *differential inhibition*. This procedure was used in a large number of experiments to examine the dogs' sensory abilities, an area of research that was later called *animal psychophysics*. For example, *easy-to-hard* training could start with a large difference in frequency of the clicks from a metronome that served as the source of stimuli; once the dog was vigorously salivating to A+ but very little to B–, then the difference between the two frequencies was progressively reduced session after session until the dog failed to respond differentially to the two stimuli. Experiments using lights of different wavelength failed to detect any ability of dogs to see colors.

Other experiments studied *stimulus generalization*. "For instance, if a tone of 1,000 d.v. is established as a conditioned stimulus, many other tones spontaneously acquire similar properties, such properties diminishing proportionally to the intervals of these tones from the one of 1,000 d.v. Similarly, if a tactile stimulation of a definite circumscribed area of skin is made into a conditioned stimulus, tactile stimulation of other skin areas will also elicit some conditioned reaction, the effect diminishing with increasing distance of these areas from the one for which the conditioned reflex was originally established."[19]

Another important effect first identified in Pavlov's lab and then extensively studied in the 1970s (see Chapter 9) was *second-order conditioning*. Such experiments start with first order conditioning of a previously neutral stimulus, say A, and then a second neutral stimulus, say B, is paired with A; thus, B– > A, in the absence of the original UCS. If this results in B now evoking a CR, second-order conditioning is said to have taken place. Neither Pavlov nor anyone since has been successful in demonstrating third-order conditioning.

Pavlov's Theories of Brain Processes Underlying Conditioning

Many of the phenomena summarized in the previous section and the experiments in Pavlov's lab that studied them were described in American textbooks of psychology by the late 1930s. However, there was at best limited mention of a long series of experiments that Pavlov's students and co-workers carried out in the early 1920s. These centered around Pavlov's theory of the processes in the brain that gave rise to conditioning. For him, as a physiologist, the main point of studying conditioning was to understand the "workings of the higher nervous system." He was dismissive of approaches that relied on cutting out parts of the cerebral cortex. "Imagine that we have to penetrate into the activity of an incomparably simpler machine fashioned by human hands, and that for this purpose, not knowing its different parts, instead of carefully dismantling the machine we take a saw and cut away one or another fraction of it, hoping to obtain an exact knowledge of its mechanical working!"[20] This argument was, however, followed by lectures that reported conditioning experiments on dogs that had had part of their cortex removed.

No neuroscientist today would quarrel with Pavlov's basic idea of how sensory inputs are processed by 'analyzers' and represented in cortical 'centers' or with the idea that a 'UCS center' sends out signals to such organs as the salivary glands. However, what seems strange are Pavlov's ideas on what happens across the cerebral cortex. He proposed that conditioned responding depends on conflicting waves of excitation and inhibition moving out from cortical centers and then receding. To measure the progress of such waves, a large number of experiments in Pavlov's lab varied the interval from, say 1 to 15 minutes, between the presentation of stimuli that were reinforced, CS+s, and other others that were non-reinforced, CS-s.

Few Western researchers have ever followed up on this kind of experiment. Partly, one can suppose, this is because Pavlov's ideas on events in the brain had been outmoded since early in the twentieth century. He seems to have known about Cajal's demonstration that the nervous system was made up of discrete neurons that made contact with each other via synapses and he quoted Sherrington who had embraced this idea. Yet Pavlov clung throughout his life to a view current in 1900 whereby "nerve cells were joined in a kind of three-dimensional netting along which nervous messages could flow roughly spherically .... it is as though, in 1900 or thereabouts, he stopped listening to what was going on elsewhere. After all, his own experiments were enough to keep him busy."[21]

At another level of theory Pavlov's claims of what happens during simple excitatory conditioning remained very clear. A connection is established between the CS center and the US center, such that when the CS center is stimulated excitation passes to the UCS center and this tends to evoke the UCR. Thus, a process of *stimulus substitution* takes place, whereby the CS now elicits the same response as elicited by the UCS, a *conditioned reflex* is formed. What was unclear – and Pavlov continued to change his mind on the topic until he died – was the relationship between the various types of inhibition he described.

Experimental Neuroses

Pavlov and his collaborators treated their dogs very well. They had individual names and were treated almost as pets. Most lived in the institute kennels for several years and served in many experiments. A calm, well-socialized dog provided good data. Consequently, when a dog started to behave in an atypical way, this stood out.

An early example was in 1911, when one of the earliest female students to work in his lab, Maria Erofeeva, tested the effects of delivering a brief electric shock as a signal for the delivery of food. Some dogs showed what became known as *counter-conditioning* in that their initial response to the shock was replaced by salivation. However, two dogs subjected to such training subsequently showed prolonged disturbed behavior.[22]

Ten years later such behavior was displayed by a dog serving in a very different kind of experiment. Nadezhda Shenger-Krestnovikova used discrimination training in which the projection of a circle onto a screen was followed by food delivery, while presentation of an ellipse was not. Once the dog was consistently salivating to the circle, but not to the ellipse, the ellipse was gradually made more circle-like up to the point – the ratio of the axes reached 9:8 – when the dog no longer responded differentially to the two stimuli. Pavlov reported that "the dog which formerly stood quietly on his bench now was constantly struggling and howling."[23] Influenced by Freud's case histories – despite firmly rejecting the latter's theories – Pavlov labeled this change "experimental neurosis."[24]

Pavlov's belief that he could provide a rigorous scientific basis for the study of abnormal behavior, including human neuroses and psychoses, was further strengthened by a natural calamity that hit St. Petersburg in 1924. A huge storm led to flooding of the institute kennels. The dogs were saved by making them swim out in small groups. They were then housed on the first floor of the laboratory until the flood had receded. When normal experimental routines were resumed, some of the dogs showed "disturbance of their conditioned reflexes" for a considerable time.[25]

For the final decade of his life Pavlov became increasingly concerned with such topics. He envisaged a normal dog as one for which the processes of excitation and inhibition were balanced within its brain. Neurosis results from a 'collision' between these processes. By no means all dogs showed disturbed behavior after serving in an experiment where a shock signaled food or in one that required discrimination between a circle and a circle-like ellipse. Similarly, only a few dogs were traumatized by the flood. This contrasted with the consistency with which dogs responded in more standard experiments. Consequently, Pavlov's long interest in the different personalities of his dogs developed into his theory of 'types.' Thus, for example, referring to the effect of the flood, he wrote that "dogs with a weak nervous system, having a predominant inhibitory process … (that) … reproduces the etiology of a special traumatic neurosis."[26]

His developing interest in mental illness led Pavlov to visit psychiatric wards several times a week when he was in his 80s. His analysis of two cases of schizophrenia, a girl in her early 20s and a 60-year-old man, in terms of "isolated inhibition of the motor region of the cerebral cortex" seems now at best quaint.[27] He commented on a case of "hysterical psychosis" in a woman who, after her husband deserted her, later taking away their child, "sank into dotage": "a closer examination of the patient shows that everything seems to be accounted for exclusively by the absence of the analytical inhibition which always accompanies our behavior, our movements, words and thoughts, and which distinguishes the adult from the child."[28] Not much empathy on display.

It is something of a puzzle that Pavlov does not seem to have been concerned with how his analyses of various kinds of human psychopathology might lead to effective therapies. Even for a dog showing symptoms of experimental psychosis, the only treatments considered were either to give it a 'bromide,' a sedative popular in that era, or sleep and prolonged rest.[29]

The development of therapies based on conditioning theories were, however, developed in the English-speaking world. Pavlov's interest in psychopathology had attracted the attention of the most influential American psychiatrist of his generation, Adolf Meyer. An earlier colleague of Meyer's was John Watson, whose *Behaviorism* of 1924 had promoted the idea that Pavlov's conditioned reflex theory would provide the cornerstone for a truly scientific psychology. Meyer visited Pavlov in 1925 and later a research-only position in his psychiatric institute was offered to an American who was working Pavlov's lab.[30]

W. Horsley Gantt (1892–1980)

Gantt was born in Virginia and had studied at the University of North Carolina, before enrolling for a medical degree at the University of Virginia. On obtaining his MD in 1920, Gantt joined the American Relief Administration in St. Petersburg to study the impact of the war and resultant famine. In 1922, knowing of Pavlov's work on digestion, he made a visit to Pavlov's lab; see Figure 1.6. There Pavlov explained his research on conditioning to the 30-year-old who had hitherto known nothing about the topic. Gantt reported that this meeting with Pavlov "stirred me emotionally, immediately made to believe that here was the method of studying mentality and its disorders." After completing further medical training in the UK, he returned to Russia as soon as he could and spent the next four and a half years working in Pavlov's laboratory.[31]

Adolf Meyer learned of Gantt's work and invited him to set up a Pavlovian laboratory within the Henry Phipps Psychiatric Clinic at the Johns Hopkins Medical School in Baltimore.[32] From 1929, when Gantt took up his appointment, he remained the

Figure 1.6 W. Horsley Gantt around the time when he first visited Pavlov. Reproduced with permission of the Alan Mason Chesney Medical Archives. Johns Hopkins Medical Institutions.

most prominent advocate of Pavlov's ideas in the USA. In 1930 Gantt established the Pavlovian Laboratory within the Psychiatric Clinic and continued to conduct research and supervise the students who worked in the lab until he was retired in 1967. Despite retirement, Gantt was able to continue his experiments until shortly before his death in 1980, leaving only one other of Pavlov's students still alive.[33] Perhaps because his return to St. Petersburg occurred a few months after the great flood, a time when Pavlov's main enthusiasm was to explain mental illness in terms of his conditioning theory, a major focus throughout Gantt's long research career was on psychopathology.

Gantt's book on the *Experimental basis for neurotic behavior*[34] reported a replication of Shenger-Krestnovikova's experiment on experimental neurosis produced – in this case – by training a dog to discriminate between ever-closer frequencies of a metronome. It also included the description of a traumatizing event akin to that of the St. Petersburg flood; in this case, his dogs had escaped one night from their enclosure and roamed the building, until discovered by a night watchman who clubbed them back into their kennels. Subsequently some dogs performed poorly in their conditioning experiments; "the inhibitory reflexes were much more unstable."[35] By contrast, some seemed unaffected; for example, Billy, who was judged to have "a strong, well-balanced nervous system."[36] The bulk of this book, however, was devoted to recounting the life and experimental treatments of a single dog, Nick, over a period of nine years starting in February, 1932. What to conclude from this mass of detail is hard to make out.

Like Pavlov, Gantt appears to have had little interest in developing conditioning methods to use as therapy for any kind of mental illness. Gantt's one paper that referred to treatment of experimental neurosis reports that a dog recovered after being given a few mild shocks.[37]

When I met Horsley Gantt in 1975, he expressed some bitterness that his work had not received the recognition that he felt it deserved, together with sadness that what he referred to as the "Pavlovian school" had never taken root in America. Perhaps he had discovered that only one of his papers was cited in an authoritative review of research on Pavlovian conditioning published the previous year.[38]

Gantt's coauthor of this paper was William Brogden.[39] In the long run Brogden was to become better-known than Gantt as, allegedly, the first person to investigate *sensory preconditioning*.[40] As described more fully in a later chapter, in experiments Brogden carried out in Gantt's Pavlovian Laboratory dogs were first repeatedly exposed to pairings of two neutral stimuli, A->B, and then B was paired with shocks; when A was then presented in a test, the dogs showed fear, even though A had never been paired with shock. This result implied that the dogs had learned that B follows A in the first stage of the experiment.

Although textbooks ever since have credited Brogden's 1939 report with being the first account of sensory preconditioning, in fact the phenomenon had been subject to a series of experiments in Pavlov's lab in the early 1930s,[41] as noted below. They were not cited in Brogden's report. In 1928, Gantt had published his translation into English of a set of lectures that Pavlov had given over a 25-year period, starting with the 1903 lecture in Madrid.[42] It seems that either Gantt no longer maintained in touch with what was happening in Pavlov's lab or he failed to tell Brogden about the Russian experiments on sensory preconditioning.

Many of Gantt's experiments were an extension of the kind of research he had carried out in Pavlov's laboratory. However, unlike Pavlov, Gantt mainly studied conditioning of cardiovascular responses and of leg flexion. Some experiments involved the conditioning of responses to drugs.[43] Later in his career, he and his students were the first to suggest that Pavlovian conditioning provided an explanation of placebo effects.[44]

Gantt can be seen as someone who fell between two disciplines. When he addressed psychiatrists and told them about his research,[45] it is unlikely that many found that this helped them to solve the clinical problems they confronted. And not many psychologists, even psychobiologists, are likely to have found talk of "collisions between excitatory and inhibitory processes" an appealing way to think about psychological processes.

In 1955 Gantt founded the Pavlovian Society and its annual meetings continue to provide an important forum for researchers studying Pavlovian conditioning. However, it is very doubtful that Gantt would have approved of the content of most of the papers presented at this meeting since 1980, let alone of the language used.

Howard Liddell and Conditioning Research on the Behavior Farm

The second important proponent of Pavlov's ideas in the USA was Howard S. Liddell (1895–1962). Like Gantt, Liddell also visited Pavlov in the early 1920s in the city recently renamed as Leningrad. However, unlike Gantt, Liddell had both an undergraduate and Master's degree in psychology and the purpose of his visit was to learn more about Pavlov's experiments on conditioned reflexes.

Liddell was born in Cleveland, Ohio and raised in Erie, Pennsylvania. After obtaining his Master's degree from the University of Michigan in 1918, he decided to enroll for a Ph.D. at Cornell University, but in physiology instead of psychology. His advisor there, Sutherland Simpson, specialized in the study of the thyroid gland in farm animals, especially sheep.[46]

In view of Liddell's undergraduate training in psychology, he was persuaded by his advisor to test whether the cretinism produced by removing a sheep's thyroid reduced its ability to learn simple tasks. To investigate the idea Liddell constructed a series of mazes to test both normal sheep and those that had had their thyroid glands removed on their ability to learn their way through a maze to access a food reward. The experiments were not a success. "From our point of view we did not have adequate control of the experimental situation. The sheep and the goats in the maze were doing what *they* wanted to do … It was not pleasant on a cold winter's day to watch a lethargic sheep lie down in the snow at the junction of the alleys and remain there for almost an hour before proceeding on its way."[47] Liddell looked for a better way to study his animals' learning abilities. In 1923, a lecture given at Cornell by a former student of Pavlov, Gleb Anrep, led Liddell to the idea that conditioning methods could provide the answer. Hence, his first visit Pavlov in 1926.[48]

Liddell started to carry out conditioning experiments with sheep in the Physiological Field Station that Simpson had established. Thus, the Field Station became the first laboratory in the USA for the study of Pavlovian conditioning. Liddell's research attracted generous grants from both the Macey and the Rockefeller Foundations and as a result his research program flourished throughout the 1930s. In 1937 Cornell University acquired a 110-acre farm that became Cornell's Behavior Farm Laboratory; its cafeteria, known as the 'Home Dairy,' was where Liddell did most of his writing and where he interacted with research assistants and students. As first Professor of Physiology and eventually Professor of Psychobiology, he could devote most of his time to research.[49]

For the majority of Liddell's experiments the UCS was delivery of a brief, mild electric shock to the foreleg of the animal – "a shock so weak that we could scarcely feel it on the moistened finger tips"[50] – and the unconditioned and CRs were a brief leg flexion. Apparatus for testing lambs is shown in Figure 1.7 and for testing a goat in Figure 1.8.

The first case of experimental neurosis in Liddell's new laboratory occurred by accident. During a routine experiment in 1927 Liddell was impatient to collect more data to report at an upcoming meeting and decided to increase the number of metronome-shock pairings given to a pair of sheep from 10 to 20 each session. The

Figure 1.7 Arrangement for establishing a conditioned reflex in a lamb in Liddell's laboratory. From Liddell (1938). Public domain.

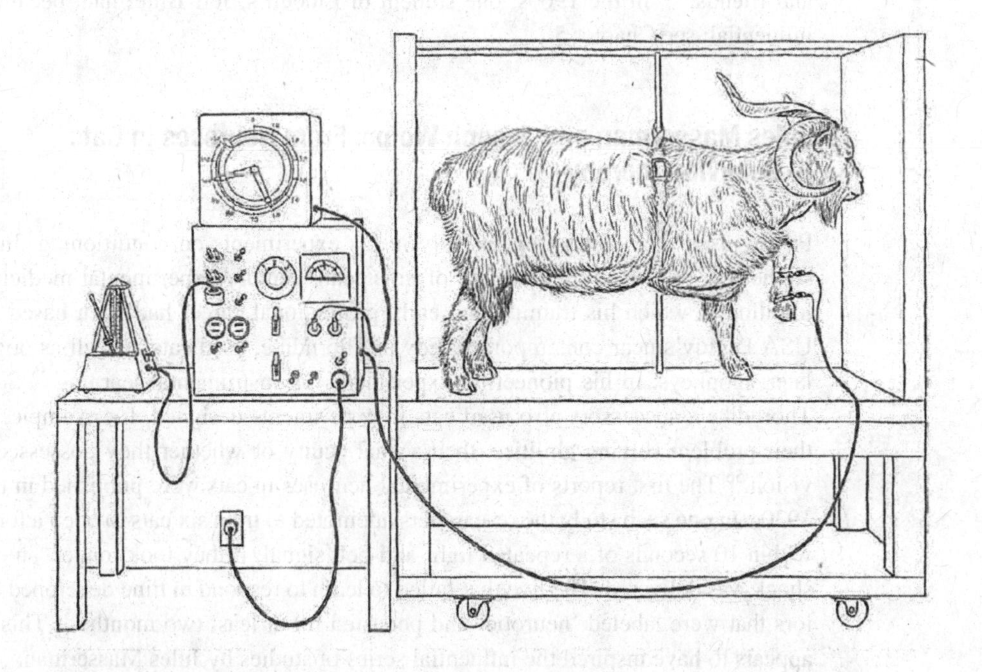

Figure 1.8 Classical conditioning of a goat in Liddell's Behavior Farm Laboratory. From Liddell (1938). Public domain.

normal sheep suddenly began to display agitated behavior and this persisted even when on subsequent sessions no signals and no shocks were given. Meanwhile, the thyroidectomized twin continued to produce precise conditioned leg flexions and "never exhibited the slightest sign of emotional excitement or alarm."[51] The normal sheep turned out to have acquired a chronic emotional disorder. Subsequent experiments established that the same outcome could be reliably produced in sheep and goats by several other methods. On a visit to Cornell one of Pavlov's assistants confirmed that the symptoms were similar to experimental neuroses seen in dogs.

This topic became the central concern of Liddell's research for the next 25 or more years. He saw the aim of his conditioning experiments on animals to be that of increasing understanding of emotions, rather than of learning processes: "The primitive forces of man's emotions are more dangerous and more devastating than nuclear fission. Who can doubt that the central scientific problem of our time is the problem of emotion."[52]

Like Pavlov, Liddell believed that the study of disturbed behavior in animals would provide a scientific basis for understanding human neuroses and what later become known as post-traumatic stress disorder (PTSD). This belief was strengthened when at 57 years of age Liddell visited Korea during the height of the Korean War to interview soldiers who had returned from the front.[53] He appears to have been widely liked and respected by his students and peers; "No one who knew Howard Liddell is apt soon to forget his ebullient, bustling manner, his earthy anecdotes … and the man who responded with sensitiveness and perceptiveness to the needs of his colleagues and friends."[54] In the 1960s, one student of Liddell's, Jeff Bitterman, became very influential; see Chapter 5.

Jules Masserman and Joseph Wolpe: From Neuroses in Cats to Behavior Therapy

Pavlov used dogs almost exclusively in his experiments on conditioning. In doing so, he was following the tradition of nineteenth century experimental medicine, the tradition in which his training and early professional career had been based. In the USA Pavlov's near contemporary, Edwin Thorndike, used cats as well as dogs, and later, monkeys, in his pioneering experiments on instrumental learning.[55] Some of Thorndike's successors also used cats in experiments designed, for example, to test their problem solving abilities, their visual acuity or whether they possessed color vision.[56] The first reports of experimental neuroses in cats were published in the late 1930s. In one such study the researchers attempted to train six cats to open a food-box within 10 seconds of a repeated light-and-bell signal; if they took longer, an electric shock was delivered. The five that failed to learn to respond in time developed behaviors that were labeled 'neurotic' and persisted for at least two months.[57] This report appears to have inspired the influential series of studies by Jules Masserman.

Masserman (1905–1989) was a young boy when his family emigrated from Poland to the USA prior to World War I. Rejecting the prospect of working in the family

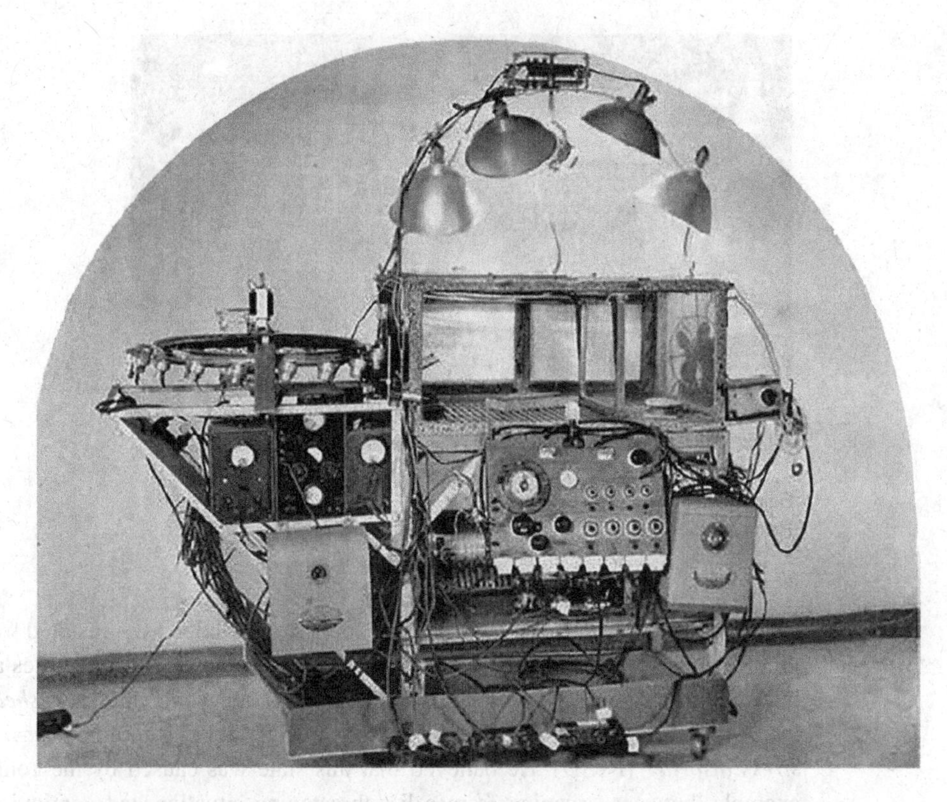

Figure 1.9 Apparatus used by Masserman to establish experimental neuroses in cats. From Masserman (1943). Public domain.

tailor shop in Detroit, he spent 10 years studying medicine, later specializing in neurology and psychiatry, bcforc obtaining a rcsidcncy in 1932 with Adolf Meyer in the latter's Psychiatric Clinic. It was during his three years there that Masserman learned about Gantt's work on neuroses in dogs. Although impressed by Pavlov's and Gantt's experiments, Masserman was highly critical of Pavlovian theory.[58]

In that era training in psychiatry was strongly influenced by Freud and other psychoanalytic theorists. Nevertheless, there was widespread belief that psychiatric practice needed to be based on better evidence than that provided by Freudian analysts; in other words, it needed to be more 'scientific.' Masserman shared this belief and became convinced that the study of neuroses in animals would contribute to satisfying this need. His ambition became "to demonstrate that all animals adapted to their environments in ways that demonstrated psychoanalytic principles, particularly the internalized response to stresses and conflicts."[59]

It was not until 1942 that Masserman began to realize this ambition. By then he had become a member of the Department of Psychiatry at the University of Chicago and had undergone training in psychoanalysis. He set up a lab in which he could run experiments that were intended to produce neuroses in cats and test ways of then curing them. His apparatus is shown in Figure 1.9.

Figure 1.10 One of Masserman's cats.
From Masserman (1943). Public domain.

Masserman believed that neuroses – or 'nervous breakdowns' – resulted from conflict. During World War II he treated many soldiers relieved of their duties and sent back to America because of what in World War I had become known as *shell shock,* then *war neurosis* in World War II, and very much later was termed *post-traumatic stress disorder* (PTSD). He believed that this state was caused by the conflict, for example, between escaping from a life-threatening situation and continuing to do one's duty on behalf of comrades and one's country.

Many of Masserman's hungry cats were faced with a conflict between obtaining the food they had earned by depressing a switch several times and being subject to a blast of air as they lifted the lid of the food box. For others the conflict was between obtaining an earned food pellet and receiving an electric shock as they crossed a grid to reach the food box.

Such treatments were regarded as successful in producing a neurosis if a previously quiet cat subsequently "tended to show restlessness or agitation" or a previously active cat "developed marked restriction of activity and a tendency to passivity." Another cat commenced "a fidgety, incessant" movement. Masserman also identified "phobic," "counterphobic," and "regressive" behavior patterns; see Figure 1.10. He claimed that the therapy he provided to such cats was based on the psychoanalytic concept of *transference*. Sometimes this consisted of simply providing a cat with a lot of attention and petting. A less humane treatment – but an interesting one in the light of subsequent developments – was to force the cat – for example – to cross the no longer live grid to obtain a food pellet and confront its fear.[60] Later, Masserman tried electro-convulsive shock, alcohol and the sedative, sodium pentothal, as potential therapies.[61]

As noted earlier, Masserman's ambition was to provide a more secure scientific basis for psychoanalytic-based treatments, but his experiments were poor science even by the standards of the 1940s. In terms of later standards, the lack of any details in his reports of methods or results, the absence of criteria for deciding whether some behavior was neurotic, the lack of an observer who was blind to the treatment a cat had received, these and further weaknesses would have made his research unpublishable. Nevertheless, Masserman's career flourished. Grants from the US Public Health Service allowed him to expand his lab, first in Chicago and then at Northwestern University, and to include monkeys as subjects in addition to his cats. For example, hungry rhesus monkeys served in experiments on 'altruistic' behavior, in which most animals chose to avoid making a response that would have given them access to food, if this also resulted in the delivery of an electric shock to a second monkey that could be seen through a window looking onto an adjoining compartment.[62]

Masserman's professional career also blossomed as his research became widely known, helped by the films of his animal experiments that he circulated and by his appearances on radio and television, and as his ideas on 'biodynamic psychiatry' were increasingly accepted. In 1979 Masserman was appointed President of the American Psychiatric Association. But in the same year, his career suddenly ended in disgrace; he had treated one of his patients with sodium pentothal and, when she woke up from her trance, she discovered that "he was having sex with her."[63]

In the long run the most influential by far of researchers inspired by Pavlov's reports of experimental neurosis was a South African, Joseph Wolpe (1915–1997). He was born in Johannesburg, where he undertook medical training at the University of Witswatersrand. During the Second World War he served as a doctor in the South African Army. His efforts to treat soldiers suffering from PTSD guided his decision to specialize in psychiatry and study experimental neuroses when the war was over. His doctoral project began with a comprehensive review of research on experimental neuroses, defining these as "unadaptive responses that are characterized by anxiety, that are persistent and that have been produced experimentally by behavioural means."[64] The review was followed by a report of a year's worth of experiments starting in June, 1947, that were inspired by Masserman's early research on the topic. Unlike Masserman, Wolpe rejected Freudian theory after finding during the war that therapies based on such theories were ineffective.[65]

One method used by Wolpe was similar to one that Masserman had used. Six cats were first trained to approach a food cup, in which minced beef pellets were delivered whenever a buzzer was sounded; once a cat was consistently producing the conditioned approach response, Wolpe now delivered an electric shock each time the animal was about to seize the food pellet. As Masserman had found, this produced changes in the cats' behavior that generalized well beyond the experimental setting. One important new result emerged from Wolpe's inclusion of a control procedure whereby six cats were exposed to a simple conditioning procedure in which the sound of a hooter was paired with shock delivery in the absence of any conditioning of a feeding response. This simpler procedure produced many of the same symptoms as the Masserman procedure. The less surprising ones included anxiety when placed in the cage and "refusal to eat meat pellets anywhere in the cage even after 1, 2 or 3 days' starvation."[66] Even introducing the cats into rooms other than the experimental room could produce signs of anxiety; its intensity was found to vary with the degree to which the new room resembled the experimental room.

Wolpe was interested in what "curative measures" would be effective in relieving his cats' neurotic behavior. "The fact that the neurotic reactions of the cats were associated with inhibition of feeding suggested that under different conditions feeding might inhibit the neurotic reactions: in other words, that the two reactions might be reciprocally inhibitory."[67] One method was to feed a cat in the room that was least similar to the experimental room and, once this was successful, to feed it in the second least similar room and so on, until the cat behaved quite normally even in the experimental room.

Wolpe reported that his "experience of the past 4 years has encouraged belief in the hypothesis that experimental and clinical neuroses are parallel phenomena. … The human subject is not often forced to undergo his conflicts or his traumata in physically confined space. He is usually kept in the anxiety-producing situation by the force of habits previously learned. For instance, a woman entangled in a humiliating marriage may be unable to get out of it because her earlier training has given a horror to the idea of divorce. Besides confining her within the marriage, this feeling of horror, being in conflict with escape tendencies, makes possible the development of a high level of emotional tension (anxiety). This tension becomes increasingly conditioned to contiguous stimuli through the drive reductions that follow every partial escape from the causative situation."[68]

Wolpe found that a curative measure similar to the one he had used with his cats could be effective in treating patients suffering from anxiety disorders.[69] He knew that this procedure had been first used in New York in the 1920s by Mary Cover Jones, a graduate student supervised by John Watson. This time the procedure was not forgotten. *Systematic desensitization* has been routinely and successfully used very widely for the treatment of phobias and other kinds of anxiety disorders ever since. Thus, it took over 40 years from Pavlov's report of experimental neurosis to the widespread adoption of a therapy that Pavlov's experiments inspired.

Following the publication of his 1952 paper, Wolpe's animal research, his Pavlovian-sounding theory of *reciprocal inhibition* – neither a cat nor a human being can be both anxious and relaxed at the same time – and the form of therapy that his research inspired became very widely known. He spent 1956/57 as a visiting fellow at the Center for Advanced Studies in the Behavioral Sciences at Stanford University, where he had considerable contact with another fellow, the philosopher, Karl Popper. The latter persuaded Wolpe of the need to make theories testable. In 1960 Wolpe returned to the USA and remained there for the rest of his life.[70]

Chimpanzees and Problem Solving: Pavlov's Final Years

The great passion of Pavlov's final years was the creation of a scientific center in the village of Koltushi, some 22 kilometers east of Leningrad. In the early years of its development Pavlov and his assistants traveled there by train. Later, he was driven there and back in the Lincoln limousine that, together with a chauffeur, was permanently on call, just one more element of the increasing generosity which he enjoyed from Stalin's government.

In Koltushi, "the Soviet state erected to his specifications a science village that combined two of his great loves, science and rural dacha life. This Institute of Experimental Genetics of Higher Nervous Activity housed a grand project to study the relationship of heredity and environment to constitution and temperament, and, eventually, to turn this knowledge to practical use in the upbringing of children and the breeding of an improved human type."[71]

Soviet generosity was accompanied by increasing influence over the research carried out in Pavlov's multiple labs. There was a steady increase in the proportion of coworkers, assistants and students who were committed communists. One or two were influential in their arguments, based on dialectical materialism, that Pavlov needed to study brain processes in primates and not confine himself to dogs. Pavlov's death was shortly followed by Stalin's 'great purge,' when many of Pavlov's former coworkers, whether communist party members or not, together with several of his friends and influential contacts in the government, were either shot or sent to the gulag.[72]

Returning to Pavlov's research in the early 1930s, a new interest was in Gestalt psychology and its challenge to associationism and American behaviorism. He and Wolfgang Koehler, whose 1919 book on *The mentality of apes* Pavlov had at least partly read, were the main invited speakers at the Psychology Congress held in New Haven in 1929.[73] At the age of 84, Pavlov began to study the two chimpanzees, Roza and Rafael, that arrived in Koltushi in 1933.[74] The two were given many of the problem-solving tasks that Koehler had described. Their behavior convinced Pavlov that they were more intelligent than his dogs, but he attributed this to the dexterity they enjoyed as a result of having effectively four 'hands.' Pavlov had no time for Koehler's claim that chimps' problem-solving involved insight – a "fictitious muddle" – and instead believed that it could be explained in terms of associations.[75] His work with the two chimpanzees and the experiments on what became known as *sensory pre-conditioning*, as described earlier, led to Pavlov to decide by 1935 that associations and conditioned reflexes were not different labels for the same processes.[76]

Many years passed before these changes in Pavlov's ideas became known in the English-speaking world. In July 1935 Pavlov began a 10-day trip to address the International Neurological Congress in London. Instead of reporting experimental results from his multiple labs, the topic of his address was "Types of higher nervous activity in connection with neuroses and psychoses, and the physiological mechanism of neurotic and psychotic symptoms."[77]

Pavlov died in February 1936.

Jerzy Konorski: Type 2 ('Motor') Conditioning and His Critique of Pavlov's Theories

Like Gantt, Jerzy Konorski (1903–1973) was a non-Russian whose life was completely changed on learning about Pavlov's research. Unlike Gantt, Konorski's huge respect for Pavlov was not uncritical. He rejected both Pavlov's theories of the neural processes underlying conditioning and Pavlov's long-standing claim that the conditioned reflex was the only form of learning.

Konorski and his friend, Stefan Miller (1903–1940), were medical students in Warsaw, Poland in the mid-1920s when they first read about Pavlov's experiments on conditioning. "The starting point for our investigations was the fact, known from everyday life, from training experience and from the evidence of behaviourist psychology, that if an animal's motor reaction leads to a 'satisfying state of affairs' (Thorndike) then it tends to be repeated in the same situation, but if it leads to an 'annoying state of affairs,' then it tends to be avoided. We subjected these generally known and universal facts of animal behaviour to conditioned reflex investigation, analysed them in regard to their structure and, to some extent, in regard to their physiological mechanism."[78]

Their first experiments involved a single dog, Bobek. In an early experiment Bobek stood facing an electric lamp. As soon as the light was switched on, Bobek's forepaw was raised by the experimenter's hand and the dog was fed. After several repetitions of this procedure, Bobek's forepaw raised without human aid as soon as the light was on. The results from a series of more complex experiments convinced them that Pavlov's theory of the conditioned reflex was unable to explain such learning, describing the reasons for this conclusion both in an article published in a French journal in 1928[79] and in a letter to Pavlov sent in the same year. They referred to their form of conditioning as a 'Type 2 Conditioned reflex.' Pavlov initially rejected their arguments and the idea of a second kind of conditioning. However, he was sufficiently interested and impressed to instruct his favorite research assistant, M.K. Petrova, and two students to replicate the Warsaw experiments. He also invited the two Poles to spend time in his lab. They arrived in 1931. Miller was newly married and stayed only a few months. Konorski stayed for almost two years.[80] During that time Pavlov failed to convince Konorski that he was wrong, while Konorski became an expert on the experimental results that had been obtained in Pavlov's lab over the previous two decades and on Pavlov's theoretical explanations for those results. Konorski also acquired the skills needed to run experiments on 'Type 1' conditioning. Pavlov eventually came to accept Konorski's account of 'motor conditioning.'[81]

After they returned from Leningrad, Miller and Konorski obtained positions in Warsaw's Nencki Institute of Experimental Biology, where they set up a conditioning lab to run experiments both on Pavlov's Type 1 and on their Type 2 conditioning. Their work on the latter became known in the English-speaking world in 1937, when they published a paper on the topic in an American journal[82] and became even better known because of the resulting exchange with the American psychologist, B.F. Skinner (see Chapter 8).

Following Pavlov, Konorski and Miller also began experiments on problem solving involving children and monkeys.[83] But these ended in 1939 with the German invasion of Poland. Miller was killed in a Nazi concentration camp and the Nencki Institute was destroyed. Konorski wrote most of a book. With the end of the war, his lab was rebuilt and in 1948 Konorski was able to visit the UK and make the final arrangements for the translation into English and publication of his book; the title was *Conditioned reflexes and neuron organization.*

Konorski's second hero was the British physiologist, Charles Sherrington, whose studies of spinal cord reflexes eventually earned him a Nobel Prize in 1932. Much earlier Sherrington had introduced the term 'synapse' and had published a book called *The integrative action of the nervous system.* It was this book of 1906 that inspired Konorski, to the extent that in 1967 he gave an almost identical title to his second book, namely, *The Integrative activity of the brain.* For Konorski, like Pavlov, the. point of studying conditioned reflexes was to discover how the brain works. What he set out to do was to make sense of what was known about conditioned reflexes in terms of Sherrington's concepts, such as excitatory and inhibitory synaptic connections between neurons, rather than the concept of the nervous system held by Pavlov, which – as noted earlier – was already generally known to be invalid by 1900. The model of conditioning that Konorski proposed in 1948 anticipated developments decades later.[84]

The neural processes underlying conditioning are beyond the scope of the present book. However, it seems appropriate to provide here one example of how Konorski's explanation of an effect that was first documented in Pavlov's lab differed from Pavlov's own account. As noted earlier, *Stimulus generalization* refers to the finding that, once a CR is established to a CS, other stimuli that are similar to the CS can also elicit the CR, but to an extent that decreases with increasing dissimilarity to the CS. According to Pavlov, the occurrence of a novel test stimulus evokes excitation in its cortical center that spreads out across the cortex with diminishing energy. When this wave reaches the CS center, the resulting excitation evokes the CR. If the wave has traveled a long way – because the test stimulus differs considerably from the CS – then the CR will be weak. Konorski's account was essentially the one that has been accepted ever since. Presentation of the CS excites a large number of cortical neurons, while presentation of a test stimulus excites a proportion of the same neurons that is related to how similar the test stimulus is to the CS. In Konorski's words: "It must be assumed that the cortical centres of particular stimuli represent complex and widely dispersed formations, that they overlap, and that this partial overlapping is the cause of generalization."[85]

To criticize Pavlov during the Stalin era could end a scientific career. "Pavlov, although he was never a member of the communist party, was a representative of progressive science; his views were considered 'errorless,' and no change in his theory could be tolerated."[86] Pavlov's successors in Russia, such Asratian, were careful not to depart from their teacher's theories. In contrast, Konorski wrote: "the above-mentioned defects, they seem so obvious to us that it is difficult not to perceive them. And yet even today many of Pavlov's disciples and followers, hypnotized by his theory, either do not see, or else do not want to see them."[87] Somehow Konorski survived public attacks on his views and personal threats. He re-established the small research institute that he had founded just before the war in order to continue experiments on conditioning.[88] With the death of Stalin his career and the Nencki Institute flourished, as described in Chapter 9.

2 Developing Habits

Clark Hull and the Hullians

William James' stirring and beautifully written chapter from 1890 on the role of habits in everyday life persuaded many American psychologists of the next generation that understanding the development of habits should be a central aim of psychology. James' student, Edward Thorndike, believed that the results from his puzzle box experiments on trial-and-error learning in cats and dogs were a major step towards achieving this aim. Habits consisted of connections between stimuli (S) and responses (R) that, according to his Law of Effect were 'stamped in' by any 'satisfying' event that followed shortly after the response.[1] Watson's studies of rats learning to navigate mazes convinced him too of the important of habits, conceived as S-R connections.[2] However, he rejected the idea that 'satisfiers' were needed for the development of habits – this was too subjective a concept – and instead proposed that habits grew stronger as a result of repetition alone, the *Law of Frequency*.[3]

The absorption of Pavlov's research into this S-R tradition began with Watson's proposal that the conditioned reflex should become the theoretical core of behaviorist psychology.[4] At the time, like anyone else who was unable to read Russian, Watson's knowledge of Pavlov's research was limited. In 1927 the publication by Anrep of a translation into English of Pavlov's lectures on the conditioned reflex meant that the non-Russian-speaking world in general, and American psychologists in particular, could at last learn about the huge number of experiments on conditioning that had been undertaken by Pavlov and his students for the past three decades.

Among the first American psychologists to study Pavlov's work in careful detail was Clark Hull. This led him to develop a general theory that combined James' emphasis on habits, Thorndike's Law of Effect, and Watson's insistence that psychology should become a science of behavior based on the concept of conditioning. Hull's ideas and those of the many bright, productive, and eventually highly influential scientists that he inspired – the Hullians – dominated American psychology until the late 1950s.

Outside of the USA behaviorism in general and Hull's particular version had far less influence. Nevertheless, when I was an undergraduate in the UK, studying psychology at Cambridge University, in 1962 a whole course of lectures was devoted to assessing the strengths and weaknesses of Hull's theories. In 1989 I arrived in Australia and found that Hullian theory was still taught at the University of Sydney as the basis for understanding both animal and human psychology.

Clark Hull's Youth and Early Career

Hull (1884–1952) had a tough childhood and adolescence. He was born in a log farmhouse near Akron in NY State, where from an early age he shared in the work around the farm. The family then moved to Michigan where Hull's formal education began in a one-room rural school and continued for a year at a high school in West Saginaw. He later enrolled in Alma College to begin training as a mining engineer. However, his education was interrupted by two long bouts of very serious illness. The first was typhoid; it left him with impaired memory for names that lasted for the rest of his life. The second was poliomyelitis; this paralyzed one leg and permanently affected his health. While convalescing from polio, Hull read William James' *Principles of Psychology* and this convinced him to study psychology.[5]

Hull majored in psychology at the University of Michigan and then carried out research on concept formation at the University of Wisconsin, where he obtained his Ph.D. in 1918 and then stayed on as a member of the Psychology Department for the next 11 years. During this period he studied a variety of topics, including the behavioral effects of tobacco, aptitude testing, and hypnosis. He became recognized as a leading expert on these last two topics.[6]

Hull also developed a strong interest in Gestalt psychology. He persuaded the University of Wisconsin to offer a one-year visiting research position to Kurt Koffka, a leading Gestaltist who had emigrated to the USA in 1924. Hull was convinced by Koffka's criticism of the inadequacy of Watson's behaviorism but not by Koffka's claims for Gestalt psychology. Hull became highly critical of this approach, claiming that it "utterly fails of the true deductive quality."[7]

Presumably because Hull was seen as one of the most promising psychologists of his generation, in 1929, at the age of 45, he was "called to the Institute of Psychology at Yale University as a research professor."[8] To become a research professor in that era was a rare privilege and to have no formal teaching duties was a luxury denied to most of his contemporaries in psychology. Hull remained in this position at Yale until his death in 1952 just short of his 68th birthday.

With the publication in English of Pavlov's lectures in 1927, conditioning became Hull's major interest. He published his first paper on the topic in the year he moved to Yale. Its first paragraph reveals that, following Watson, he proposed to assimilate Pavlov's work within the American S-R tradition. He described the conditioned reflex "as an automatic trial-and-error mechanism which mediates, blindly but beautifully, the adjustment of the organism to a complex environment."[9] Intriguingly, Hull sketched a hypothetical conditioning experiment in which a bell is paired with the delivery of food. "Let us suppose that a conditioned alimentary reflex has been set up to a bell of a certain pitch," he wrote,[10] when almost certainly knowing that Pavlov never used a bell as a conditioned stimulus (CS). Maybe he believed his readers would be more comfortable with the example of a bell than with a metronome of adjustable frequency that Pavlov actually used? This passage appears to be the forerunner of hundreds of subsequent accounts and cartoons that have Pavlov teaching dogs to salivate when a bell sounds.

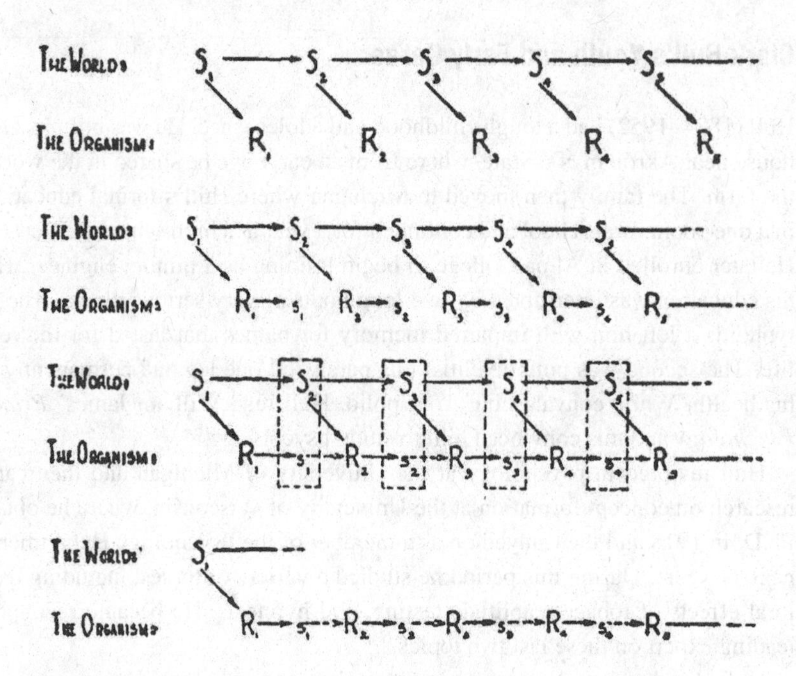

Figure 2.1 Going from top to bottom these figures illustrate Hull's theory of how, as a result of reinforcement, a system could start by simply reacting to environmental stimuli, then to produce internal stimuli ("fractional stimuli"), followed by the development of an integrated chain, and ultimately the equivalent of a "train of thought."
From Hull (1931). Public domain.

In a short note also published in 1929 Hull made an even greater claim for the importance of conditioning. "It is believed by increasing numbers of students of human and other mammalian behavior that the conditioned reflex, with its power of substituting one stimulus for another, is the basic mechanism not only of ordinary habits but of the entire mental life."[11] The note described an electromechanical device that could simulate a dozen properties of a conditioned reflex. The idea was Hull's; the implementation was by a Wisconsin engineering colleague.[12]

In a series of papers Hull described how to go beyond "the typical undergraduate behaviorist's glib explanation of the more complex forms of habit phenomena by saying of each that it is a case of stimulus and response"[13] and to explain, step-by-step, how some phenomenon can be deduced from general principles. One topic treated in this way was trial-and-error learning.[14] More ambitiously, another was the analysis of "knowledge and purpose as habit mechanisms."[15]

The flavor of Hull's theorizing at this time is suggested by the following example. As shown in Figure 2.1, Hull's treatment of 'knowledge' starts with the proposal that a series of stimuli (S_1, S_2, S_3, etc.) elicit a series of overt conditioned responses (R_1, R_2, R_3, etc.). "Now a high-grade organism possesses internal receptors which are stimulated by its own movements. Accordingly, each response (R) produces at once a characteristic stimulus complex and stimuli thus originated make up to a large extent the internal component of the organism's stimuli complexes. Let these internal stimulus components be represented by *s*'s."[16] These internal stimuli and the equally undetectable internal responses that they were assumed to elicit came to play a major role in Hull's later theorizing and that of his followers.

The sequence of figures shown in Figure 2.1 shows the transition from the succession of external stimulus-response pairs via the integration of this sequence to the state in which the occurrence of S1 alone triggers a long sequence of responses triggered by internal stimuli. These transitions were said to involve 'the principle of redintegration' which Hull does not explain. He summarized his argument by stating that "the world in a very important sense has stamped the pattern of its action upon a physical object. The imprint has been made in such a way that a functional parallel of this action segment of the physical world has become a part of the organism. Henceforth the organism will carry about continuously a kind of replica of this world segment. In this very intimate and biologically significant sense, the organism may be said to know the world."[17]

The Lure of Equations: Hull's *Principles of Behavior*

Throughout the twentieth century, experimental psychologists strove for recognition of their discipline as being a 'real' science. By the middle of the century, this was to a large extent successful in the USA, at least as indicated by the number of psychologists elected to membership of the Academy of Science. In the UK it was not until the last decades of the century that more than a handful of psychologists were elected to the Royal Society. In Australia, only two psychologists had been elected to its Academy of Science by the end of the twentieth century.

One argument that Watson used to support his plea that psychologists confine themselves to objective data was that only then would psychology "make its place in the world as an undisputed natural science." In the first half of the twentieth century the natural science that psychologists most envied was physics. Physics was built on mathematical equations. Therefore, psychologists should strive to express their findings in terms of equations.

In an autobiographical essay Hull reported that he "came to the definite conclusion around 1930 that psychology is a true natural science; that its primary laws are expressible quantitatively by means of a moderate number of ordinary equations; that all the complex behavior of single individuals will ultimately be derivable as secondary laws … (as also) all the behavior of groups … *i.e.,* strictly social behavior as such, may similarly be derived as quantitative laws from the same primary equations. With these and similar views as a background, the task of psychologists obviously is that of laying bare these laws as quickly and accurately as possible, particularly the primary laws. This belief was deepened by the influence of my seminar students, notably Kenneth W. Spence and Neal E. Miller."[18]

Hull's Wednesday Evening Seminar was held every week and was open to all graduate students and staff members in psychology, but on condition that they had read Newton's *Principia* of 1687 (*Mathematical Principles of Natural Philosophy*).

Figure 2.2 The Institute of Human Relations at Yale University was housed in the building on the left of the main entrance.
Reproduced with permission from the Human Relations Area Files, Yale University.

Not just Hull's salary but also those of several more junior researchers and research assistants were paid out of a grant by the Rockefeller Foundation. A further sum of nearly two million dollars made possible the construction of a building to house many members of the Institute of Human Relations; see Figure 2.2. This grant reflected the effective influence of the president of Yale, James Angell, a psychologist who had been Watson's Ph.D. advisor at Chicago three decades earlier.[19]

As a result of the huge grant, the facilities for behavioral experiments, involving both animal and human participants, that Hull and his associates had at their disposal far eclipsed those available for such research almost anywhere else in the world. It was in this context that Hull developed the ideas that were laid out in 1943 in his major work. The title, *Principles of Behavior*, was deliberately modeled on Newton's *Principia*, since Hull aspired to become the Newton of psychology.

What Hull sought to do was extraordinarily ambitious. His aim was to construct a theory of behavior that would cover all existing experimental data on learning and motivation, a theory that was based on a set of clearly defined postulates ('primary laws') and corollaries ('derived laws'), and one sufficiently detailed to generate quantitative predictions for future research.

The core assumption regarding learning was that all learned behavior is based on stimulus-response connections – habits, as measured in units of *habs* – that were formed whenever a stimulus was followed by a response that in turn was followed by some reinforcing event. To this point, Hull was simply adopting Thorndike's Law of Effect of 1911. What he added was a definition of 'reinforcement,' namely, any event that reduces a drive stimulus, symbolized by D. Hull's core assumption regarding motivation was that drive stimuli were produced by a limited set of biological needs, such as the need for food, water, or a cooler or a warmer temperature. Thus, for example, a full bladder produces a distinctive drive stimulus that is reduced as 'micturition' (Hull's term for urination) takes place and this reduction reinforces whatever sequence of stimuli and responses preceded it; drive stimuli both energize behavior and their reduction reinforces stimulus-response connections.

The combination of a drive stimulus, D, and a habit, an S-s→r-R connection, produces overt behavior via a *reaction potential*, $_{S}E_{R}$, measured in *wats* (as a nod to John Watson). However, the latter is subject to modification by a number of factors. These included both *conditioned inhibition*, measured in *pavs* (as a tribute to Pavlov), and *reactive inhibition*. Hull proposed that reactive inhibition accrued whenever a response was made and that its magnitude was a function of the effort involved in making the response. The *law of less work (or least effort)* refers to the unsurprising finding that in general, when given a choice of responses that all produce the same outcome or of alleyways that lead to the same goal box, animals prefer the least effortful response or the shortest route. According to Hull, this is because previously the preferred response or route produced the least reactive inhibition. Hull's treatment of extinction was problematic, as discussed later.

According to Hull, the expression of a habit in overt behavior is also modified by 'oscillation.' In the final chapter of his 1943 book he proposed three reasons why the science of behavior had taken so long to reach the stage that physics had attained three centuries earlier. One was the complexity of the subject matter. Another was the corrupting influence of anthropomorphism; that is, the failure to adopt a purely mechanistic analysis of the behavior of animals, including humans, and to regard "the behaving organism as a completely self-maintaining robot."[20] The final reason, he suggested, was that behavior is inherently variable; repeating the exact same stimulus fails to evoke the exact same response. Oscillation reflects the variability of neural responses. As an aside, Hull noted that "it is presumably responsible for many of the phenomena grouped by the classical psychologists under the head of 'attention'."[21] For Hull to include attentional processes within his theory would have been to relapse into anthropomorphism.

For many of the topics that Hull analyzed the only data he could examine were those reported by Pavlov. For other topics, there were also experimental results produced over the previous two or so decades by various American researchers. Hull tended to highlight data obtained within the Institute of Human Relations from research in which he had been involved, if only as advisor. As discussed in Chapter 1, *stimulus generalization* refers to the finding that, once a response has become conditioned to a particular stimulus, it will also be evoked by other stimuli to the extent that they are similar to the original stimulus. Important results regarding stimulus generalization were obtained by a Ph.D. student, Carl Hovland, who later became a pioneering and highly influential researcher into the psychology of persuasion. Hovland's human participants were subjected to a conditioning procedure in which a tone was paired with delivery of an electric shock to the wrist and the measured response was a change in conductivity across the fingers, the *galvanic skin response* as it was known for many years and more recently as the *skin conductance response*.

Such skin conductance conditioning experiments posed a potential threat to Hull's belief that all learning was based on reinforced stimulus-response connections. Just as in Pavlov's experiments, Hovland's procedure was one in which the unconditioned stimulus (US) – the shock in this case – occurred following the CS no matter what the participant did. To squeeze such examples into a Law of Effect straitjacket Hull needed to claim that, for example, the arrival of food reinforced Pavlov's dogs' tendency to salivate or, for example, the termination of the shock reinforced the change in skin conductance exhibited by Hovland's participants.

In rejecting Pavlov's theory of conditioning, Hull made the patronizing comment that "it is not difficult to understand how Pavlov could have made such an error. His mistaken induction was presumably due in part to the exceedingly limited type of experiment which he employed."[22] However, Hull does not seem to have been completely confident about applying the Law of Effect to Pavlovian conditioning: In the final chapter of the *Principles of Behavior* he compared "primitive trial-and-error learning" with "conditioned reflex learning" and suggested that it is "very likely they are at bottom the same process."[23] He was aware of the powerful arguments that there are two distinct forms of conditioning, instrumental (or operant) conditioning and Pavlovian (or classical) conditioning put forward a few years earlier by the British psychologist, Grindley,[24] by the Polish researchers, Konorski and Miller[25] (see Chapter 1), and, above all, by Hull's younger American contemporary, Skinner,[26] whose research was well known to Hull (see Chapter 8). However, none of these were cited by Hull in 1943.

Challenges to Hull's Claims

For many contemporaries, the *Principles of Behavior* was "one of the most important books published in psychology in the twentieth century."[27] Others believed that the magnificent edifice that Hull constructed was based on foundations that were at best shaky. As already noted, interpreting Pavlovian conditioning as simply another form of habit formation was questionable. More generally, was Hull's claim correct that all forms of learning are based on the S-R-Reinforcement model? Could learning occur in the absence of reinforcement? Do all reinforcing events involve drive reduction? Do rats in mazes learn *where* food is to be found as well as – or instead of – what responses are required in order to reach the food?

These questions were addressed by various researchers, as described in later chapters. This section focuses on a further question: Is it true that animals fail to anticipate the outcomes of their actions during instrumental conditioning – not knowing that pressing the lever will be followed by a food pellet – but just acquire a blind habit? Surely human participants given a number of tone-shock pairings learn that each time the tone is sounded, it will be followed by a shock? According to Hull in 1943, all that they learn is to withdraw their hand to the tone. Experiments examining such questions that are described next were a major reason that Hull introduced changes to his theory, as described in his final book, *A behavior system,* of 1952.

As noted earlier, in 1924 John Watson had also claimed that all learning consists of the acquisition of S-R connections, habits. This was then challenged partly on the basis of experiments carried out in Tolman's laboratory on the Berkeley campus of the University of California showing – at least to many people's satisfaction – that animals learn to expect particular outcomes. Thus, after being shown a piece of lettuce being hidden under a large cup and then released to retrieve this reward, a monkey would contentedly eat the lettuce; in contrast, after being shown a much more highly prized banana being hidden that was then surreptitiously exchanged for a lettuce leaf, when the same monkey discovered the lettuce, it looked all around the cup, shrieked at the observers and walked away leaving the lettuce untouched.[28]

This fairly informal example of an *incentive shift* experiment was accompanied by more systematic examples using rats. Thus, in one study rats were first trained to run down a straight alley for a reward consisting bran mash and, once their speeds had reached a fairly constant level, the mash was exchanged for less-preferred sunflower seeds; this resulted in an abrupt decreased in their running speed to a level below that of rats that had been trained on sunflower seeds from the start;[29] (see Chapter 3).

Hull's hypothetical robot was based on the analogy of an automated telephone exchange in which 'expectancies' could have no place. To meet the challenge of results suggesting that animals learn what to expect in various situations, in 1931 Hull proposed an explanation for such data in purely stimulus-response terms. His account was similar to the one he proposed to explain the acquisition of knowledge in S-R terms, as described earlier; see Figure 2.1.

When I approach the door of my house or office, I usually pull the relevant key out of my pocket well before I reach the door. This is an example of what Hull termed a *fractional antedating goal reaction*, which I will refer to here as a FAGR. It is 'fractional' in the sense of being only part of the response I make as I approach the door and 'antedating' in that, after many repetitions, it occurs well before I reach the door. The examples Hull gave included human *ejaculation praecox* and salivation by Pavlov's dogs.[30] He suggested that, when a series of stimuli and responses leads to a reinforcing event, a 'goal,' then – as a result of repeated reinforcement of this series – those fractions of the goal response that are compatible with responses in the series become elicited by ever earlier stimuli in the series.

Furthermore, as noted earlier, as early as 1931 Hull proposed that FAGRs generate proprioceptive feedback in the form of *fractional antedating goal stimuli* that have secondary reinforcing properties. The latter function to integrate the series S-R connections into a single action. So, once one of Elliot's rats has run down the alley and found bran mash at the end of each of many such trials, the "mouth movements of a masticatory nature" appropriate to bran mash are elicited as soon as the animal is placed in the runway. However, when eventually it runs down to find sunflower seeds instead of mash, the integrated stimulus-response sequence is disrupted so that on the next trial the rat runs more slowly. Hull concluded that "the anticipatory goal reaction seems also to constitute the physical basis of the somewhat ill-defined but important concept of purpose, desire, or wish, rather than the drive stimulus as has sometimes been supposed … this hypothesis also renders intelligible the 'realization of an anticipation' by an organism."[31]

Somewhat surprisingly the FAGR received only limited attention in Hull's 1943 book. In contrast, it appeared in full glory as a fundamental concept throughout Hull's final book, *A behavior system.*

Incentives Transform Underlying Habits Structures into Reactions

In 1951 Hull wrote that "during the last eight years the fundamental hypotheses of my system have matured considerably."[32] This suggestion of a curiously passive process, evoking thoughts of good wine or cheese, was an understatement with regard to the profound changes in his theory. One entirely new development was to include the concept of an acquired drive. This was prompted by the work of his close colleague, Neal Miller. In these experiments, rats were first shocked in a distinctive chamber and then, in the absence of any shock, learned to operate a wheel to open a door that allowed them to escape from the chamber.[33] Previously, the only animal-based experiments that Hull had taken into account used food or some other positive reward. Now he appealed to *fear* as a secondary drive and its termination as a reinforcer to explain the growing body of research on fear conditioning and avoidance learning (see Chapter 4).[34]

The more fundamental change to Hull's system was prompted by further research on incentive shifts. As described above, early experiments in Tolman's lab had studied the disruption caused by *qualitative* shifts from a highly valued reward to a less valued one. Research on *quantitative* shifts in reward appears to have started with experiments on chickens. Gwilym Cuthbert Grindley was one of the few researchers outside of North America and Russia with an active interest in how animals learn. In the United Kingdom he provided the only surviving link with nineteenth century British comparative psychology; as a research student and then tutor at Bristol University he was in contact with Lloyd Morgan, now Vice-chancellor, and on the latter's behalf Grindley gave lectures on animal intelligence.[35] In 1929 Grindley reported an experiment in which he gave from one to six grains of rice as a reward to groups of young chickens for reaching the goal box at the end of an alleyway or of a simple maze. He found that the greater the number of grains, the more rapidly the chickens learned their way through the maze and the more rapidly they ran at asymptote.[36]

Grindley's data were variable to the extent that a respectable journal a few years later would have rejected his paper on the grounds that the group differences did not reach statistical significance. This weakness in Grindley's data provided Leo Crespi with the justification for carrying out far more extensive and more carefully designed experiments on magnitude of reward. Crespi had graduated from the University of California in 1937 and then moved to Princeton University where his rat experiments earned him his Ph.D. in 1942. His later, highly successful career was in the new area of public opinion research.

As a graduate student, Crespi constructed a 20-foot long runway that ended in a large square 'food box.' The time a rat took to reach the food box was automatically recorded. Rats were given one trial a day and reward consisted of a number of 0.02-g pellets that Crespi made from dog food. And, very unusually for that time, the data were analyzed using analysis of variance. The first of the key findings from his three experiments was to confirm that running speeds increased more rapidly and reached a higher steady level with larger rewards; thus, in Experiment 3 rats were rewarded with 256 pellets ran faster than those rewarded with 64 pellets, while the latter ran faster than those given 16 pellets. The second finding was what subsequently became known as the *Crespi effect* or *successive negative contrast.* In Experiment 2 all rats were initially trained on 16 pellets but later three groups were shifted to a smaller reward, either 0, 1, or 4 pellets, whereupon they displayed various disruptive behaviors – such jumping and biting the apparatus – that Crespi described as expressing frustration and ran far more slowly. Furthermore, in Experiment 3 the groups trained initially on 256 or 64 pellets were shifted to 16 pellets, whereupon for the next six or more trials their speeds remained below those of the group maintained on 16 pellets throughout; as shown on the right of Figure 2.3. In addition to this 'depression-effect,' he also found weaker evidence for an 'elation-effect,' whereby an upward shift in amount of reward almost immediately resulted in an increase in running speed; as shown on the left of Figure 2.3.[37]

In his *Principles of Behavior* of 1943 Hull cited data on salivary conditioning from a single dog, Billy, reported by Gantt,[38] as well as Grindley's results from groups of 10 chickens in each condition, and decided that large rewards produce stronger habits than small rewards. When he did get around to reading Crespi's report, Hull cited it as "the best study available in this field"[39] and made a fundamental change to his theory, one that distinguished between learning – in the form of increases in habit strength – and performance. It was a point that Tolman had long argued and Crespi[40] strongly echoed. Now, according to Hull, habit strength is simply a function of how many times an S-R sequence has been followed by reinforcement, no matter how large or small that reinforcing event. The major factor now governing performance is *incentive* value, represented by *K*, which is largely a function of the quality and quantity of a reward. This could now explain the depression and elation effects that Crespi had reported. It did not explain why, for example, a rat shifted from 256 pellets to 16 pellets ran more slowly for several trials than a rat maintained on 16 pellets throughout; Hull dismissed such results as representing the impact of transient emotional reactions.

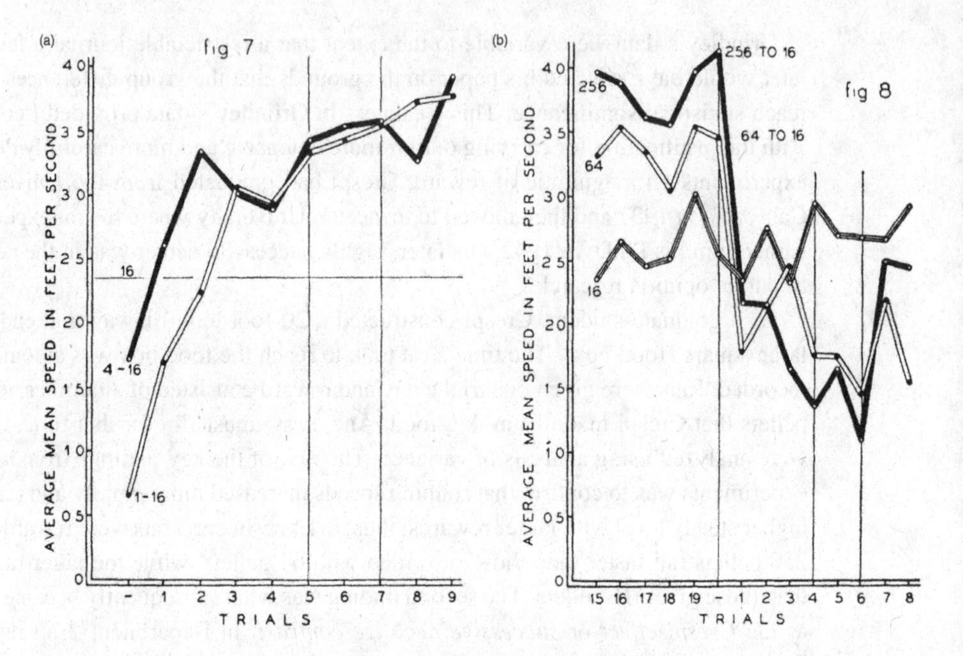

Figure 2.3 (a) shows data indicating an 'upshift,' whereby groups of rats initially trained on 1 or 4 pellets ran faster down a runway when shifted to a 16-pellet reward than a group trained throughout on 16 pellets. (b) shows data indicating a 'downshift' – or *Successive Negative Contrast* effect – in groups shifted from 256 or 64 pellets to a reward of only 16 pellets. From Crespi (1942). Copyright by the Board of Trustees of the University of Illinois. Reproduced with permission from the University of Illinois Press.

Hull now extended his 1931 concept of a fractional antedating goal response – the FAGR – to account for shifts in the magnitude of reward. He ignored Crespi's claim that this does not work: "First, one is forced to assume that there are different goal-response to different quantities of incentive as well as to different qualities. This ad hoc extension greatly strains the concept. Second, it does not explain the *elation* effect, i.e. increases in amount of reward."[41] Thus, now central to Hull's theory is the idea that incentive is represented by FAGRs; after receiving a large number of pellets as reward on several trials, the vigor of a rat's 'masticatory movements' and presumably amount of salivation at the start of a trial is greater than if it had previously received only a small number of pellets and this results in greater energization of the habit, which is now a sequence of S-R connections integrated by the FAGR into a single action, that of running down the runway.

In Hull's 1943 book, the basic equation expressing how vigorously an animal would make a learned response to a stimulus (the "reaction potential" of a habit, ${}_s\boldsymbol{E}_r$) was, according to Postulate 7[42]:

$${}_s\boldsymbol{E}_r = f({}_s\boldsymbol{H}_r) \times f(\boldsymbol{D})$$

where ${}_s\boldsymbol{H}_r$ represents "habit strength," the strength of the connection between the stimulus and the response, and $\boldsymbol{D}$ represents the strength of the current drive.

Nine years later, the equivalent equation had become more complex, in that the reaction potential, ${}_{s}E_{r}$, is now a function of two further factors. ***V*** represents the general finding that responses to intense stimuli are stronger than those to weak stimuli, *stimulus intensity dynamism*. More importantly, ***K***, represents *incentive motivation*. Thus, Postulate 8 of 1952 reads:

$${}_{s}\boldsymbol{E}_{r} = \boldsymbol{D} \times \boldsymbol{V} \times \boldsymbol{K} \times {}_{s}\boldsymbol{H}_{r}$$

In the Preface to his 1942 book Hull singled out Kenneth Spence as making the greatest contribution to the development of Hull's *Principles*. Nine years later Hull paid a similar tribute to Spence: "through his unfailing interest in and understanding of the problems here discussed, and through criticisms, suggestions, and relevant experiments which he and his students have performed, (Kenneth Spence) has contributed to a degree feebly recognized by these few lines."[43] In recognition of Spence's major contributions, it became common to refer to the 'Hull-Spence' theory.

One issue that divided the two theorists was whether the relationship between drive and incentive was multiplicative, as Hull suggested in his Postulate 8, shown above, or whether, as Spence suggested, the relationship was additive, *D+K*. The attempt to decide between these claims generated many runway studies. As late as 1965, this issue was the concern of a review paper in a top journal. As the author noted, a problem preventing any easy resolution was how to relate *D* and *K* to actual experimental manipulations.[44] For example, is the level of a hunger drive (*D*) to be measured terms of how many hours since an animal was last fed or in terms of the percentage of body weight that the animal has lost?

The hypothetical, never-measured FAGRs were then used to explain a range of phenomena that for many had previously demonstrated the fundamental inadequacy Hull's system. These ranged from: Koehler's account of animals' ability to solve detour (*Umweg*) problems and of chimpanzees' solutions to various problems on the basis of insight[45]; later experiments showing that rats can work out how to reach a goal box by integrating two separate learning experiences[46]; the finding that chimpanzees need earlier experience of poking objects with sticks before they could spontaneously use a rake to reach a banana that was otherwise out of reach[47]; and even the report that rats would maintain responding on a ratio schedule that was progressively increased from a stage in which the rat needed to make just eight lever presses to obtain a food pellet to a stage at which 193 such responses were required.[48]

An important example of the scope of Hull's late theories is provided by the latent learning effect. Using the complicated maze shown in Figure 2.4, Hugh Blodgett placed his rats in a start box and removed them as soon as they reached the designated goal box. One of his three groups always found food in the goal box, whereas a second group never did. The third and critical group was given no food for the first 10 trials but then found food in the goal box for the remaining trials. Whereas this last group had previously taken as long a time to reach the goal box as the never-rewarded group, within a trial or two from the introduction of reward they were reaching the goal box and making as few errors, that is, entries into blind alleys, as the group that had been rewarded throughout.[49] Blodgett and, his advisor, Tolman, viewed the latter result as

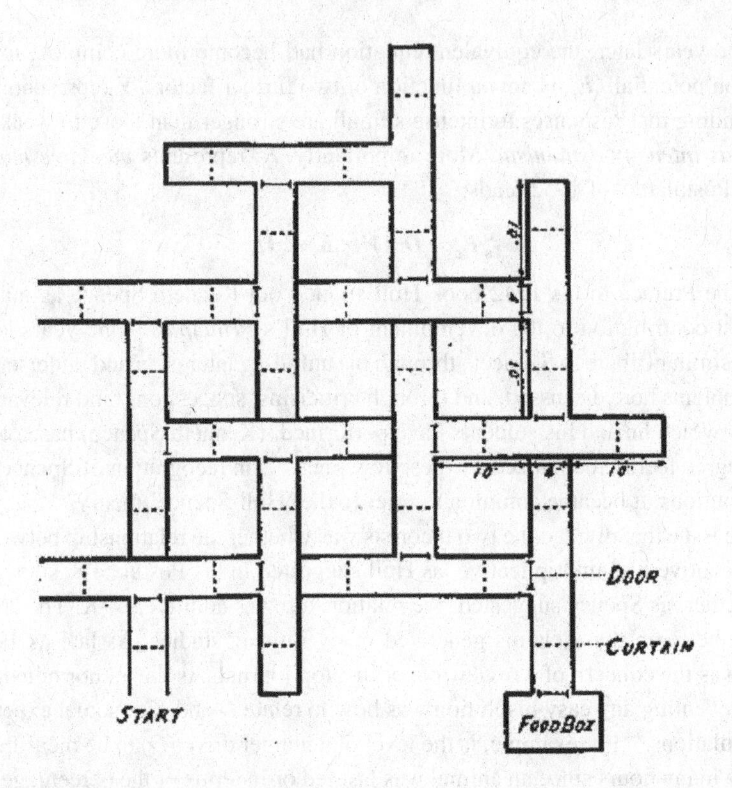

Figure 2.4 Maze used in Tolman's Berkeley lab by Blodgett to study latent learning and by Elliott to study incentive shifts.
From Elliott (1928). Public domain.

an example of learning without reinforcement and believed that the introduction of a food reward simply motivated this third group of rats to utilize what they had already learned about the spatial features of the maze.[50] Following an earlier suggestion by his collaborator, Kenneth Spence,[51] Hull instead claimed that rats in the shifted group had acquired the habit of running from the start to the goal box on the basis of the weak reinforcement provided by being removed from the goal box and returned to their home cage; the provision of food provided incentive based on a FAGR to change from a weak reaction potential to a strong one.

What impressed many of his contemporaries was that in many cases Hull's analyses of such phenomena made testable predictions, whereas that was rarely true of appeals to insight or to 'cognitive maps' within an animal's head. On the other hand, it was not easy to make direct tests of the claims made for fractional antedating goal responses and stimuli or for the appeal to very weak reinforcing events that were seemingly just as effective in establishing habits as strong reinforcers. Furthermore, as detailed by the Oxford-based psychologist, Tony Deutsch, there were major problems with Hull's claim that FAGRs can explain the results of reasoning and latent learning experiments.[52] Subsequently, Deutsch replaced Hull's S-R model with his own S-S model to explain the same set of phenomena.[53]

The Partial Reinforcement Extinction Effect: Amsel vs. Capaldi

Thorndike had always rewarded his dogs and cats for escaping from their puzzle box and never tested what happened to the habits they had acquired if reward was consistently withheld. For many years no study of maze learning examined the effect of no longer rewarding a rat when it reached the goal box. In contrast, the study of extinction had been a central topic for Pavlov from the very start of his research on conditioned reflexes. On reading Pavlov's 1927 book, Hull knew his system had to contain a theory of extinction.

As noted earlier, in his *Principles* of 1943 Hull proposed two mechanisms to account for extinction. One was called *reactive inhibition*, I_r, this supposed that, whenever a response occurs, it generates a process that tends to stop it recurring. The other was termed *conditioned inhibition*, ${}_sI_r$, and was described by Hull as a "habit of not responding." Experiments on what was termed *latent extinction* suggested that performance of a response was not needed for extinction of a maze habit; simply exposing a rat to an empty goal box several times reduced the speed with which it ran through a maze.[54] Much more serious for Hull's account of extinction was a conceptual analysis that concluded that the combination of Hull's two types of inhibition predicted that "not only should the learning curve inevitably decline to its starting point with continuous reinforcement but, in fact, learning should be impossible altogether."[55]

A further confronting challenge to Hullian theory was posed by the *partial reinforcement extinction effect (PREE)*. To take a standard example, if a hungry rat is trained to run the length of a straight runway for 20 trials but receives a food reward on only 10 of those trials, when extinction is introduced by removing all rewards, the rat will persist in running for many more trials than a rat that was rewarded on every trial. A simple S-R-reinforcement theory might suppose that the strength of the S-R connection – the *habit strength* of running down the runway – would be greater following 20 reinforcements than after only 10.

Various theories of the PREE were put forward from the 1950s onwards and for several years it became one of the most popular topics among those studying learning. Many theories were shown either to be wrong or to apply only to a limited range of situations. This left a general theory termed *generalization decrement*. This claims that the degree to which animal continues to respond under extinction depends on the similarity between the extinction conditions and those experienced during training. "A partial reinforcement schedule … establishes associations or responses that are less disrupted by the conditions of extinction."[56] Two rival accounts made quite different claims about what the critical "associations or responses" might be. The battle between them dominated research on the PREE during the 1960s. One influential version of generalization decrement theory was firmly within the framework of Hullian theory. It was proposed by Abram Amsel.

Amsel (1922–2006) was born in Montreal. He majored in psychology at Queens University in Kingston, Ontario, and then obtained a Masters degree in 1946 from McGill University that was based on a study of rote learning of nonsense syllables. He then enrolled as a Ph.D. student at Iowa University, with Kenneth Spence as his advisor. There he switched from studying human learning to learning by rats. In 1948 he was awarded his Ph.D.; his thesis reported experiments that supported Hull's theory of drive stimuli. On obtaining his Ph.D., Amsel obtained an academic appointment at Tulane University in New Orleans where he remained until moving to the University of Toronto in 1960. He remained there for nine years, an important period when Toronto's Department of Psychology was particularly vibrant, notably in human memory research. Meanwhile, in 1964 Spence had left Iowa to become a professor at the University of Texas where he died in 1967. Amsel was appointed a full professor there in 1969, a move that could be seen as recognition as one of Spence's most important successors. Amsel remained at Texas until he retired in 1999. Like many prominent learning theorists of his generation, he was combative and hypercritical, at least in his academic life.[57]

In 1958 Amsel gave a full account of his frustration theory, as it become known. The first paragraph stated: "This paper is based on the proposition that an adequate theory of instrumental behavior must involve three types of goal event: (a) Rewarding events – usually the presence of stimuli which evoke a consummatory reaction appropriate to some condition of deprivation; (b) Punishing events – noxious stimulation at the termination of a behavior sequence; and (c) Frustrative events – the absence of or delay of a rewarding event in a situation where it had been present previously." He pointed out how little attention had been given to the third category, "frustrative events," or to "frustration as a primary, aversive motivational condition."[58]

In fact, he had been giving such events a great deal of attention in experiments he carried out with his students at Tulane. Many of those were concerned to show that frustration energized behavior. His 1958 paper summarized these results and also suggested how his theory could explain phenomena such as the PREE. In brief, this idea was that during partial reinforcement training, the animal experiences frustration on non-rewarded trials but nonetheless continues to respond and is occasionally rewarded for doing so; this results in frustration becoming a stimulus for responding; during extinction, frustration is again experienced and again evokes responses, whereas for animals that have previously experienced only consistent reinforcement the frustration they experience in extinction is new in this context.

This was not the language that Amsel used. Instead, he used the language of fractional anticipatory responses, defining a new category as *fractional anticipatory frustration* (r_F-s_F). "When Ss are switched to extinction, partially reinforced Ss have been trained to respond in the presence of antedating frustration stimuli, whereas the consistently reinforced Ss have not."[59]

Particularly convincing evidence for the basic idea came from a series of studies from Amsel's Toronto lab in the late 1960s. These 'slow response' experiments were mainly run by his student, Michael Rashotte. Thus, in one of the first experiments of this kind five hungry rats were trained to run down both a 5-foot black runway and a 5-foot white runway for food pellets delivered in the goal boxes. They were given five trials each day in each runway. In the white runway, they were trained to take their time to reach the goal box by delivering pellets only if they took more than five seconds to arrive. This was a difficult task for the rats; the proportion of trials in which they received a reward ranged from only 6% to 50% of trials and they acquired “idiosyncratic (superstitious) behaviors” when in the white runway. In the black runway, there was no restriction on timing and the rats were rewarded on every trial. The most interesting results were obtained when extinction conditions were introduced for both runways. Then the rats began to display their individual superstitious patterns of behaving in both runways. According to Rashotte and Amsel, “it could be argued that, for each S, the low percentage reward in the (white) runway promoted conditioning of s_F to a specific form of the approach response and that, in extinction, r_F-s_F was evoked in the CRF runway and acted to elicit that form of the approach response to which it had earlier been conditioned.”[60] In other words, the rats had learned a specific response to frustration in the white runway that was expressed when they experienced frustration in the black runway for the first time.

Following this early experiment, several slow-response studies were carried out using improved equipment and with control, groups added. These confirmed the initial results.[61] In the meantime, a serious rival to frustration theory had been developed by John Capaldi.

Egidio John Capaldi (1928–2020), was born in Philadelphia, majored in psychology at La Salle College and in 1952 enrolled for a Ph.D. at the University of Texas at Austin, long before either Spence or then Amsel arrived there. Capaldi’s advisor was Hugh Blodgett, who as a student in Tolman’s lab had run the influential latent learning experiments described earlier (see Figure 2.4). After obtaining his Ph.D. in 1955, Capaldi remained as an assistant professor at the University of Texas until 1969, the year Amsel arrived there. Capaldi was then appointed as a professor in the Department of Psychological Sciences at Purdue University where he remained until retiring in 2011.

A 50% partial reinforcement condition can be arranged in various ways. The simplest is to arrange that rewarded trials (R-trials) alternate with non-rewarded trials (N-trials). Alternatively, runs of, say, three N trials might alternate with runs of three R trials. Or a more random sequence of R and N trials can be used. Capaldi’s early experiments demonstrated that the sequence of R and N trials during training was an important factor in determining resistance to extinction.

One of Capaldi’s early experiments compared three groups of rats that were trained to run down a straight-alley runway to end up in either a goal box that contained reward or an identical box that never contained any reward. The N_0 group was rewarded on every trial, the N_1 group was rewarded on alternate trials, and in the N_3 group sequences of three rewarded trials alternated with three non-rewarded trials. Although both these last two groups received 50% reinforcement, only the N_3 group showed the PREE over the three-day extinction stage;[62] see Figure 2.5.

In 1966 Capaldi described his *sequential hypothesis* based on many experiments of this kind: “The central assumptions of the hypothesis are that nonreinforced trials occasion a specific and distinctive internal stimulus and that this stimulus is progressively modified by successive nonreinforcements.”[63] Capaldi was later hailed by some as a pioneer of the later ‘cognitive’ wave in animal learning research. His theory can be described as claiming that during partial reinforcement training, the memory of a sequence of non-rewarded trials accompanies a response that is at last rewarded; thus, early in extinction, the animal expects that the sequence of non-reward trials will similarly be followed by a rewarded trial. This would be consistent with the Tolmanian background of his Ph.D. advisor. This was not, however, the language that Capaldi used in 1966. Instead, his sequential hypothesis was couched in Hullian-resonant terms such as 'habit strength.'

Capaldi did not suggest that the sequence of non-rewarded trials was just one more factor contributing to the PREE. He insisted that this was by far the most important factor. His theoretical paper contains only a passing reference to Amsel, implying that frustration made at best a minor contribution to the PREE under some conditions.

In 1957 Allan Wagner had reported experiments in his Master's thesis from the University of Iowa that Amsel cited as supporting his frustration theory. In 1964, now at Yale University, Wagner was co-author of a report of an experiment showing that the effect of non-reinforcement could be similar to that of electric shock, a result that was beyond the scope of Capaldi's theory.

In the first stage of this experiment, three groups were trained to run down a straight runway. Group C ("consistent") were given a food reward on every trial. Group N (for "non-reinforcement") was the traditional partial reinforcement group that was rewarded on 50% of the trials, using varying sequences of rewarded and non-rewarded trials. The unusual group, Group P for "punishment," were rewarded on every trial but in addition, they received a shock when they arrived in the goal box on 50% of the trials, with the intensity of the shock starting low and gradually increasing over days. After eighteen days of training, six days of testing were introduced. For half the rats in each group, the test conditions were simple extinction. This revealed the typical PREE, whereby Group N extinguished much more slowly than Group C. The new result was that Group P also extinguished more slowly than Group C. This last effect was small.

A much stronger effect was found in testing for resistance to punishment. This involved delivering both food and a shock on every trial. Rats that had been trained with 50% reward continued to run quickly down the alley, while those trained with 100% reward steadily slowed down. The authors commented: "If there is more than a conceptual similarity between the emotional responses of fear and anticipatory frustration, it would be reasonable to expect some degree of transfer of behaviors learned in the presence of one to occasions when the other is aroused.... In this context, the present findings of a partial transfer between the learned resistances to punishment and extinction would argue for a degree of commonality between the two emotional responses."[64]

Analyses of the PREE in the late 1960s could be seen as a battle between supporters of Capaldi and those of Amsel. Years later one of Amsel's graduate students during that period described the atmosphere. "In the lab, Amsel often characterized the competition with Capaldi in terms that we imagined were a carryover from his being educated at Iowa in the postwar years. Theoretical 'camps,' 'skirmishes,' 'attacks,' 'defeats,' and 'victories' were all part of the discussion. It seemed that the stakes were high. It was a great motivator, and great fun for the students."[65]

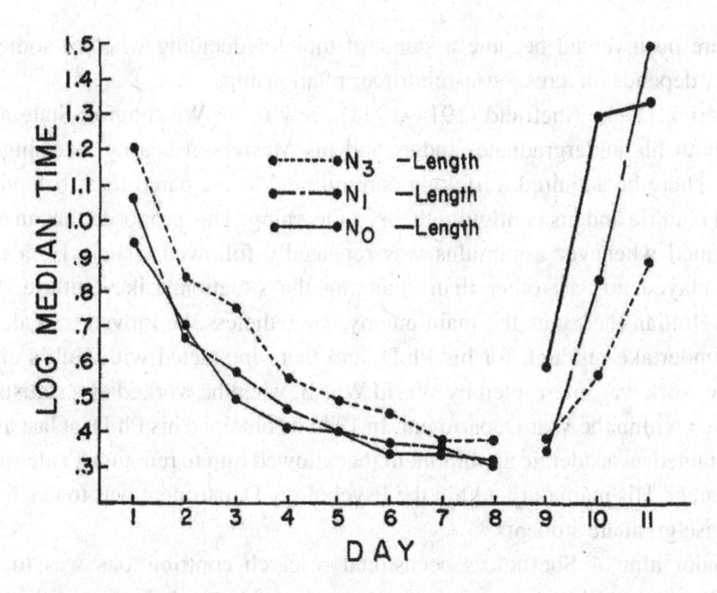

Figure 2.5 In Capaldi's experiment rats were given eight days of training in a straight runway, followed by three days of extinction. Rats given exposure to runs of three non-reinforced trials during training (Group N_3) were slowest to extinguish.
From Capaldi (1964). Reproduced with permission from the American Psychological Association.

Mackintosh's 1974 detailed review of research bearing on theories of the PREE concluded that Capaldi's sequential theory provided the best account of the bulk of the evidence but that frustration also played some role.[66] From the mid 1970s research in Amsel's lab at the University of Texas concentrated on developmental studies of learning using very young rats. A postdoctoral researcher, Jaw-Sy Chen, was the primary contributor to these experiments.[67]

The Omission Procedure and Visceral Learning

As noted at the start of this chapter, when in 1929 Hull described his "functional interpretation of the conditioned reflex," he launched two false beliefs. One was of little importance: the implication that Pavlov used a bell as a CS. The other belief was of much greater consequence, namely, that Pavlov's dogs were salivating because they were rewarded for doing so, just as Thorndike's dogs and cats had obtained food when they escaped from a puzzle box.

Within a few years convincing arguments for treating Pavlovian conditioning and instrumental learning as distinct types of learning had been made by Konorski, Skinner, and others (see Chapters 1 and 8). These were widely accepted, except by followers of Hull. Indeed, believing that Pavlov's dogs salivated because they had been rewarded for doing so became almost a mark of a true Hullian. It took until 1965 before a dedicated anti-Hullian provided direct evidence that this was not true. The procedure he invented became a standard tool for deciding whether some learned behavior depends on a response-reinforcer relationship.

Frederick Duane Sheffield (1914–1994) grew up in Washington State and completed both his undergraduate studies and his Masters at nearby Washington University. There he acquired a lifelong commitment to the pared-back behaviorism of Edward Guthrie and his contiguity theory of learning. This proposed that an S-R habit was formed whenever a stimulus was repeatedly followed closely by a response; reward played no role other than changing the situation. Like Guthrie, Sheffield viewed Hullian theory as the main enemy. Nonetheless, he moved to Yale University to undertake research for his Ph.D. and there interacted with Hull's circle. His graduate work was interrupted by World War II, when he worked as a statistician and researcher within the War Department. In 1946 he obtained his Ph.D. at last and a year later obtained an academic appointment that allowed him to remain at Yale for the rest of his career. His main duty within the Psychology Department was to teach statistics and advise graduate students.[68]

A major aim of Sheffield's occasional research contributions was to disprove Hull's theory that all primary reinforcers were based on the reduction of drive stimuli, which in turn were based on some physiological need. One set of experiments verged on the bizarre: male rats that ran down a runway and climbed a hurdle to gain access to females were allowed intromission but were then prevented from copulating. The results suggested "that elicitation of a prepotent consummatory response, rather than drive reduction, is the critical factor in reinforcement of instrumental responses."[69] Other experiments were straightforward: Access to a saccharin solution worked as an instrumental reward even though it reduced no drive and no biological need.[70]

Sheffield's idea for an experiment involving what he named as an *omission procedure* came out of studies he had carried out on avoidance behavior. Almost as an afterthought, he realized that "an additional appeal of the experiment, however, was that it offered an indirect opportunity to test the single-principle law of effect interpretation of classical conditioning, which suggests that hidden Thorndikean operations account for Pavlovian phenomena."[71] The procedure consisted of a modification of Pavlov's salivary conditioning preparation in which a tone signaled that a dog would shortly gain access to food, as long as the dog did not salivate when the tone was present; or – to put it another way – food was omitted on trials when salivation was detected. Thus, the dog was not rewarded – in a 'Thorndikean' sense – for salivating.

The first problem was the technical one of setting up Pavlovian salivary conditioning as almost no American learning researcher had done. Once two students, including David Williams, had acquired the necessary surgical skills and other technical challenges had been met, the experiment could, at last, be started. The results were patchy. Those obtained from one dog, Vicki, were what Sheffield had hoped for: "We regarded the fact that she maintained a fairly stable level of 50% responding during 800 trials as strong evidence against the operation of the law of effect in the case of the salivary response."[72] A second dog, Belle, was not so obliging: she did not show any conditioned salivation to the tone used as the CS even when trained in a standard Pavlovian manner.

In the chapter in which Sheffield described the omission procedure, he referred to tests with more dogs. But their data were never reported. While Sheffield himself seems to have lost interest, perhaps dismayed that the huge effort had yielded so little, others soon ran with the idea.

Setting up Pavlovian conditioning of licking by rats proved a lot easier than working with salivation by dogs. Unlike Sheffield, Patten and Rudy introduced the omission condition from the very start: their rats continued to lick on the majority of trials in which a 3-second stimulus was followed by access to a sucrose solution only if they did not lick.[73] An even more important result was obtained by David Williams, when he and his wife, Harriet Williams, applied the omission procedure to what became known as autoshaped pecking; if grain is delivered soon after a response key is illuminated, pigeons will start to peck at the illuminated key. The Williamses discovered that pigeons continued to peck at a response key, when it was illuminated, on a sizeable number of trials even though the delivery of grain was omitted on trials when the pigeons did peck the key;[74] see Chapter 6.

In a few ways, Neal Miller's (1909–2002) career was similar to that of Sheffield. He also obtained his bachelor's degree at the University of Washington where he was strongly influenced by Guthrie. In 1932, after obtaining a Masters degree in experimental psychology from Stanford University, he enrolled for a Ph.D. at Yale where he joined the group around Hull in the Institute of Human Relations. Here, unlike Sheffield, he abandoned any adherence to Guthrie's theories and instead embraced Hull's theories. In 1936 Miller was appointed as an instructor at Yale and in 1941 as an Associate Professor. During World War II, as a Captain in the Army Air Corps, he contributed to improvements in the selection and training of pilots, before returning to Yale, where he became a full professor in 1950. In 1966 he moved to Rockefeller University.

Miller was enormously productive from his Ph.D. years onwards. For over 15 years he maintained a fruitful collaboration with the sociologist John Dollard that resulted in a series of books that interpreted both human behavior in general and Freud's theories of abnormal behavior in terms of Hullian drive theory. In the late 1930s, Miller's collaboration with Hobart Mowrer in research on avoidance learning led to the idea of fear as an acquired drive and the invention of the shuttle box, a device for studying escape and avoidance learning that led to important theoretical advances (see Chapter 4). In the 1950s Miller became increasingly involved in what later became known as behavioral neuroscience, exploring the effects of various drugs and of various kinds of brain lesions. The latter work led to important discoveries concerning the neural basis of feeding. By the time he moved to Rockefeller University Miller had become one of the leading neuroscientists of his era and probably enjoyed what was then among the greatest incomes from grants to support the experiments in his lab carried out by postdoctoral staff and numerous Ph.D. students. Newly arrived postdoctoral researchers learned to call him 'Mr. Miller.'[75]

Miller did not drop his early interest in learning as new areas of research were explored within his lab. He was not impressed by the two-factor theory of avoidance learning that his previous collaborator, Hobart Mowrer, proposed. This theory was based on the assumption that Pavlovian conditioning and instrumental learning involve different processes (see Chapter 4). In 1952 Miller saw no need to abandon Hull's claim that all learning takes the form of habit acquisition via 'stimulus-response-reinforcement.' "Until rival theories of learning are developed to at least the primitive level of current reinforcement theory, they are not likely to displace it."[76]

Ten years later Miller reported a particularly interesting study of the conditions under which a stimulus becomes a secondary reinforcer. This succeeded in confirming the hypothesis "that in a situation in which there is more than one stimulus predicting primary reinforcement, e.g., food, the more informative stimulus will be the more effective secondary reinforcer."[77] This experiment anticipated a number of studies reported a few years later that were inspired by the same basic idea and that laid the groundwork for associative learning theories in the 1970s (see Chapter 9). However, Miller did not follow up on this study. Instead, in the period before he moved to Rockefeller University, the conditioning of visceral responses became Miller's main interest.

The then general view was one that, for example, Skinner had proposed in 1938, namely, that such responses are subject only to Pavlovian conditioning. This view went with the claim that the few studies claiming to have obtained instrumental conditioning of such responses were flawed; a standard criticism was that the researchers had not ruled out the possibility that the conditioning procedure had changed some skeletal response which had an unlearned tendency to affect the recorded visceral response.[78]

Miller believed this to be an important issue. "The problem of whether or not visceral responses are subject to instrumental learning (also called operant conditioning or trial-and-error learning) has basic significance for both the theory of psychosomatic symptoms and the theory and neurophysiology of learning."[79] He also had thought of a way of overcoming the standard criticism leveled at previous research of this kind. Part of the solution was to paralyze the rats by injecting curare to prevent any skeletal responses from occurring, while keeping them alive by artificial ventilation.[80] But how to deliver positive reinforcement to a paralyzed animal? The answer was to use a recent discovery made in another lab but one that Miller's lab had then investigated: Electrical stimulation of certain brain structures that acted as a powerful positive reinforcer.

The first experiment to use this combination of paralysis and brain stimulation was carried out by a Yale Ph.D. student, Jay Trowill. While his rats were paralyzed for about three hours, he arranged that some of them were rewarded by brain stimulation for decreasing their heart rate and others received such stimulation when their heart rate increased. The effects were small but significantly different from changes in heart rate recorded in yoked control subjects. Trowill concluded "that fairly strong evidence for, but not definitive proof of, the instrumental learning of a visceral response in a curarized animal has been secured."[81]

In 1965 Leo DiCara obtained his Ph.D. from New York University and obtained a postdoctoral position in Miller's lab. A year later he joined Miller in the newly set up lab at Rockefeller University. The project he was assigned was to improve on

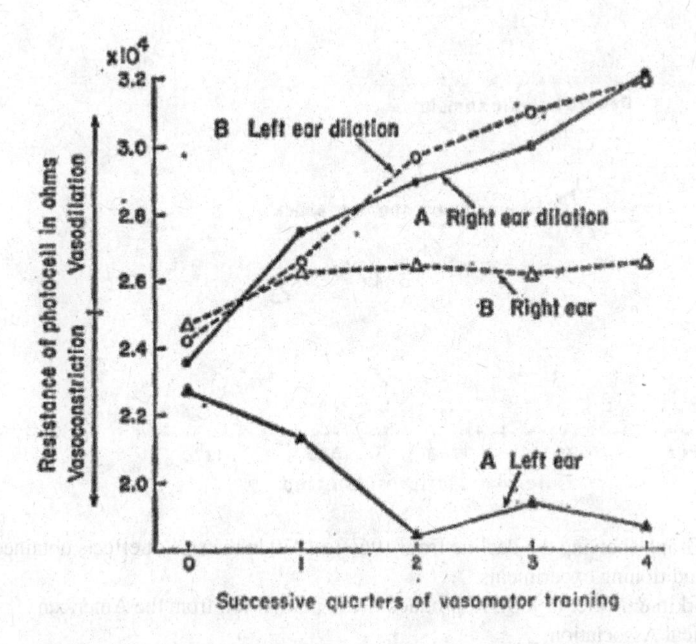

Figure 2.6 Data from an experiment on vasomotor conditioning, in which paralyzed rats received positive brain stimulation. In Group A this was given after an increase in blood flow through the right ear and for Group B after such an increase in the left ear.
From DiCara & Miller (1968). Reproduced with permission from the American Association for the Advancement of Science.

the procedures used by Trowill and obtain more definitive evidence for instrumental conditioning of a variety of visceral responses. A key person in Miller's Yale lab was a senior technician, Dr. Eleanor (Ellie) Adair. At Yale, DiCara relied on her help to operate and maintain the complex equipment used in his first experiments. Ellie did not move to Rockefeller.[82] Nevertheless, during his five years there DiCara's project blossomed.

He began by obtaining more substantial changes in heart rate by introducing a shaping procedure – reinforcing first small changes, then ever larger ones – and other refinements in the methods that Trowill had used.[83] Subsequent experiments expanded the range of visceral responses that were subject to instrumental conditioning to include colon motility, gastric motility, gastric blood flow, and urine output.[84]

A particularly dramatic example was conditioning of an increase in blood flow into one of a rat's ears, while no change occurred in the other; as shown in Figure 2.6. DiCara and Miller concluded from these and other results that "the specificity of learning of a variety of visceral responses demonstrated in these experiments makes it difficult to salvage the strong traditional belief that the instrumental learning of visceral responses is impossible by trying .... to explain the visceral changes produced by training as the indirect effects of somatic learning mediated by the cerebrospinal nervous system."[85]

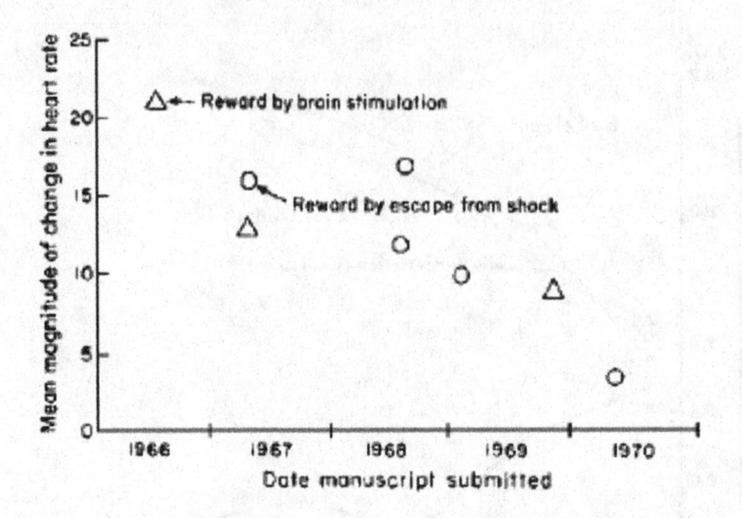

Figure 2.7 Graph showing the decline from 1966 to 1970 in the size of effects obtained from visceral conditioning experiments.
From Dworkin & Miller (1986). Reproduced with permission from the American Psychological Association.

Positive results of this kind continued to be published until a new Ph.D. student, Barry Dworkin, failed to get the results he expected. It prompted him to examine the size of the effects obtained in the many experiments on conditioning of heart rate that had been obtained in Miller's lab in the period from 1966 to 1970. He constructed an unusual graph showing how these sizes changed over these years. As shown in Figure 2.7, the graph showed a steady downward trend towards no effect at all after 1970.

After obtaining his Ph.D. from Rockefeller in 1973, Dworkin obtained an appointment at Pennsylvania State University. There, in collaboration with Miller, he carried out experiment after experiment, using over 2,000 rats, in an attempt to pinpoint what procedural details of the early experiments had contributed to their success that may have been changed in later experiments. To no avail. "We have been unable to reliably demonstrate visceral learning in acutely paralyzed rats in spite of an intensive replication effort, and in the 12 years since the last of the original publications appeared, there has been no report from any laboratory of consistent replication of any one of these experiments .... it is concluded that the original visceral learning experiments are not replicable and that the existence of visceral learning remains unproven."[86]

Well before this report was published, in 1976 DiCara, at the age of 39, shot himself, and in 1981 Trowill, at the age of 44, also died suddenly. His death was also rumored to result from suicide.[87]

As for Miller, his involvement in conditioning of visceral response had been related to his strong interest, dating back to his youth, as to how psychological states affect the body. That interest had first taken him to Freud's theories. In the 1960s there was great fascination in the West with the reported feats of Indian yogis in being able to control bodily functions such as heart rate. Miller became increasingly interested in biofeedback, defined as "the process of gaining greater awareness of many physiological functions of one's own body, commercially by using electronic or other instruments, and with a goal of being able to manipulate the body's systems at will."[88] In 1980 Miller retired from Rockefeller University but, unwilling to end his involvement in research, obtained a voluntary position as a Research Associate at Yale. He remained active, mainly in studies related to health psychology and medical treatments, until he died in 2002 at the age of 91.

Miller and Sheffield were colleagues within the same department for many years. Consequently, Miller must have been aware of Sheffield's use of an omission contingency and the data that appeared to undermine Miller's belief that instrumental learning provided the basis for salivary conditioning. However, I have never seen any reference that Miller made to Sheffield's or others' use of an omission contingency. Perhaps Miller was skeptical of data from a single dog?

Hull's Legacy

In 1975 a review of research on incentive motivation concluded that "it has become increasingly apparent that .... the theory of incentive motivation proposed by Hull and Spence was fundamentally unsatisfactory."[89] As for the concept of *fractional anticipatory goal responses (FAGRs)*, for Spence, these remained potentially observable responses such as mouth movements, while by 1965 for Miller these had become largely unobservable central events; as such, they were indistinguishable from Tolman's expectancies.[90] By the 1970s, FAGRs were essentially forgotten by researchers investigating animal learning.

So, from the movement that Hull launched and that dominated psychology in the USA for three decades, what has been learned about how animals learn, what they learn, and the conditions under which learning takes place? Not a great deal from Hull himself. He never saw himself as an explorer who would make new discoveries about learned behavior. He tried to make sense of what was known and, where there were gaps, suggested that others might run the necessary experiments. On the other hand, many of those he inspired directly in the 1940s and 1950s, notably Kenneth Spence and Neal Miller, were highly productive experimenters. Spence's contributions and those of his students to the study of discrimination learning are described in Chapter 7.

Miller's major contributions were in areas of behavioral neuroscience that are not recounted here; the failure of his project on instrumental conditioning of visceral learning was an unusual exception in his career. According to Google Scholar in 2021, the long list of Miller's publications shows the 20 most cited to include four of the books written when he was still a member of the group around Hull in the late 1930s and 1940s. One was *Frustration and aggression*, published in 1939.[91] Of the experimental papers, only two in the top 20 were related to animal learning: both were on his early studies of fear as an acquired drive. The remainder were on topics in behavioral neuroscience or medical science.

Both Miller and Spence trained and inspired a large number of researchers who became equally productive. Amsel and Wagner, whose research on partial reinforcement and frustration is described in this chapter, provide just two examples from this generation.

One could argue that more new phenomena were discovered and in the long run greater understanding of animal learning obtained by those who set out to prove that Hull was wrong than by his followers. Examples of this in the present chapter include the analysis of incentive shifts and contrast effects and the discovery that animals can remember previous sequences of rewards and non-rewards. Notably, Tolman's and his students' rejection of Hull's claim that all learning consists of the acquisition of habits led to the study of spatial learning, as described in Chapter 3.

A more abstract contribution of the Hullian movement was to increase rigor in claims about what animals learn. Rigor in terms of conceptual thinking; it was no longer acceptable to make vague appeals to common sense when criticizing some Hullian claim. Also, rigor in terms of well-designed experiments and appropriate statistical analyses of the data they produced.

No one since Hull has attempted a general theory of both learning and behavior, certainly not one that aspired to emulate Newton's transformation of physics. It seems almost certain that no one with a shred of plausibility ever will.

3 Learning Where Things Are and Where Events Happen

The history of twentieth century research on spatial learning in animals consisted of four phases. The first phase took place in the two decades following the First World War, mainly in Berkeley, California, where Edward Tolman and his many students carried out a highly innovative series of experiments on maze learning in rats. Tolman's claim that this involved a different kind of learning from that studied in conditioning experiments inspired by Pavlov triggered the second phase, a furious decades-long controversy over what came to be seen as a central issue in psychology. When this controversy was widely regarded as resolved in favor of Tolman's critics by a definitive review by Restle[1] in 1957, a third phase was initiated, one of neglect of the topic. The fourth and final phase began in the 1970s when spatial learning research was reinvigorated, largely as a result of research on the hippocampus that eventually led to the award of a Nobel Prize in 2015.[2]

A different line of research blossomed in the 1970s that, although conceptually related to studies of spatial learning, was a quite separate development. These studies were concerned with the questions of what an animal learned about its surroundings – the context – when subjected to a conditioning procedure and of what role such learning played in the outcome of such an experiment.

Tolman's Early Studies

In Chicago in the very early years of the twentieth century, John Watson studied how his rats learned to find their way through a maze to the food awaiting them in the goal box. In doing so he both established a research tradition that endured well into the middle of the century and a theoretical tradition by claiming that what his rats learned were a series of stimulus-response connections. A particularly influential example was what became known as the 'kerplunk' experiment. Rats were first trained to run up the first alley and then turn left into a second alley and, once they had been well trained, the position of the left-turn was moved to further up the first alley. Watson reported that many of the rats turned to their left at the former position of the second alley and bumped into the wall.[3]

Some years later, accompanied by a new graduate student, Karl Lashley, Watson returned to the Florida Keys to continue his project on the migratory behavior of noddy and sooty terns. The two researchers made very little progress in understanding how the terns were able to find their nesting area on a particular island from a starting point many hundreds of miles away. However, once arrived in their nesting area, how the birds found their particular spot seemed less mysterious. When local landmarks were moved around, rather as Watson had previously changed the layout of a rat maze, the terns ended up in the wrong place. Watson and Lashley concluded that the terns were stimulus-response learners just like rats.[4]

In 1915 Edward Tolman obtained his Ph.D. from Harvard University. Previously he had been an undergraduate at the Massachusetts Institute of Technology, where reading William James' *Principles of Psychology* convinced him that his future lay in psychology rather than in chemistry, physics, or mathematics. During a study visit to pre-war Germany, he had become enthusiastic about Gestalt psychology. In 1918, 32-year old Tolman was appointed to what turned out to be a lifetime position at the University of California, Berkeley, where he set up what became a remarkably productive laboratory for studying how rats learned to find their way through a bewildering variety of mazes. His Gestalt background contributed both to his complete rejection of Watson's stimulus-response theories and to the way he thought about the findings that he and his many students obtained.[5]

The aims behind many of these experiments were, first, to demonstrate that rats learn *where* to go in a maze, rather than what responses to make at a given point, and, second, that they learn what to *expect* in a particular place. One example of the first kind of experiment used a T-maze that could be filled with water either to a depth that allowed rats to wade their way to the end of the arm that contained food (on a raised shelf) or filled to a depth that required the rats to swim. Half were trained under the 'wade' condition and then tested when they had to swim; for the others, this was reversed. Both groups performed well on test.[6] For Tolman, this was conclusive proof against Watson's "old kinesthetic doctrine of maze learning."[7] Instead of "muscle twitches," Tolman argued that behavior needs to be defined at a molar level – "turning left" or "approaching the light," for example – and this was an argument that was generally accepted from then on, even by those who dismissed most of his other claims.

Far more controversial were studies of *latent learning*, starting with the experiment reported by Blodgett that was described in Chapter 2.[8] To recap briefly, the experiment contained three groups of rats that were exposed once each day to the complicated maze shown in Figure 2.4. The critical group was one in which for the first 10 days there was no food, just like the second group; but from the 11th day onwards these rats found food in the goal box, just like the first group. Although for the first 10 days, the critical group took much longer to reach the goal box than the food-rewarded group and frequently entered arms that had dead ends, almost immediately after the 11th, and rewarded, trial the critical group performed as well as the group that had been rewarded from the outset. Tolman argued that the function of the food reward was to provide an incentive for the rats to convert what they had learned during the first 10 days into performance.

Experiments designed to demonstrate that rats acquire *expectancies* of what to find in a particular place were later called *incentive shift* experiments. The early studies of this kind that were carried out in Tolman's laboratory all consisted of *downshifts*, such

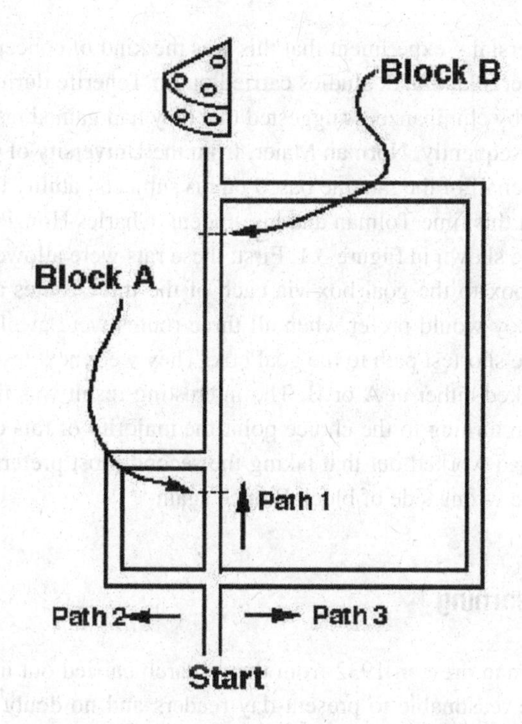

Figure 3.1 Maze used to test for insightful problem-solving in the rat. The critical evidence came from a test in which a rat, previously trained to take all three paths to the goal box, started on its path (the shortest, Path 1) but then discovered a block at position B; on returning to the "crossroads," would it then take (inappropriately) what had been the second most preferred route (Path 2) or show understanding of the problem and choose the least preferred, but now appropriate, route, Path 3?
From Tolman and Honzik (1930). Public domain.

as those in which a well trained rat, one used to find food when it reached the goal box, suddenly found nothing. The rat's immediate and vigorous searches in the place where food had previously been found were proof, as far as the experimenters were concerned, that it had expected to find food.[9]

As described in Chapter 2, these studies followed ones involving less severe down-shifts, namely, ones in which rats were trained on a highly preferred reward and then shifted to a less preferred one. In such experiments rats trained to traverse a maze to obtain bran mash as a reward found sunflower seeds instead; the seeds were ignored as the rats searched the goal box, even though normally the rats would readily eat such seeds.[10] In an incentive shift study that for once involved a species other than the rat, a monkey was trained to upturn a bucket to obtain the banana that was hidden inside; once trained, it reacted with fury when finding a lettuce leaf instead, even though the latter was an acceptable food under normal circumstances.[11]

In the 1932 book, *Purposive behavior in animals and men,* in which Tolman summarized these and many other studies, he was not entirely clear as to what kind of knowledge a rat acquired about the maze in which it was trained. He did not introduce the term 'cognitive map' until 1948. Nevertheless, it seems from one particularly impressive – and controversial – experiment that this was the kind of concept he had in mind. Wolfgang Koehler's landmark studies carried out in Tenerife during World War I of problem-solving by chimpanzees suggested that they had gained insight into the use of tools[12] and, subsequently, Norman Maier, from the University of Chicago, made a similar claim on behalf of the rat, one based on his subjects' ability to solve a spatial problem.[13] Around this time Tolman and his student, Charles Honzik, trained and tested rats on the maze shown in Figure 3.1. First, these rats were allowed to find their way from the start box to the goal box via each of the three routes and were then tested as to which they would prefer when all three routes were available; not surprisingly they chose the shortest path to the goal box. They were next tested when this direct route was blocked either at A or B. The interesting result was that when the block was at B, after returning to the choice point the majority of rats chose the longest route, as if they had worked out that taking the second most preferred route would only get them to the wrong side of block B once again.[14]

Place vs. Response Learning

The conclusions that Tolman drew in 1932 from the research carried out in his laboratory might seem quite reasonable to present-day readers and no doubt to many of his contemporaries. However, they were anathema to the neo-behaviorists led by Hull who had come to dominate the study of animal learning in the 1930s; (see Chapter 2). For them, the language Tolman used either reeked of anthropomorphism – terms like 'purposive' and 'expectancy' – or included hyphenated strings of Gestalt-influenced terms such as 'sign-gestalt-expectation' and 'means-end-field' that appeared as impediments to developing a purely mechanistic account of behavior. In a typically disarming Preface to his book, Tolman had written: "I wish now, once and for all, to put myself on record as feeling a distaste for most of the terms and neologisms I have introduced. I especially dislike the terms *purpose* and *cognition* and the title *Purposive behavior*."[15]

More to the point, researchers influenced by Hull found it easy to explain away some key findings, such as latent learning, and discovered that others were difficult to replicate. The latter problem often turned out to hinge on methodological details of a particular experiment. As Tolman himself had emphasized, whether a rat showed 'insight' when tested in the maze shown in Figure 3.1 depended on the use of elevated runways that allowed a clear view from the start box of the whole maze and its surroundings; no such behavior was shown unless these conditions were met. Many of the failures to replicate his results used enclosed mazes with high walls that allowed only a limited view of the world outside. In any case, the pigmented (hooded) rats used in the Berkeley laboratory had a clearer view of the world than the albino rats with limited vision used by many of Tolman's critics.[16]

Sometimes the Hullians set out to test what were seen as implications of Tolman's theory. A particularly interesting example is provided by a series of studies of what became known as *irrelevant incentive* learning. The first in this series, by one of

Tolman's fiercest critics, Kenneth Spence, reported results that in fact supported Tolman's ideas. Rats were first allowed to explore a Y-maze when fully sated with both food and water; during this preliminary training food was placed at the end of one arm and water at the end of the other. The only reward for reaching the end of an arm was for the rat to be picked up and returned to the company of its cage mates. Parenthetically, one suggested explanation for latent learning results of the kind reported by Blodgett in 1929 was that a rat was rewarded for reaching the goal box by being returned to its home cage. But, back to the Y-maze experiment: On test, the rats were either made hungry or thirsty beforehand. The results showed that when hungry, these rats tended to choose the arm that had previously led to food, and, when thirsty, they chose the one that had led to water. This difference was even found to be statistically significant; an early example of the application of statistical analysis to psychological data.[17]

This initial irrelevant incentive experiment was followed by several such experiments, many of which failed to confirm the original finding. In one such experiment rats were initially trained when thirsty to reach a cup of water at the end of the arms of a T-maze and at the end of one the arms they needed to climb over a pile of rat chow to reach the cup. When later tested when hungry, they showed no sign of having learned which arm contained food.[18]

In the early 1940s, while other laboratories attempted to confirm or explain away earlier results from Tolman's laboratory, Tolman himself concentrated on a phenomenon, *vicarious trial and error* or *VTE,* a term coined to describe the "hesitating, looking-back-and-forth sort of behavior that rats can often be observed to indulge in at a choice-point before actually going one way or the other."[19] A series of VTE experiments led Tolman to conclude that "the animal's activity is not just one of responding passively to discrete stimuli, but rather one of the active selecting and comparing of stimuli."[20] In terms of resolving the theoretical differences between Tolman and his critics, this VTE research seems less than critical.

The Place-vs-Response Debate

What were much more important were two experiments run by two particularly able students of Tolman, Benbow Ritchie and Donald Kalish, who shared their mentor's commitment both to the theory of cognitive maps and to the causes of civil liberties during the McCarthy era. The aim of the first of their experiments, subtitled 'orientation and the short-cut,' was to demonstrate that, even if rats were trained to make an initial left turn in order to eventually reach the goal box, as shown in Figure 3.2a, when that route was blocked and an opportunity was later offered to head directly to the food box, they would take it. The magnificent 'sun burst' maze available on test is shown in Figure 3.2b. What will be discussed later is the circle behind the food boxes marked 'H'; this designates a desk lamp with a five-W bulb that provided the only illumination during experimental sessions. Of the 56 female pigmented rats used in the experiment, 36% headed towards the food boxes on their

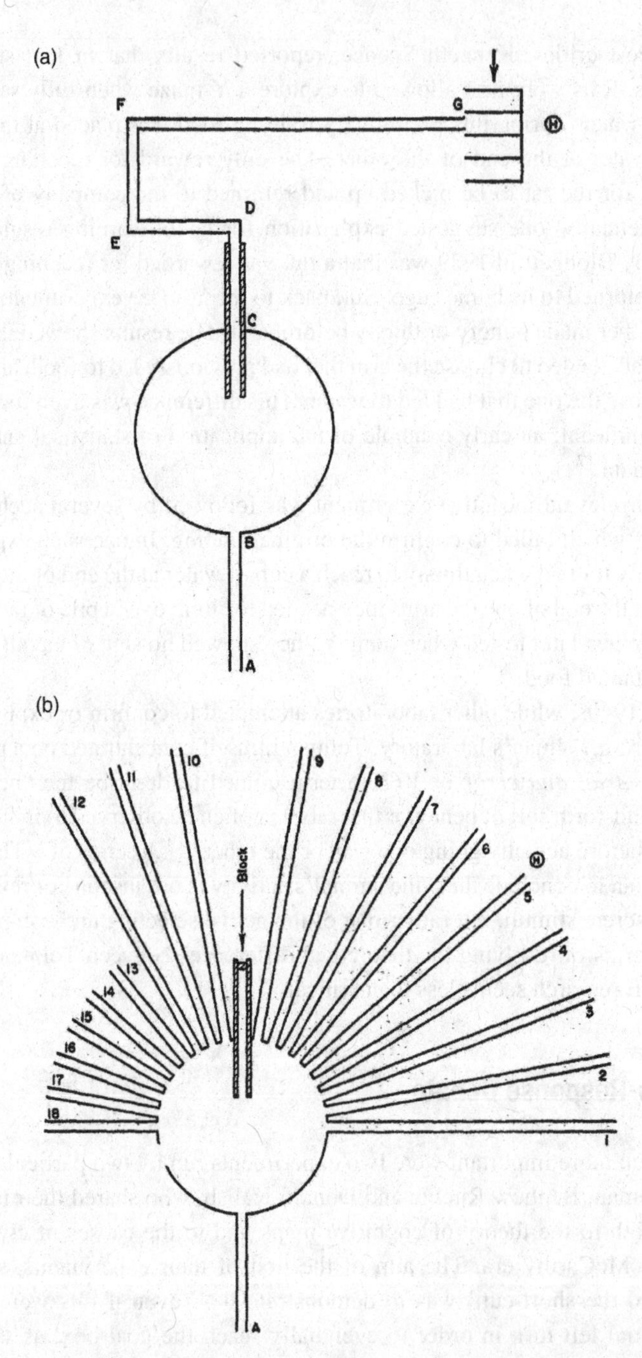

Figure 3.2 Sunburst maze. Rats were first trained in the simple maze shown in (a); this involved making a turn to the left on leaving the central platform. The rats were then tested in the full sunburst maze shown in (b). Would they show habit learning and tend to choose a left arm or place learning by choosing an arm pointing towards the goal box?
From Tolman, Ritchie, and Kalish (1946a). Reproduced with permission from the American Psychological Association.

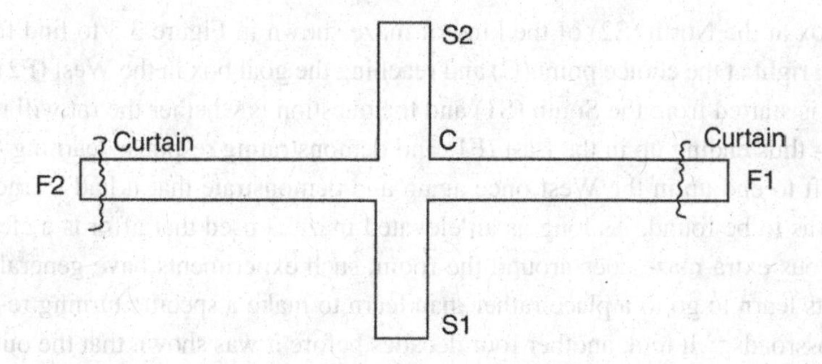

Figure 3.3 Of the many different mazes built in Tolman's lab, the only one that continues to be used is this "Cross" maze. During the training stage, a rat, when started in the "south" from S1, learns to turn right at the crossroads (C) to reach the goal box in the "east" (F1). On test, the rat is started from the north (S2), a turn to the left at the crossroads indicates "place learning," whereas a turn to the right suggests "S-R" or "habit" learning. From Tolman, Ritchie, and Kalish (1946b). Public domain.

first test trial. The authors concluded that these rats had learned the location of the food box, whereas the others had perhaps not had enough training. They were somewhat defensive, and not entirely convincing, when rejecting the possibility that all that the 'good' 36% had learned was to run towards the light, a response that was rewarded during training.[21]

It is likely that the sunburst maze shown in Figure 3.2 was the first and last to be constructed. In contrast, the much simpler maze reported next by the same researchers was used in the same essential form for decades to follow. This elevated maze took the form of a cross. As shown in Figure 3.3, goal boxes were placed at the east (F1) and west (F2?) ends of an eight-foot runway, while a rat could be started at either the south (S1) or the north end (S2) of the shorter cross arm. The eight rats in the *Response-Learning* group were given six trials each day: On three of these, they were started at S1 and could find food only if they turned right at the crossroads to end up in F1, since access to F2 was blocked, while on the other three trials, these rats were started at S2 and needed again to turn right, since food was now available only in F2. The rats in the *Place-Learning* group were also given six trials per day, with three starting at S1 and three starting at S2. However, for three of them, food was always to be found in F1, and for the other five food was always in F2. In the *Response-Learning* group after 12 days of training, a total of 72 trials, only four rats of the eight reached the criterion of 10 successive trials without errors (i.e., turning the wrong way at the crossroads), while in the *Place-Learning* group, all eight from the start box reached the criterion within eight trials. Thus, there was very strong support for the authors' conclusion that "in situations where there are marked extra-maze cues, place-learning is simpler than response-learning."[22]

This last experiment triggered literally dozens of experiments on the place-vs.-response issue over the next few years. A simpler and long-lasting version of the plus-maze procedure is one in which a rat is first trained, for example, with the start box at the North (S2) of the kind of maze shown in Figure 3.3 to find food after turning right at the choice point (C) and reaching the goal box in the West (F2); on test, the rat is started from the South (S1) and the question is whether the rat will turn right again – thus ending up in the East (F1) and demonstrating response learning – or now turn left to end up in the West once again and demonstrate that it had learned *where* food was to be found. As long as an elevated maze is used that affords a clear views of various extra-maze cues around the room, such experiments have generally found that rats learn to go to a place rather than learn to make a specific turning response at the crossroads.[23] It took another four decades before it was shown that the outcome of such an experiment also depended on how much training a rat was given: After eight training trials rats with an intact hippocampus showed place learning on test, but after sixteen training trials they tended to demonstrate response learning.[24]

One of the most influential of Tolman's students was Henry Gleitman. One of his doctoral experiments foreshadowed the *reinforcer devaluation* procedures that were developed almost three decades later as a powerful analytical tool, one that has been used extensively ever since. The hungry rats in Gleitman's experiment were first trained to find food at the end of both arms of a special T-maze. This was unusual in that the goal boxes at the end of each arm were quite different in shape and color, but both had metal grid floors and were detachable from the maze. In the following nine-day training phase, the rats were transferred to a different room where the detached goal boxes had been placed on tables; each day they received food as usual after entering one of the goal boxes, but were given an electric shock after entering the other. For the final test rats were returned to the original training room where the T-maze had been reassembled; at the choice point 22 out of the 25 animals chose to enter the arm leading to the safe goal box. Tolman and Gleitman concluded that "the animals acted as if they remembered that the bright baffle-containing box was on the left, and the dark hurdle-containing box on the right" and that the results provided a clear demonstration of latent learning.[25] However, they were forced to admit – possibly in response to a Hullian reviewer of their journal submission – that the result might also be explained in terms of "extremely hypothetical notions" based on fractional anticipatory goal responses, as proposed to explain the results of a somewhat similar experiment.[26]

After obtaining his Ph.D. Gleitman obtained a number of teaching positions, ending up in Philadelphia in 1953, where his lectures on introductory psychology at the University of Pennsylvania had an enormous impact on many of the students there. Thus, he can be seen as bringing Tolman's Californian view of psychology to the East Coast. Gleitman's lectures formed the basis for an introduction to psychology[27] that was the most impressive and influential textbook of its generation. It had reached its eighth edition by the time he died in 2015.

In 1952 a prominent Hullian theorist came to the conclusion that trying to answer the question of what a rat learns in a maze is to enter a "theoretical blind alley"; on the basis of an argument that one hopes was easier to follow in that era, he proposed that whether rats learn "cognitive maps" or "stimulus-response associations" is a pseudo-problem.[28] In an amusing reply, titled 'The circumnavigation of cognition,'

Ritchie suggested that the same kind of argument would have reduced the fifteenth century debate over whether the world is round or flat to a pseudo-problem.[29]

The most influential version of Kendler's argument came from Frank Restle's major review of this literature.[30] His 'resolution' of the place-vs-response question was to suggest that it was all a matter of what kind of cue controlled the rat's behavior. If there was a prominent extra-maze cue close to the goal box, such as the light by the goal box in the sunburst maze shown in Figure 3.3 or the rat's home cage, a particularly effective cue, then the rat would learn to approach such a cue and thus show place learning. On the other hand, in the absence of salient extra-maze cues, intra-maze cues would become associated with turning responses.

With the advantage of decades of hindsight, Restle's conclusion can be seen as a pseudo-resolution. Tolman's proposal that rats acquire a 'map' of a maze implied more than that they can learn to approach certain extra-maze cues or avoid others. However, he does not appear to have spelled this out. The general opinion among learning theorists in that still behaviorist era was that, following Restle's resolution, Tolman's cognitive maps and his claims that behavior is goal-directed and that "there is more than one kind of learning"[31] could now be forgotten. The now venerable idea, that the only kind of learning was the acquisition of habits based on S-R connections, lived on, except that now the 'R' could include 'approach S,' as Hull had suggested in 1952, the year he died. Tolman outlived his long-term theoretical opponent for a few more years. By the time he died in 1959 few researchers shared Tolman's belief that "everything important in psychology (except such matters as the building of a super-ego, that is everything save such matters as involve society and words) can be investigated in essence through the continued experimental and theoretical analysis of the determiners of rat behavior at a choice point in the maze."[32] In contrast to the Skinner box, the shuttle box, and the single runway, for almost 20 years following Restle's 1957 paper, mazes were generally abandoned as a tool for studying how animals learn.

Maps vs. Routes

The most famous neurological patient of the twentieth century was HM. He suffered from very severe epilepsy and, to treat this, brain surgery was carried out that invaded his hippocampus, a part of the brain that had previously received relatively little attention.[33] This produced amnesia of "the purest form" ever studied to that date; HM was said "to live in the eternal present" for the rest of his long life.[34] This and related studies were carried out in McGill University in Montreal, Canada, and led to heightened interest in the hippocampus, among those who studied memory in rats as well as neuropsychologists.

As a probably unique example of electrophysiological data prompting a major change in behavioral research, John O'Keefe, working at University College in London after obtaining his Ph.D. from McGill University, reported that individual cells in the hippocampus of rats placed in a small arena responded only when the rat visited a particular place within the enclosure.[35] The report was a short communication based on "preliminary evidence" obtained using the new technology of recording from single cells in moving animals, but nevertheless was given a bold and provocative title, *The hippocampus as a spatial map*. This invited the inevitable criticisms that O'Keefe was over-interpreting preliminary data obtained under conditions that were not fully controlled. However, subsequent studies confirmed that the title was justified.[36] In 1970 O'Keefe had been joined by another McGill Ph.D., Lynn Nadel, who had just completed his Ph.D. dissertation on the hippocampus. Almost immediately they started together on what became a magnificent book that reviewed philosophical, psychological, behavioral, ethnological, and physiological theories and evidence concerning perception and memory for places.

The part of the book, *The hippocampus as a cognitive map,*[37] that probably most psychologists remember is where a distinction is drawn between *routes* and *maps*. Their argument was illustrated using the map shown in Figure 3.4. "How do I get from High Beach to Upshire?" a lost rambler might ask. One type of answer is to provide a route, consisting of a list of directions as to what landmark to approach or at what point to make a right or left turn, and so on. O'Keefe and Nadel pointed out that such information has the following properties: On the positive side, it can be quickly conveyed, it demands no particular skill but, on the negative side, it is completely inflexible, that is, of no use if the rambler decides to go to Waltham Abbey instead of Upshire or if he or she gets lost somewhere along the route or if part of the instructions are damaged or for someone who wishes to get *from* Upshire *to* High Beach. On the other hand, to give a map to the lost rambler – or to draw one on the spot – provides a great deal of flexibility, in that it allows reaching Upshire from any point on the map or, for that matter, from any place to any other place, and also can be joined up to other maps; the disadvantage is that to draw a map requires much time and effort and to use a map requires special skills.

The argument is not that a rat contains in its hippocampus something closely resembling the map shown in Figure 3.4. Rather, O'Keefe and Nadel argued that, no matter how the relationships between various points in an animal's environment are encoded in the brain, this representation has the same flexible properties as everyday physical maps. In doing so, they clarified the distinction that Tolman had failed to make clear between a rat acquiring a cognitive map of the maze to which it was regularly exposed and learning a route, whether one composed of turning responses or approach responses, to get from the start box to the food in the goal box.

By the time that *The hippocampus as a cognitive map* was published in 1978, mazes were already starting to reappear in animal laboratories. One that has been used extensively ever since 1976 is the elevated eight-arm radial maze shown in Figure 3.5. This was designed by David Olton, a few years after he obtained his Ph.D. from the University of Michigan and was now a relatively junior academic at the Johns Hopkins University; later he joked that the maze was his ticket to tenure. One possible source of inspiration for Olton was a study of the spatial memory capacity of young chimpanzees by Emil Menzel at the State University of New York, Stony Brook. The unusual method used by Menzel was to carry a chimp around a large compound while allowing it to observe where Menzel hid 18 food items. When a chimp was then

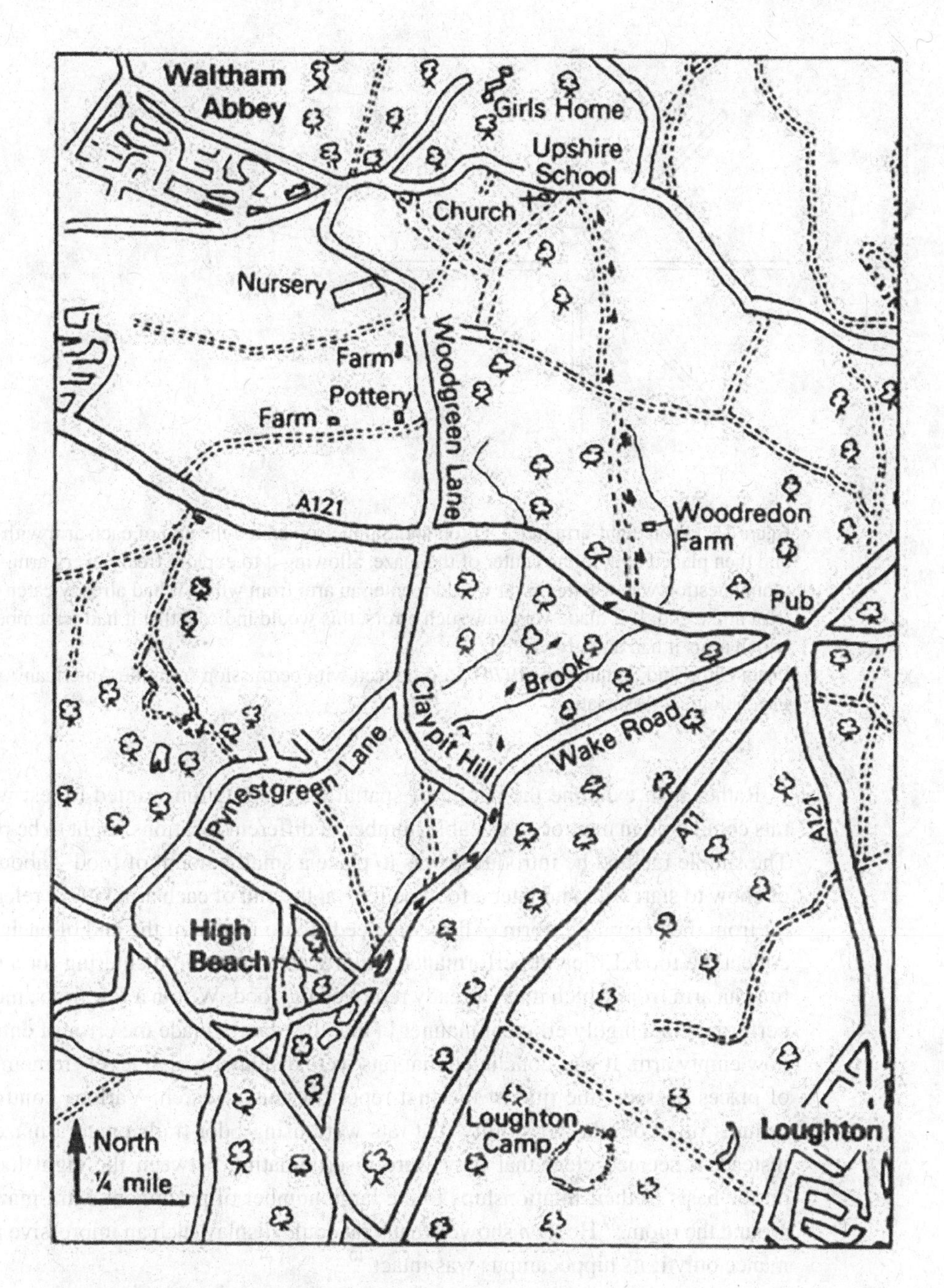

Figure 3.4 How to get to Upshire? By following a *route* defined by instructions on what to do at choice points and in relation to landmarks? Or by using a *map*?
From O'Keefe and Nadel (1978). Public domain.

released to search for these items, performance was usually impressive, locating over 12 items on average while following an efficient search pattern that suggested that it had remembered the majority of the hiding places.[38]

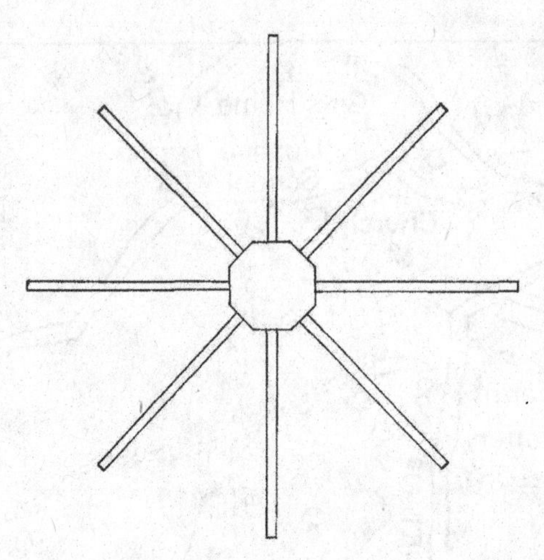

Figure 3.5 Olton eight-arm maze. Olton and Samuelson baited the end of each arm with a treat and then placed a rat in the center of the maze, allowing it to explore freely every arm. Their main question was whether a rat would reenter an arm from which it had already eaten the treat at the end. If it made very few such errors, this would indicate that it had remembered which arms it had already visited.
From Olton and Samuelson (1976). Reproduced with permission from the American Psychological Association.

Rather than examine the nature of spatial learning, Olton wanted to test whether rats could hold in memory a sizeable number of different locations, eight to be precise. The simple method he introduced was to place a small amount of food – about 0.1 g of chow to start with and later a food pellet – at the end of each arm before releasing a rat from the central platform. All the rat needs to do is to visit the end of each arm to collect the food. Efficient performance requires the rat to avoid entering for a second time an arm from which it has already removed the food. Within a few trials, most rats performed in a highly efficient manner in that they rarely made the error of entering a now empty arm. It was concluded that rats were displaying high level "remembrance of places passed," the title of the first report on this research. Various control procedures ruled out the possibility that rats were using odor trails or intra-maze cues; instead, it seemed clear that they were discriminating between the eight locations on the basis of their relationships to the large number of prominent extra-maze cues around the room.[39] He then showed that a rat could display such an impressive performance only if its hippocampus was intact.[40]

Most people who found themselves surrounded by a ring of cupboards each known to contain a $10 note would, after opening the first cupboard, move to the next one, and then the next, and so on, to move methodically round the circle. Such a response strategy requires no "remembrance for places past." When placed in an eight-arm maze some rats respond in this manner. Most zig zag to and fro across the maze in what appears to be far from a methodical response strategy. However, to rule out the latter possibility is not easy. What was still required in 1977 was a method that

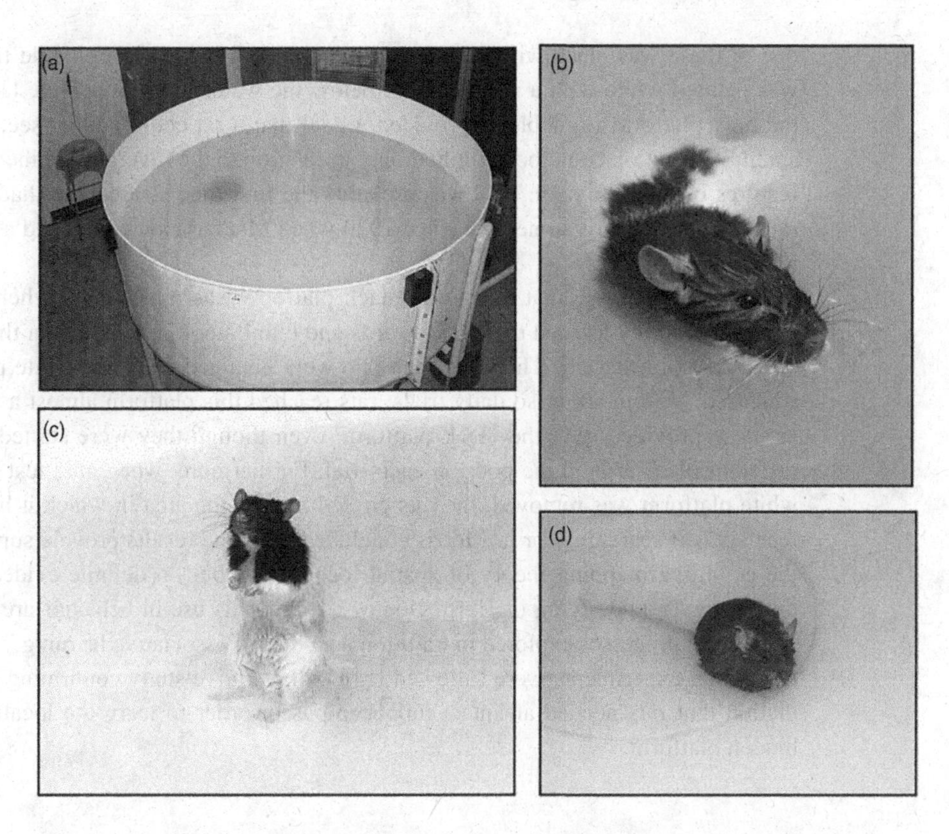

Figure 3.6 The Morris water maze tests whether a rat can locate a submerged platform on the basis of distant visual cues. As devised by Richard Morris (1981), the submerged platform cannot be seen or smelt by a swimming rat because the water is cloudy. Climbing onto the platform is rewarding. (a) The original maze used by Morris; (b) A rat searching for the platform; (c) A rat that has found the platform; (d) A mouse searching for the platform. From Morris (2008). Reproduced with permission from Richard G.M. Morris.

provided completely unambiguous evidence that a rat's behavior depended entirely on identifying a place on the basis of an array of distant cues.

One of the topics that O'Keefe and Nadel discussed in their 1978 book was the amazing ability of some Pacific islanders, Puluwatans, to travel vast distances in their canoes and arrive at their destination, some relatively tiny speck of an island. Possibly this or possibly his hobby of sailing may have influenced Richard Morris, but the immediate prompt for his development of the *water maze* were tanks containing various sea creatures in the Gatty Marine Laboratory at the University of St. Andrews in Scotland, where the newly appointed lecturer's laboratory was placed. The first water maze consisted of a circular pool, 130 cm in diameter, containing water to a depth of 40 cm that was made opaque by adding some milk. The water was heated to a temperature of about 26 degrees centigrade, a fairly comfortable temperature for a rat swimming in the pool. As suggested by Figure 3.6, a rat was unable to climb out of the pool but was able to climb onto one of two kinds of platform;

one of these was black with a surface 1 cm above the water level, while the other was painted white with a surface 1 cm below the water level. When the latter was the only platform available, it provided a goal that a rat could neither see, hear or smell; its location could be identified only in relation to the several prominent visual features beyond the pool. This was probably the first time a researcher had trained rats to swim towards some goal since 1930 when Macfarlane had flooded a T-maze with water.[41]

It was no surprise that, when the black platform was placed somewhere in the pool, rats rapidly learned to swim towards and climb upon it, whereupon they were removed and dried off. The important data were obtained when the white platform was used: Within six or so daily trials, rats reached this platform almost as rapidly as those provided with the black platform, even though they were started from a different place around the pool on each trial. Furthermore, when on a test trial the white platform was removed, the rats circled around the area in which it had been located as if searching for it. Morris concluded that "the results provide support for the cognitive mapping theory of spatial localization but no definite evidence that the processes underlying the formation of a map or its use in behavior are distinct from those processes explored in traditional studies of associative learning."[42] These two initial experiments were followed by a collaborative study confirming the prediction that rats needed an intact hippocampus in order to learn the location of a hidden platform.[43]

Concluding Comments on Cognitive Maps

Although a few researchers after 1981 questioned whether Tolman was correct in claiming that rats acquire cognitive maps of their environment, that is, a form of neural encoding that has the flexibility of a human-generated map, the evidence in support of Tolman has become overwhelming.[44] Since 1981 the eight-arm maze and the water maze have become standard tools for behavioral neuroscientists.

Learning Where Conditioning Takes Place

In his 1927 lectures Pavlov reported that, when dogs that were relatively new to conditioning experiments involving food were led from their kennels into the laboratory, they would begin to salivate. Thus, they had – in his terms – acquired a conditioned reflex to this environment. However, once some discrete stimulus had become strongly established as a conditioned stimulus, the dogs ceased to exhibit this environmental conditioned reflex.

His successors in the USA, especially those using electric shock in their experiments, must have been aware of their animals rapidly learning that conditioning took place in a particular setting. Somehow this form of place learning was rarely mentioned and, on the rare occasion when noted, it was regarded as of little interest.

Quite abruptly that attitude changed; in the 1970s there was an explosion of research on the role that the place – the environment or context – in which some conditioning procedure was carried out played in what was learned. The research and ideas described so far in this chapter appear to have had little influence on this sudden interest in the role of context. Instead, it was driven by the revolution in theoretical analyses of conditioning that took place in the late 1960s and that are later described in detail, in Chapter 9. These theories, most notably the Rescorla-Wagner model of 1972, gave an important role to associations between a context and the events that take place within that context. Several representative examples of early experiments on context conditioning are described in this section. All were carried out either as part of a doctoral research program or shortly after completion of a Ph.D. Context was a topic that seems to have had a special appeal to young researchers in the early 1970s.

A fundamental assumption in the Rescorla-Wagner model and related theories is that of competition between stimuli for associative strength, with the added assumption that a context can compete with the events that occur within it. The first direct test of this last assumption used measurement of a context's association, probably the first time this had ever been done. An ingenious idea of how to do so was that of a Ph.D. student at University College, London. John Odling-Smee placed a black chamber that could be used for fear conditioning within an open area that was painted white and brightly lit. Since the latter was mildly unpleasant for rats, when placed in the open area, they soon entered the dark chamber; in a pretest, the time a rat took to enter the chamber was recorded. During a subsequent conditioning stage that took place in the chamber when the door was closed, tones were occasionally sounded and shocks delivered. In a post-test that followed several conditioning sessions, a rat was again placed in the bright, open area with the door to the dark chamber now open once again and the rat's latency to enter the chamber recorded for a second time. The extent to which a rat's latency to enter the chamber exceeded its pretest latency was taken as a measure of its fear of the chamber, reflecting the strength of its association between this context and shock.

In these experiments, rats' fear of the chamber was found to be inversely related to how well during conditioning the tone had signaled when the shocks were delivered. Thus, when every shock was immediately preceded by a tone, rats trained under these conditions showed little subsequent avoidance of the chamber in the posttest.[45] On the other hand, when the tones had been poorly correlated with shocks or failed to signal precisely when a shock would occur, rats avoided the chamber in the posttest. Over a series of experiments of this kind, the results were entirely consistent with predictions from the Rescorla-Wagner model. Unlike the other young researchers mentioned here, on completing his Ph.D. and obtaining an academic appointment, Odling-Smee did not continue research on animal learning and instead became an evolutionary theorist.

Odlling-Smee's experiments are examples of overshadowing designs, ones that examine competition between the conditioning context and discrete – or punctate – events, like the tone, for association with some outcome. A major stimulus for the development of the new associative learning theories was Leon Kamin's discovery of the blocking effect, whereby initial exposure to, say, a tone-shock pairing subsequently produced little or no learning about a novel light stimulus when a rat was exposed to a procedure in which a compound stimulus, the tone and the light presented simultaneously, was immediately followed by a shock.[46] The first clear demonstration that prior conditioning of a context could block subsequent learning about the significance of a stimulus presented in that context was obtained from pigeons in an autoshaping experiment.

After serving in the US Army in Viet Nam, Arthur Tomie completed a Masters degree and Ph.D. at the University of Colorado that were based on a series of pigeon experiments. Tomie continued with similar research after obtaining his Ph.D. and on taking up an academic position at Rutgers University in 1975. There he expanded his study of what is known as the *US-pre-exposure effect*. The method used in his experiments took the following form. In an initial stage pigeons were given a series of unsignaled food deliveries in a conditioning chamber; in the following, autoshaping, stage these deliveries were signaled by a key-light that came on a few seconds beforehand. This procedure was one that rapidly produced vigorous pecking at the illuminated response key in pigeons with limited prior exposure to food deliveries in the conditioning chamber. In contrast, in the group that had received a large number of unsignaled food deliveries in the first stage, pecking at the key-light developed much less rapidly.

The evidence that this US-pre-exposure effect involved blocking came from experiments in which Tomie manipulated contextual cues. He did this, for example, by giving one group of hungry pigeons – the *Change* group – an initial stage consisting of 10 sessions in each of which these pigeons received 60 unsignaled presentations of a hopper full of grain in a chamber that had corrugated brown cardboard lining the walls. The next stage consisted of a standard autoshaping procedure for all groups in a chamber that was unlined: Sixty presentations of the hopper were again given in each session, but each presentation was preceded by 6-second illumination of the response key by a green light. Pecking at the green key developed as rapidly in this *Change* group as in a control group that had been pre-exposed to only a few unsignaled hopper presentations in the first stage. In contrast, a *No change* group that had received many unsignaled hopper presentations in the first stage, but in an unlined chamber, were slow to peck at the green keylight. Thus, the US-preexposure effect was found only when contextual cues in the initial stage were identical to those in the subsequent autoshaping stage. These and other, better controlled experiments by Tomie, strongly supported his claim that the US-preexposure effect is largely, if not solely, the result of blocking by contextual cues.[47]

These first two examples demonstrated how context associations can compete with excitatory conditioning to standard discrete stimuli. The next example involves inhibitory learning. Andy Baker completed a Ph.D. at Dalhousie University in 1974 and, before starting a long and productive academic career at McGill University in Montreal, spent two years as a postdoctoral researcher in the UK working with Nick Mackintosh at the University of Sussex. In the course of a series of experiments during his Ph.D. he found a result that was the opposite of what he had predicted. He discovered that *between-session negative correlations* could produce condition inhibition.

Baker arranged that rats were shocked on some days but exposed to a noise in the absence of any shocks on other days. When in a subsequent test the group given this training received a series of noise-shock pairings, these rats developed a fear reaction to the noise much more slowly than groups given various control treatments. This *retardation* test was followed by a *summation* test in which the noise was combined with a light that had previously been paired with shock; the noise was found to reduce the fear response that otherwise occurred when the light was presented on its own. Both tests confirmed that the between-session negative correlation had produced conditioned inhibition. Casually speaking, rats given this treatment learned that the noise signaled a period of safety in an otherwise dangerous environment.[48]

Baker reported further experiments that produced results similar to those found by Odling-Smee: If, during between-session negative correlation training, the shocks were signaled by a flashing light, then no inhibitory learning occurred to the noise. Thus, harking back to Pavlov's original report and consistent with Odling-Smee's results, the flashing light overshadowed conditioning of fear to the context and, because these rats did not fear the context, the noise did not become a safety signal.

Early studies of taste aversion learning by Garcia and his colleagues strongly suggested that, whereas rats readily associate a taste with subsequent illness, they are unable to associate sickness with the place in which they were made sick. In 1973, this suggestion was shown to be wrong by Phillip Best and his graduate students – one of whom was Mike Best (no relation) – at the University of Virginia. They used a version of what later become known, and widely used, as a *place preference chamber*. Their chamber consisted of two compartments of equal size, one painted black and the other white. Normally, when given a choice, their rats strongly preferred to be in the black chamber. However, a group that on two occasions had been placed in the black chamber for two minutes and on removal had been injected with an illness-inducing drug, apomorphine, avoided the black chamber in a subsequent test. Other groups that had been treated in the same way but had been allowed to drink a saccharin solution or water during the two conditioning trials, showed less avoidance of the black chamber on test. Best and his students concluded both that rats can readily associate a place with illness and that such learning can be overshadowed by taste cues.[49]

The US-pre-exposure effect described above for pigeons given an autoshaping procedure had also been found to occur for taste aversion learning: When rats are given a number of illness-inducing procedures prior to a taste – illness pairing, they display weaker taste aversions than rats given a taste-illness pairing when the illness-inducing procedure is novel. On obtaining his Ph.D. at the University of Virginia in 1971 a later student of Phillip Best, Jerry Rudy, obtained an academic position at Princeton University. Here he started to analyze the US-pre-exposure effect in taste aversion learning. In a series of experiments, he demonstrated that this effect occurred only if, prior to pairing saccharin with lithium chloride injection, his rats had been placed in a distinctive black chamber and injected with lithium. As long as the black chamber was novel at the time of the chamber-lithium pairings, subsequent saccharin-lithium pairings were less effective in producing an aversion to saccharin. Most of the results pointed to a simple explanation in terms of context blocking, similar to that proposed by Tomie (1976) to explain the US-pre-exposure effect in pigeon autoshaping. However, a major problem with this analysis was that conditioning an aversion to the black chamber attenuated subsequent saccharin aversion learning whether or not exposure to the black chamber preceded exposure to saccharin. This required a more complicated account that would be too much of a diversion to describe here.[50]

The research described up to this point established both that animals can learn that, when in a particular place, they can expect to receive food or electric shocks – a completely predictable finding for either a learning theorist or a layman – or expect to feel ill afterward – a less predictable finding – and, more importantly, that such learning can interfere with learning about the significance of events that occurred within that context or shortly after leaving the context. The final series of experiments described here led to the demonstration that contextual cues can acquire further properties.

Some of the earliest experiments on conditioned reflexes carried out in Pavlov's laboratory were concerned with extinction. After a stimulus such as the ticking of a metronome had come to elicit saliva as a result of being paired with meat powder, further presentations of this stimulus on its own led eventually to the stimulus failing to elicit any more saliva. Pavlov decided that the extinction procedure had not erased the original conditioned reflex but instead had come to inhibit its expression. Two phenomena prompted this belief: The ability of the metronome to elicit saliva was restored at least temporarily both after some time had elapsed since the extinction treatment, *spontaneous recovery,* and when some salient event occurred – such as the loud noise of a door being slammed – during an extinction session; *disinhibition* was Pavlov's term for the latter effect; see Chapter 1.

In the early 1970s Bob Rescorla was at Yale University and already recognized as one of the two foremost learning theorists of his generation. Together with two graduate students, Donald Heth and Chris Cunningham, Rescorla began the search for further factors that would *reinstate* an extinguished conditioned fear response. Their experiments were based on the idea that, during extinction, memory of the shock that had generated fear during conditioning faded to some extent, as if the rats came gradually to forget quite how frightening this experience had been. This account suggested that, if the rat was reminded of the shock, conditioned fear would reappear. A series of experiments confirmed this prediction.[51] For example, in one experiment initial training consisted of a single session in a conditioning chamber, in which two 2-minute presentations of a flashing light and two 2-minute presentations of a tone were each followed by an intense shock. Subsequent extinction and testing were carried out in a Skinner box where the rats had already been trained to press a lever for food. The extinction treatment consisted of four days in which the 2-minute tones were presented on four occasions in the absence of any shock. By the end of this treatment, there was no longer any sign of fear when the tone was presented, in that the rats continued pressing the lever for food at the same rate as when the tone was absent. The next day rats were returned to the conditioning chamber where the experimental groups received either two unsignaled shocks or two shocks each signaled by the flashing light. In the test that took place the next day these two groups displayed reinstated fear to the tones, while two control groups did not.[52]

Rescorla and his students were confident that they had ruled out the possibility that their reinstatement effect depended on context conditioning. This was challenged by a graduate student at the University of Washington, Mark Bouton, whose first series of experiments convinced him and his mentor, Bob Bolles, both that context conditioning played a major role in reinstatement of fear to a stimulus that had been extinguished and that decline in memory for events during conditioning sessions, as described by Rescorla, played an insignificant role in extinction.[53] Their second series of experiments had a greater and more enduring impact. These demonstrated what they termed the *renewal effect*. The following proved to be a key experiment. This involved three distinct contexts. The rats first received fear conditioning to a tone in one context, A. Fear of the tone was then fully extinguished by repeatedly presenting the tone on its own in either of the other two contexts, B and C. Finally, half the animals that had been given the extinction treatment in B were tested in C and vice versa, while the remaining rats were tested in the context in which their fear of the tone had been extinguished. On test, fear of the tone was found to be renewed in the rats tested in a novel context. Thus, extinction proved to be context-specific: The rats had learned that in Context B they had nothing to fear from the tone but, when shifted to the novel context, this could not inhibit their fear.[54]

In 1985, an edited book was published that surveyed the huge amount of research on the role of contextual cues that had been published over the previous 15 or so years.[55] As detailed in its first two chapters, this research had uncovered several ways, in addition to those described here, in which context learning could affect what an animal learned about events within that context and how such learning was expressed in the animal's behavior.[56]

4 Fear, Avoidance, and Punishment

The major theme of this chapter is the attempt to understand avoidance learning. In an experiment of this kind animal will experience some unpleasant event, usually a mild electric shock, unless it makes the designated response that will prevent this event from occurring. The first experiments of this kind were performed in St. Petersburg by Vladimir Bekhterev. The latter viewed such learning as equivalent to Pavlov's conditioned reflexes. It took two decades before it became clear that avoidance learning is a form of instrumental learning and quite distinct from Pavlovian conditioning.

When an animal makes a successful avoidance response, no event follows the response. Therefore, the question arose as to what provides reinforcement of the response. An answer to this question was provided by Hobart Mowrer whose two-factor theory is still influential today, both in laboratories and in clinics where clients with phobias receive therapy. Nonetheless, ways in which the theory is inadequate became clear from the 1950s onwards, notably following experiments performed by Richard Solomon and his many students. It was not until well into the 1970s that there was at last an account of how animals learn and maintain an avoidance response that has been broadly accepted ever since.

Mowrer's theory included the claim that fear could become a Pavlovian conditioned response. One of the minor themes of this chapter is how the study of fear conditioning blossomed in the 1960s, with major contributions again being made by Solomon's students. Some of them went on to study what became known as *learned helplessness*. One-way of measuring how afraid an animal is of a particular stimulus is the extent to which the occurrence of the stimulus interrupts the performance of some positively reinforced behavior or increases the vigor with which an avoidance response is performed. Such interactions between Pavlovian and instrumental behavior were first studied by B.F. Skinner and one of his first doctoral students, Bill Estes. Their work was in the context of a further minor theme of this chapter, punishment; this can be seen as the opposite of avoidance learning in that the aversive event is delivered only if the animal makes the target response.

From Bechterev's Dogs in a Harness to Guinea Pigs in a Circular Cage

In the early years of the twentieth century, Pavlov's major rival was Bechterev, a scientist who differed from Pavlov in many different ways. First, and most important in the present context, Bechterev was not trained as physiologist with surgical skills; instead, his general training in medicine in St. Petersburg was followed by specializing in psychiatry. After many years studying both psychiatry and psychology in France and Germany, he returned to Russia as professor of psychiatry at the University of Kazan. Bechterev eventually arrived back in St. Petersburg in 1893, when he was appointed as Professor of Psychiatry at the same Military-Medical Academy that provided Pavlov with an endless stream of students to run his experiments. While Pavlov's highly routine life was concentrated around his laboratory, Bechterev committed himself to a range of different activities, professional and political as well as academic, and had only sporadic direct involvement in his research projects. Despite their differences, Pavlov and Bechterev shared two strong beliefs: One was that psychology needed to be an objective science and the other was that the conditioned reflex provided the key building block for such a science.[1]

On learning of the conditioning experiments being carried out in Pavlov's laboratory, in 1907 Bechterev enlisted the help of V.P. Protopopov to set up his own research program on conditioning. Instead of measuring drops of saliva, as in most of Pavlov's experiments, in Bechterev's laboratory what was measured was whether or not a dog flexed a hind leg to a signal paired with shock to that leg. A crucial aspect of the procedure was that, since the electrodes were attached to the leg, shocks were delivered whether or not the dog flexed its leg to the signal. On the other hand, in similar experiments with human participants starting some three years later, if a subject made the designated response *before* the shock was due, no shock was delivered. Maybe this version of the procedure made it easier to find volunteers to take part in such experiments? It took till 1926 before a Russian researcher using dogs compared the two procedures, that is, unavoidable as opposed to avoidable shock, and found the latter procedure to promote faster and more stable learning of the flexion response.[2] Since the report of this study was in Russian, it is unlikely that it influenced similar studies undertaken a decade later in the USA.

Given the bitter rivalry between Pavlov and Bechterev, it is somewhat ironic that, when Pavlov's loyal disciples in the USA, Horsley Gantt, and Howard Liddell, set up conditioning laboratories, they generally used Bechterev's shock-based procedure. Their example was followed by an increasing number of researchers, predominantly psychologists, who, although inspired by Pavlov, lacked the surgical skills needed to perform experiments that measured salivation.

One early development by such a researcher that had long-term consequences was the invention of the *shuttle box* to study shock-based learning. Lucien Warner, a researcher at Columbia University, set out to measure "the association span of the white rat" using a method that "should be solvable by the humblest animals."[3] His equipment consisted of a chamber containing a floor through which an electric shock could be delivered and divided into two compartments, separated by a low fence, as shown in Figure 4.1. The procedure consisted of presenting a warning signal (WS) that was followed by a shock as long as a rat remained in its current compartment; by jumping over the fence to the other compartment the rat could either escape the shock or avoid it altogether. Intriguingly, his experiments were "carried out in the writer's

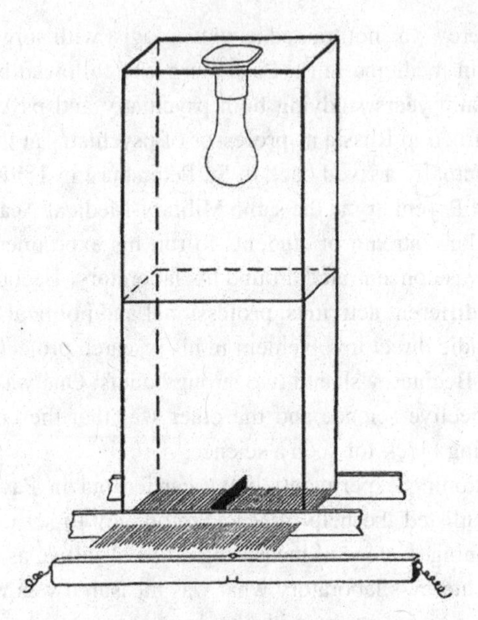

Figure 4.1 In the first study of avoidance learning to use a shuttle box a rat needed to learn to jump over the barrier when the light came on in order to avoid a shock.
From Warner (1932). Public domain.

private laboratory, located in the country at a distance from highways, tramcars, and other sources of vibration and sound."

Warner measured the 'association span' of his rats by arranging that the interval between presenting the WS, a 1-second buzzer, and the onset of a shock was varied across different groups of rats. Those given the shortest interval, 1-second, learned very rapidly to both escape and then avoid the shock; in contrast, not one of the 10 rats given the longest interval, 30-s between the buzzer and the shock, learned to escape, let alone avoid the shock. Warner noted that the response rats made to avoid the shock was qualitatively different from that involved in escaping the shock; he suggested that this presented a challenge to current theories of conditioning but did not seek to meet this challenge.

For researchers interested in sensory processes in animals, shock-based conditioning procedures could be very useful. Elmer Culler ran an animal laboratory at the University of Illinois that concentrated on measuring the hearing abilities of various mammals, including guinea pigs and dogs, as well as rats. As an improvement on the methods previously used in this laboratory, an enterprising graduate student, William Brogden, developed a new method for training their animals. He adapted an activity wheel, so that movement of the wheel could be measured when an animal placed within the wheel reacted to a shock or to a sound signaling that a shock was shortly to arrive.[4]

Using this equipment, a key experiment was later carried out using guinea pigs. For Group A the standard procedure for this kind of shock-based learning was employed, in that these guinea pigs could both escape the shock by turning the wheel a short distance and even completely avoid the shock by making this response during the

(a)

(b)

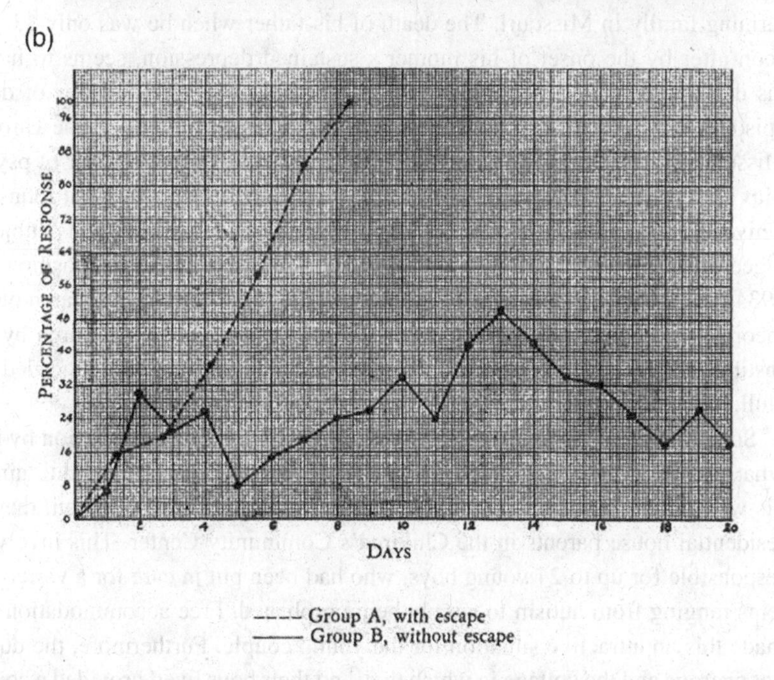

Figure 4.2 A guinea pig from Group A, when placed in the activity wheel (a), could avoid ("escape") a signaled shock if it ran in time. Guinea pigs in Group B ("without escape") were unable to either avoid or escape the shock. As shown in (b), the percentage of trials in which Group A ran increased steadily. This did not happen with Group B.
From Brogden, Lipman, and Culler (1938). Public domain.

2-second 1kHz tone that was used as an immediate WS. For Group B a strictly Pavlovian procedure was used, in that the WS was always followed by the shock, no matter how these animals responded. Twenty-five trials were given each day. The averaged results are shown in the figure below; within 5–13 days of such training the guinea pigs in Group A were performing perfectly, in the sense that they avoided the shock on every trial. In contrast, Group B made the specified response – turning the wheel a certain distance – on only a small percentage of trials, and their behavior essentially remained unchanged over 500 trials, as shown in Figure 4.2.[5]

Even though turning the wheel a short distance was both the Pavlovian unconditioned response and the avoidance response, Culler argued that these results demonstrated the inadequacy of Pavlov's claim that a conditioned response always resembled the unconditioned response: "The CR, in brief, is nature's way of getting ready for an important stimulus and its form depends on the 'incentive' provided by the unconditioned stimulus."[6] The full implications of this experiment later become much clearer, mainly as a result of the work of O. Hobart Mowrer.

Mowrer and the Early Development of Two-Factor Theory

Orval Hobart Mowrer (1907–1982) – usually known as 'Hobart' – was born to a farming family in Missouri. The death of his father when he was only 13, followed soon after by the onset of his mother's sustained depression, seems to have led to his decision to study psychology and also to have triggered a series of depressive episodes that recurred throughout his life. As an undergraduate at the University of Missouri Mowrer worked as a research assistant to its only professor of psychology, Max Meyer, an early behaviorist. Mowrer went on to obtain a Ph.D. at Johns Hopkins University, where his dissertation was on spatial orientation and his subjects were pigeons. He continued research in this area during two postdoctoral positions. Then in 1934 he obtained a fellowship at Yale University and started the research on learning theory that defined the rest of his career. Two years later he was hired by the Yale Institute of Human Affairs, where he joined the high-powered team headed by Clark Hull; see Chapter 2.

Soon after arriving at Yale Mowrer made his first major contribution by inventing what proved to be a highly effective application of Pavlovian conditioning. When his wife, Holly, also a psychologist, came to join him in New Haven, they became residential house parents in the Children's Community Center. This involved being responsible for up to 24 young boys, who had been put in care for a variety of problems ranging from autism to simply being orphaned. Free accommodation no doubt made this an attractive situation for the young couple. Furthermore, the duties were not onerous and the cottage in which they and their boys lived provided a comfortable and 'beautifully modernized' environment. There was just one problem. The place stank of urine. Around half of the boys were persistent bed-wetters.

Whereas the staff had become resigned to this problem, the Mowrers decided to find a solution. Both were undergoing psychoanalysis at the time and had a positive interest in Freud's theories, particularly his claim that maladaptive behavior is normally a symptom of some deep psychological malaise. According to the Freudian theory of symptom substitution, getting rid of one symptom would simply lead to the development of another that could be even worse. However, instead of being tempted to psychoanalyze the bed-wetters and instead of trying out one of the traditional, mainly barbaric, and largely ineffective treatments for enuresis, Mowrer turned for advice to his new colleagues at the Institute for Human Affairs. These included "Clark Hull, Neal Miller, Don Marquis, Kenneth Spence, Ivor Hovland and others, who were very much interested in the psychology of learning, a relatively new field to me … and (I) became convinced that the enuresis problem at the Children's Center could be solved by simple, Pavlovian conditioning."[7]

The idea was to arrange that, when a special pad became wet with urine, this would ring a bell that was loud enough to wake up the boy. In Pavlovian terms, the bell served as the unconditioned stimulus. Mowrer assumed that the internal stimulus produced by bladder distension could serve as a conditioned stimulus. After a few pairings of this internal stimulus with the bell, the former should evoke the response of waking up in time for the boy to get to a bathroom to urinate. Mowrer and his wife made up pads, ones comfortable to sleep on, that contained wire mesh arranged in such a way that, when the pad become wet, the bell was sounded. When first tested out on the champion bed-wetter, an 11-year-old, after four nights of being woken by the bell, the boy started to get up before urinating. Apparently, he retained a "dry-bed habit"' forever more, as did almost all of the 38 other bed-wetters in the Center who were subsequently given the pad-and-bell treatment. The treatment has been widely and very successfully used to treat enuresis ever since.

Finding out a short time later about the key experiment by Brogden, Lipman, and Culler could have made Mowrer realize that the success of the pad-and-bell method was not just based on Pavlovian conditioning but also involved avoidance learning. Their study became the starting point for the series of experiments and theoretical developments that ensured that Mowrer's contribution has been remembered ever since.

In the second half of the twentieth century Freudian theory and the practice of psychoanalysis were generally seen as fundamentally opposed to behaviorist-oriented learning theories, in terms of underlying philosophy, scientific approach, and the treatment of mental illness. It is therefore intriguing that Mowrer, along with contemporaries such as B.F. Skinner, took Sigmund Freud's theories very seriously. The 1936 English version of Freud's *The problem of anxiety* was particularly important in the development of Mowrer's analysis of avoidance learning. In what appears to be his last presentation to the weekly meeting of researchers from the Institute of Human Relations, the 'Monday Night Group,' he favorably compared Freud's analysis of anxiety with that of Pavlov. Mowrer's "stimulus-response analysis of anxiety" concluded: "(i) that anxiety, i.e., mere anticipation of actual organic need or injury, may effectively motivate human beings and (ii) that reduction of anxiety may serve powerfully to reinforce behavior that brings about such a state of 'relief' or 'security'."[8]

In 1940 Mowrer left Yale to take up an Assistant Professorship in the Department of Education at Harvard University. Within a relatively short time, he and his students were running experiments with rats that used a form of shuttle box. They addressed the question posed in the following quote. "It had been previously taken for granted by various writers that it is in some sense rewarding to an experimental subject to avoid a noxious unconditioned stimulus. It is easily seen that is rewarding to escape from such a noxious stimulus. But how can a shock which is not experienced, i.e., which is avoided, be said to provide a source of motivation or of satisfaction?"[9]

The answer Mowrer gave to this question became known as *2-factor theory*. The first factor was the Pavlovian one of forming an association between some event, the danger signal, and an aversive event – electric shock in Mowrer's experiments – so that the danger or warning signal came to elicit anxiety. The second factor involved instrumental learning, whereby responses that decreased anxiety were strengthened, following the Law of Effect. An early experiment provided important support for the theory. On each trial a buzzer was sounded to warn the rat of an imminent shock; the rat could escape the shock once this began by moving to the other side of the shuttle box and, furthermore, could avoid the shock completely if it made this response before the shock was delivered. The conditions for three groups of rats differed only in whether or not the response of moving to the other side also terminated the buzzer. The key result was that the two groups that had no control over this WS learned more slowly to avoid the response. According to the theory, the superior performance of the group that could switch off the buzzer was because this stimulus had come to elicit anxiety and terminating it reduced anxiety.[10]

For some years Mowrer oscillated between using *anxiety* and *fear*, treating these terms as synonymous. Eventually, he settled – as researchers have done ever since – for the term 'fear' to describe the state of an animal anticipating an electric shock. In doing so, he essentially accepted Freud's distinction on the grounds that "fear has a consciously perceived object and anxiety does not."[11] Thus, in 1946 the title of a second experimental report by Mowrer and his student, Ross Lamoreaux, was "Fear as an intervening variable in avoidance conditioning."[12]

Up to this point, the procedures used in experiments on avoidance learning were ones in which the avoidance response was the same as – or at least very similar to – the response the animal made when shock was delivered. Thus, the guinea pigs used by Brogden, Lipman, and Culler both reacted to a shock by moving their wheel a short distance and could prevent the delivery of shock by making the same movement. Similarly, the rats used in the 1942 experiment by Mowrer and Lamoreaux both reacted to a shock by moving to the other side of their shuttle box and could prevent the delivery of shock by doing so. In their 1946 experiment, Mowrer and Lamoreaux showed that their rats could acquire an avoidance response that was quite different from the response they made to the shock itself.

The 1946 experiment used a highly innovative, multi-factorial design that allowed the researchers to apply Fisher's analysis of variance to their data, a statistical approach that did not become widespread in psychology until many years later. In discussing their data, the authors criticized Pavlov for ignoring motivational factors in

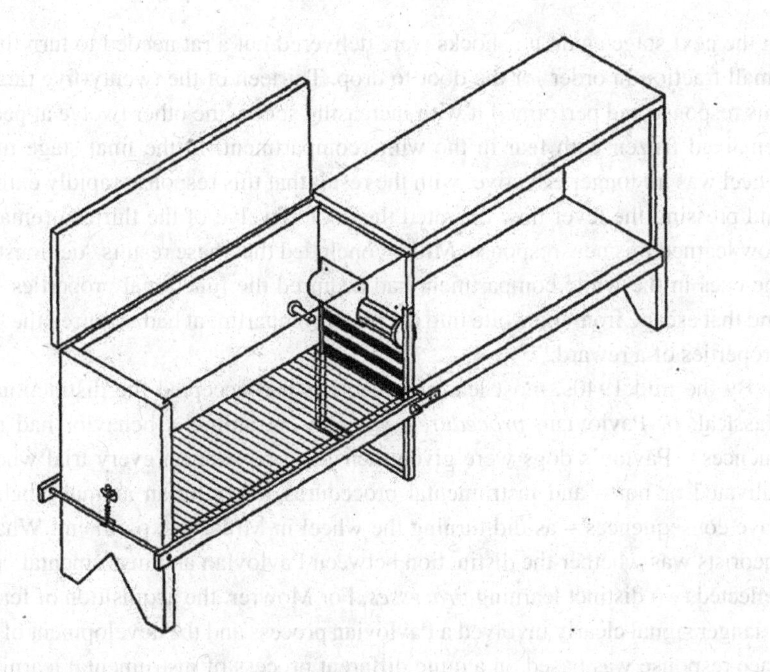

Figure 4.3 This one-way shuttle-box was used to train rats, when placed in the left-hand compartment containing the electrifiable grid floor, to escape or avoid a shock by turning a wheel that would open a gate to allow them into the other ("safe") compartment.
From Miller (1948). Public domain.

his theory of the conditioned reflex but praised Freud: "Pavlov's attempt to extend the same narrow formulae into which he forced his experimental results with animals over into the field of psychiatry is eloquent testimony of their inadequacy. One of the keys to Freud's success is that from the outset he saw fear, lust, and anger – or, more accurately, fear *of* lust and anger, i.e., 'anxiety' – as the central problem in the psychoneuroses."[13]

The shuttle-box apparatus used by Mowrer in his early research at Harvard was designed in collaboration with his former colleague at Yale, Neal Miller.[14] Miller constructed a variation on the shuttle-box to use in a classic experiment that provided much stronger evidence for the arbitrary nature of an avoidance response than that reported by Mowrer and Lamoreaux. As shown in Figure 4.3, this consisted of two compartments separated by a door that could be dropped out so that the rat could get from the white compartment containing an electrifiable grid to the safe compartment painted black; the door could be operated either by the experimenter or by the rat turning a wheel mounted above the door or pressing a lever to the left of the door.

In his experiment, Miller first gave his rats 10 trials, each containing 12 shocks, 5 seconds apart, in the white compartment – an excessive amount by later standards. At the end of each trial, the door was dropped so that a rat could escape into the safe compartment from the continuous shock that followed. Then followed five non-shock trials in which the experimenter dropped the door as soon as a rat approached it.

In the next stage again no shocks were delivered but a rat needed to turn the wheel a small fraction in order for the door to drop. Thirteen of the twenty-five rats acquired this response and performed it with increasing speed; the other twelve appear to have remained frozen with fear in the white compartment. In the final stage turning the wheel was no longer effective, with the result that this response rapidly extinguished, and pressing the lever now operated the door. Twelve of the thirteen remaining rats now learned this new response. Miller concluded that these results "demonstrated that the cues in the white compartment had acquired the functional properties of a drive and that escape from the white into the black compartment had acquired the functional properties of a reward."[15]

By the mid 1940s, most learning theorists had accepted the distinction between classical, or Pavlovian, *procedures,* whereby an animal's behavior had no consequences – Pavlov's dogs were given their meat powder on every trial whether they salivated or not – and instrumental procedures, whereby an animal's behavior did have consequences – as did turning the wheel in Miller's experiment. What divided theorists was whether the distinction between Pavlovian and instrumental procedures reflected two distinct learning *processes*. For Mowrer, the acquisition of fear towards a danger signal clearly involved a Pavlovian process and the development of an avoidance response was based on a quite different process of instrumental learning; hence the term '2-process theory' applied to his analysis of avoidance learning. In contrast, as described in Chapter 2, Miller remained loyal to Hull's single process theory and argued that the acquisition of fear can also be explained in "stimulus-response, law-of-effect terms," although not specifying how this could be done.[16]

Anxiety, Fear, and Punishment in the Skinner Box

In a 1946 lecture Mowrer distinguished between, on the one hand, behavior that is under voluntary control, involves the skeletal musculature, is mediated by the central nervous system, and is subject to the Law of Effect, and on the other behavior that is involuntary, involves the smooth muscles and glands, is mediated by the autonomic nervous system, is often experienced as an emotion, and is subject to Pavlovian conditioning. Later that year Mowrer met B.F. Skinner at the annual meeting of the American Psychological Association and learned that Skinner had already drawn the same distinction; Mowrer's excuse for not knowing that Skinner had published these ideas eight years earlier was that they were expressed in terms that did not further the "interests of clarity and quickness of communication" and in a book that it seems few had read.[17]

As detailed in Chapter 8, as a postdoctoral fellow at Harvard University Skinner developed what became known as the Skinner box or operant chamber. The initial experiments he carried out using this new piece of behavioral equipment were reported in his 1938 book, *The behavior of organisms.*[18] Positive reinforcement, in the form of food pellets, was used in almost all of Skinner's Harvard experiments and, with one exception, the only aversive events were the withdrawal of food, as in extinction. The exception was an experiment on punishment that is discussed later. When Skinner left Harvard in 1936 to take up an academic post at the University of Minnesota, he fitted his new laboratory with Skinner boxes that had grid floors through which electric shocks could be delivered. He was joined by a particularly brilliant Ph.D. student, Bill Estes. Between them, they carried out a series of experiments that can be seen as producing the first study of an interaction between Pavlovian and instrumental conditioning. The importance of this study is hardly revealed by the title of the 1941 report, "Some quantitative properties of anxiety."[19]

In these experiments, rats were first trained to press a lever for food using what the researchers referred to as *periodic reconditioning*. This meant that, following some initial training, every four minutes a single lever-press would produce a food pellet; this schedule would later be called a *Fixed Interval (FI) 4-min* schedule. It ensured that during a 1-hour session, the rats would respond steadily in the intervals between pellet deliveries, thus providing an "orderly baseline that made it possible to follow with ease the development of the 'anticipation' (the authors' scare quotes) of the shock during subsequent repetitions of the situation."[20] Rats were able to anticipate intermittent brief shocks because each shock was preceded by a 3-minute tone and these tone-shock pairings were presented twice per session over six sessions. After a few such pairings the rats stopped responding in the presence of the tone. This 'conditioned anxiety state' was no longer elicited by the tone after this had been presented several times without ending with a shock.

These results were described in the idiosyncratic manner that Skinner had developed in his earlier research. Instead of comprehensive reporting of the data set, the Estes and Skinner just pointed to samples of cumulative records – some mechanically averaged over several animals – in support of their description of what was found. The style is reminiscent of the way Pavlov reported his experimental results, pointing in his case to sample records showing how many drops of saliva a single dog had produced. However, just as subsequent comprehensive experiments confirmed almost all the claims that Pavlov made, the very many experiments that followed Estes and Skinner in examining *conditioned anxiety* – or, as the phenomenon was later termed, the *conditioned emotional response* (CER) or simply *conditioned suppression* – fully confirmed all the claims in this original report.

In 1911 Thorndike proposed that, just as a 'satisfying' event strengthened a connection between a stimulus and the response that produced the satisfying event, the occurrence of an aversive event following a response, as in a punishment procedure, weakened such a connection. In 1932 Thorndike changed his mind on this topic. On the basis of some unimpressive data from experiments on human learning, he argued against the idea that reward and punishment had symmetrically opposite effects; instead, he proposed that punishing a particular response led to the strengthening of some other response that competed with the punished response.[21]

Several of Skinner's early Harvard experiments addressed this issue. Using the term *negative reinforcement* instead of *punishment*, Skinner – who was always fond of gadgets – introduced a mechanism for "delivering a sharp slap to the foot or feet used (by the rat) to press the lever," delivered by the lever itself in the course of being depressed. The apparatus consisted of "an electrically-operated double hammer striking upward against the two shafts of the lever behind the panel."[22] Not surprisingly, during extinction of lever-pressing previously reinforced with food pellets, slapping the rat whenever it pressed the lever reduced the response rate. In discussing the recovery of lever-pressing when the slaps ceased, Skinner suggested that such negative reinforcement should not be seen as the opposite of positive reinforcement – thus agreeing with Thorndike – but that the slap-produced transient reduction in responding results from "an emotional state of such a sort that any behavior associated with feeding is temporarily suppressed." He noted that these slaps were only mildly aversive and that experiments using shocks might lead to a different conclusion.[23]

Experiments of this kind were reported in Estes' Ph.D. thesis which was submitted in 1943. They constituted the first systematic study of punishment. The methods Estes used were similar to those that he and Skinner had previously developed to study anxiety. Rats were given preliminary training designed to produce a stable rate of lever-pressing even when such responses produced a food pellet only every four minutes and to produce prolonged responding in extinction when food pellets were withdrawn altogether. Over a series of eleven experiments, Estes introduced periods in which either every lever press or only some lever presses were immediately followed by a shock; he recorded rates of responding both during such periods and after the shock had been discontinued. Across experiments such punishment of lever-pressing varied according, for example, to whether it involved weak or strong shocks or to whether it was administered while lever-pressing was still being reinforced with pellets every four minutes.

The conclusions that Estes drew from this substantial set of experiments were similar to those reached earlier by Skinner. The effects of punishment were seen as resulting from arousal of an emotional state that had a transient, but no long-term, effect on behavior. Thus, Estes appealed to the same process of response suppression that he and Skinner had reported two years earlier. In summarizing his results, he claimed to have found "no evidence that the correlation of the punishment with the response per se in the role of a 'negative reinforcement' is important."[24]

In reaching this conclusion Estes downplayed some data from his final, and particularly innovative, experiment. This involved first training the rat to respond on two retractable levers, mounted two inches apart to the left and right of the food aperture. Preliminary training ensured that the rats responded at an equal steady rate on both levers. Subsequently, a single experimental session included a 10-minute period in which shocks were delivered whenever a rat pressed the left lever. Estes then tested how a rat would respond when only one of the levers was present. Initial tests revealed complete generalization of the effects of such punishment to the unpunished lever, a result he emphasized in his general conclusions. However, repeated testing revealed more rapid recovery of responding on the lever that had not been punished, thus showing "that a part of the effect of punishment is quite specific to the punished response."[25] Many years later several studies revealed that the response-shock contingency was far more important than Estes had previously suggested.[26] Years later he admitted that he had been wrong to reject a symmetry between the effects of positive reinforcement and punishment.[27]

As a Ph.D. student, Estes had reached a conclusion wholly in line with the views of his supervisor. This was despite using experimental methods and data analyses that were abhorrent to Skinner. Thus, Estes used group designs, counterbalanced conditions across groups, reported means and statistical analysis of differences between means, and rarely introduced samples of cumulative records to illustrate the behavior of his rats. His occasional attempts to fit data to mathematical equations foreshadowed his later development as a pioneering mathematical psychologist.

When WW II ended, Skinner wrote a utopian novel, *Walden Two.*[28] A major theme in this book, and continued in his highly influential, *Science and human behavior,*[29] and in his articles on educational practice[30] was that traditional controls of human behavior by societies, institutions, or in the classroom rely too much on aversive methods such as punishment or avoidance that are ineffective, relative to positive reinforcement, and involve "the anxieties, the boredom, and the aggressions which are the inevitable by-products of aversive control."[31] Unlike Estes, Skinner retained his view that punishment only produced a transient emotional state.

In a graduate seminar in 1965, I listened as a graduate student, Howard Rachlin, challenged Skinner as to why he would not accept the considerable evidence from animal experiments of symmetry between positive reinforcement and punishment. Skinner simply ignored the challenge.

Mowrer's Subsequent Career

At Harvard, the main focus of Mowrer's interests shifted towards understanding human neuroses and finding effective psychotherapies. He retained his respect for Freudian and related theories, but not for the practice of psychoanalysis, a form of treatment that had done nothing for his own psychological problems.[32] In 1947 he gained wide popular attention from a lecture at a major scientific meeting in which Mowrer argued that "sin was responsible for nearly all mental illness" and that "what was needed was a return to religion."[33] A year later he left Harvard to take up a research professorship at the University of Illinois.

In view of this shift in Mowrer's beliefs, his new research interests that included vocal imitation in birds and the acquisition of human language and, above all, his recurrent bouts of depression (he was hospitalized during a particular serious episode in 1950), it is remarkable that he continued to undertake experiments on avoidance learning[34] and to publish two substantial books.[35]

Mowrer's 1960 *Learning theory and behavior* is mainly important for the expansion of his two-factor theory from an analysis of avoidance learning to a general theory of learning; he argued for the importance of the emotions *hope* and *disappointment* in conditioning involving positive rewards to complement the role of *fear* and *relief* in aversive learning. Two further points are worth noting about this book. First, the analysis of research and theories published by his contemporaries is accommodating

Figure 4.4 O. Hobart Mowrer, aged 67. Unlike all other contributors to a series of autobiographies, he insisted that the photograph should include his wife and three children. From Mowrer (1974). Public domain.

in style in contrast to the combative stance adopted by most other learning theorists of his generation. Second, the book's intended readers appear to be the several hundred or so psychologists already convinced of the central importance of the large number of experimental studies that Mowrer discusses in great detail; for most readers of the era without such a conviction it must have seemed a very tedious book.

During the 1960s Mowrer showed a decreasing interest in learning theory. Moving on from his religious period, his main preoccupation became the development of a form of group therapy, he called *integrity therapy*, that was related to the encounter groups that spread out of California in the 1970s. It has been suggested that repressed homosexuality was a major contributor to his recurrent bouts of depression.[36] His ever-supportive wife and collaborator, Molly Mowrer, died in 1979 and in 1982 Hobart Mowrer took his own life at the age of 75;[37] see Figure 4.4.

Richard Solomon and His Many Impressive Students

Before Mowrer left Harvard in 1948, it seems that he may have either lent or donated his shuttle box equipment to a colleague, Richard Solomon.[38] If true, then this was almost a physical symbol of handing over a baton. In 1950 Solomon reported his first avoidance experiment at a conference[39] and for the next two decades, his work and that of his students dominated the study of fear and avoidance.

In background and in personality Solomon could not have been more different than Mowrer. Richard Solomon (1918–1995) was born in Boston into what appears to have been a very stable family background that emphasized achievement, intellectual debate, and personal responsibility. Despite a mixed performance as a high school student, he proved to be an outstanding undergraduate at Brown University, where the quality of the academics in the psychology department led him to focus on psychology. After graduating in 1940 he decided to stay on at Brown to undertake a Ph.D. in the laboratory of Harold Schlosberg. A spell of war-related work was followed by completion of his Ph.D. in 1947 and – remarkably – in the same year Solomon was appointed an Assistant Professor in the Department of Social Relations at Harvard.

By tradition, most advisors of Ph.D. students, then as now, are named as co-authors of research papers reporting work by their students, even when the supervisor's involvement may have been limited. Solomon was different. Even when he had been closely involved in experiments carried out by a student and urged by the latter to add his name to the report, he would usually refuse: "I have plenty of publications and an established career, but you are just beginning. You need the authorship much more than I." It was one of the many characteristics that made him so influential as a teacher and a mentor.[40]

Having Schlosberg as an advisor must have influenced Solomon's later involvement in research on animal learning. However, for several years he was involved in some very different areas of research. There is no evidence that Solomon or any of his students ever used the shuttle box for rats loaned by Mowrer. Instead, when Solomon eventually ran experiments on avoidance learning, the subjects were dogs. This choice was made following advice from a postdoctoral researcher, Lyman Wynne, that far more was known about the physiology of the dog's fear response than about that of rats.

Mowrer had claimed that "informed and forward looking psycho-clinicians today are generally agreed that the most important and exciting future advances in the understanding of personality and in psychotherapy will come from the mutual modification and blending of learning theory and psychoanalysis." He pointed to Freud's analysis of the way that traumatic experiences – as suffered by shell-shocked combatants in World War I – could lead to neurotic behavior.[41] These suggestions may well have prompted Solomon to run his first experiments on avoidance learning. He wanted to study the persistence of behavior that was acquired under conditions of a traumatic experience.

These experiments used the equipment shown in Figure 4.5 and a procedure that became known as the two-way shuttle box procedure. A dog is first placed in, for example, the left chamber and, when a WS starts, it has 10 seconds to jump over the shoulder-high barrier to avoid the shock. The WS comprised both the offset of the lights in the chamber in which the dog had been placed and raising of the gate that otherwise prevented a dog from jumping into the other chamber. Once the dog has reached the other chamber, the gate is closed during the eight minutes until the next trial starts, when the dog now needs to learn to jump from the right-hand chamber back into the left-hand chamber. And so on.

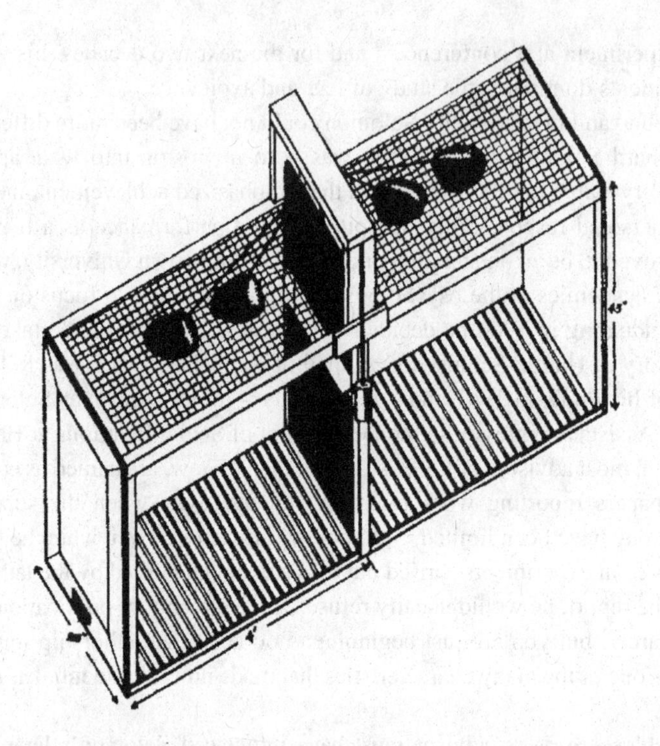

Figure 4.5 This two-way shuttle box was used extensively in Solomon's and other labs. Dogs were trained to avoid a shock when the warning signal (WS) commenced by jumping over the barrier to the other compartment; when the WS was later repeated, they needed to jump back to the original compartment; and so on.
From Solomon and Wynne (1953). Public domain.

Solomon's students were later renowned for the elegance of their experimental designs and careful choice of control conditions. In contrast, this first experiment was quite crude. He and Wynne decided that to obtain non-extinction of avoidance in a large number of dogs, they would need to use an intense enough shock to produce "traumatic avoidance learning." Instead of checking the assumption that higher levels of shock were needed to obtain avoidance behavior that would be highly persistent, they simply used a single group of thirty-three dogs. Their experiments would not have been approved by any animal ethics committee in later decades when such committees were introduced to monitor animal research in the USA. To obtain "an intense fear reaction … the dog is suddenly stimulated through the steel grid floor by a high-voltage electric shock, of an intensity just below the tetanizing level … (with the result that) … the dog will scramble rapidly and vigorously around the compartment, slamming into walls, perhaps, or leaping up against them; he will simultaneously emit a high-pitched screech, will salivate profusely, will urinate and defecate in a manner which could be called 'projectile elimination' and will roll his eyes rapidly, and jerkily …."[42]

There was a dramatic contrast between the dogs' early reaction to such shocks and their behavior once they were regularly jumping over the barrier before any shock arrived. The use of a large number of dogs revealed a wide range of individual differences, particularly in what a dog did after it had made its first avoidance response. Thus, three out of the thirty-three avoided all potential shocks over the next thirty trials; to achieve a similar level of success, a further seven needed only one subsequent experience of a shock when they failed to jump in time; and a further eight dogs failed to avoid the shock on two occasions before they showed persistent avoidance behavior.

According to Mowrer's two-factor theory, there should be periods in which an animal successfully performs an avoidance response, albeit becoming somewhat slower as trials are repeated, but interrupted occasionally by at least one trial when the animal does not react fast enough and receives a shock. This prediction follows from the idea that, as successful avoidance responses are repeated, Pavlovian conditioned fear of the WS will extinguish and a refresher WS-shock pairing is needed to restore this fear, and thus restore the motivation to perform the avoidance response. Avoidance behavior of the eighteen dogs described above failed to meet this prediction; as well as consistently avoiding the shock, their response to the WS tended to become quicker rather than slower. Solomon and Wynne noted that "the instrumental act (i.e., jumping over the barrier) occurs before, or simultaneously with, any emotional reactions elicited by the CS. When this occurs, it appears that the animal is acting cognitively."[43] What they meant by 'acting cognitively' was not spelt out till later.

The surprising persistence of avoidance behavior in the first study raised the question: How could such avoidance behavior be extinguished? The answer could be useful to clinicians in treating clients with persistent non-adaptive neurotic behavior. A series of experiments, in which – as in the initial study – dogs were first trained to jump the barrier to avoid a traumatic shock, revealed that a combination of two treatments was required to obtain extinction of this avoidance response for a majority of dogs. One consisted of a 'reality check,' consisting of presenting the WS while a dog was confined to a compartment and then failing to deliver a shock. The other was to punish the avoidance response by delivering a shock once the dog had landed in the other compartment. In the absence of such 'treatment' dogs continued to respond ever more rapidly for 10 trials a day over a 20-day period.[44]

From a perspective of over 60 years later, these experiments can appear needlessly cruel. There is nothing to indicate that the researchers were unusually insensitive to the suffering of their dogs. However, in that era, there was generally far less concern for the well-being of animals; the animal liberation movement hardly existed. Furthermore, any qualms that the researchers may have had about the procedures their dogs had to endure were likely met with the firm conviction that the results would convey very direct benefits to the treatment of mental illness. In this context, it is noteworthy that the extinction experiments were reported in the *Journal of Abnormal and Social Psychology* rather than in the journals in which studies of animal learning were normally reported.

One of Solomon's first Ph.D. students was a co-author on the extinction paper. Leon Kamin (1927–2017) was from a similar background to that of his advisor: From

Massachusetts and of Jewish origin. He obtained his BA from Harvard University in 1949 and remained there to undertake graduate work, obtaining a Ph.D. in 1954. As detailed in Chapter 9, in 1954 Kamin was interrogated by the McCarthy's Un-American Activities Committee. This led to him being *persona non grata* at universities in the USA. However, he was very welcome in Canada, where he was appointed in 1954 as a Research Associate at McGill University in Montreal and then as an Assistant Professor at Queens University in Kingston, Ontario.[45]

The widely cited experiment that Mowrer and Lamoreaux reported in 1942[46] was justifiably recognized as providing strong support for Mowrer's claim that avoidance behavior was reinforced by termination of the WS. As described earlier, the key finding was that a condition in which the avoidance response – running to the other side of the shuttle box – both prevented the shock and terminated the WS produced faster acquisition than control conditions in which the avoidance response had no effect on the WS and its only consequence was to prevent the shock being delivered. What received little attention was that the control rats also acquired the avoidance response, even if more slowly than the warning-signal terminators. As Kamin noted,[47] almost all other experiments on avoidance learning had confounded the two factors: Termination of the WS and avoidance of the shock. He pointed out that theories such as Mowrer's predicted that a response should be learned if reinforced by termination of the WS, even if it failed to avoid the shock. He designed a factorial experiment to determine whether both termination of the WS and avoidance of shock had main effects on acquisition of the behavior of running to the other side of a shuttle box and to determine whether these factors interacted.

The experiment was run in the new laboratory that had been set up for Kamin at Queens University. The apparatus was a modified shuttle box that had no barrier separating the two compartments and the WS was "an ordinary buzzer." Instead of the dogs that he had learned to work with at Harvard, Kamin now used rats. His design generated four groups: The *Normal* group, for which running to the opposite compartment when the buzzer was sounded both terminated the buzzer and avoided the shock; the *Avoid-US* group, for which response avoided the shock but had no effect on the buzzer; the *Terminate-CS* group, for which the response switched off the buzzer but did not prevent the arrival a shock; and the *Classical* group, for which a shock always followed the buzzer no matter what a rat did.

As shown in Figure 4.6, it turned out that both factors – termination of the WS and avoidance of shock – independently could produce acquisition of the shuttling response. As entirely expected, the *Normal* group, for which the factors were combined, learned more rapidly and maintained a far higher level of performance than the other groups. It turned out, however, that the duration of the CS was a critical factor. When this was only five seconds long, responding developed as rapidly in the *Terminate-CS* condition as in the *Avoid-US* condition; shown in Figure 4.6a.[48] In a subsequent experiment that used exactly the same design and methods, but lengthened the CS to 10 seconds, responding developed no more rapidly in the *Terminate-CS* condition than in the *Classical* control condition where the rats' responses had no consequence; shown in Figure 4.6b.[49]

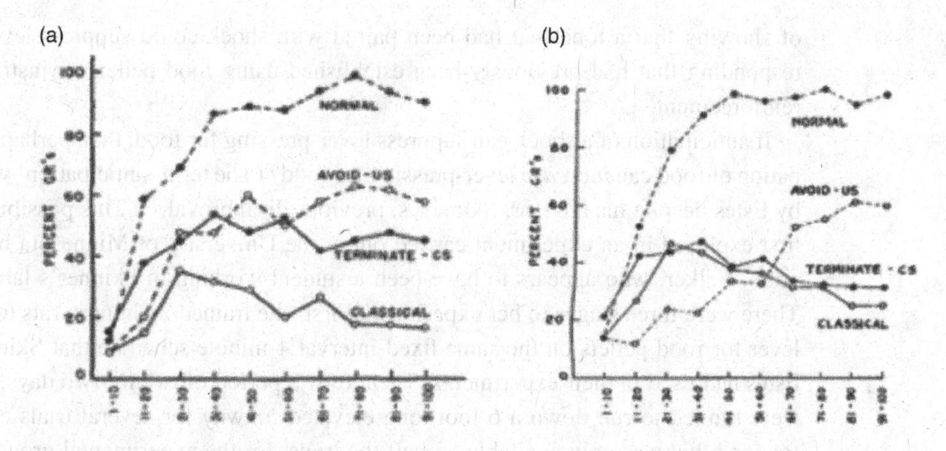

Figure 4.6 Acquisition of avoidance learning by four groups of rats in a two-way shuttle box. Data shown in the top graph (a) is from an experiment in which for the "Normal" group responding both avoided the shock and terminated the 10-second CS; for the "Avoid-US" group responses avoided the shock but did not terminate the CS; for the "Terminate-CS" group responding terminated the CS but did not avoid the shock; and for the "Classical" group responding had no consequences. In this experiment, the CS duration was only five seconds (Kamin, 1956). (b) Data shown in the bottom graph (b) is from a follow up experiment using an identical design and method, except that the CS lasted for 10 seconds unless terminated by the response of moving to the other compartment.
A from Kamin (1956) and B from Kamin (1957). Public domain.

The explanation for the importance of CS duration offered by Kamin was that under the *Terminate-CS* condition, the punishing effect of the shock that always follows the response was more effective with the 10-second CS duration.[50] Inspection of these two figures strongly suggests what became a widely held view of 'normal' avoidance learning: During early trials termination of the WS is the consequence that provides the more effective reinforcement of the avoidance response, while later the avoidance contingency – response means no shock – becomes increasingly important in maintaining this behavior.

Kamin did not draw any firm conclusions from these two experiments. He simply noted that 'cognitive theory' was unable to account for the data obtained from the *Terminate-CS* groups, while data from the *Avoid-US* groups pose "a real problem for S-R interpretations of avoidance learning."[51]

The Development of Pavlovian-to-Instrumental Transfer Methods as a Major Research Tool

Pavlovian-to-Instrumental Transfer (PIT) refers to any demonstration that a stimulus that is established as a Pavlovian conditioned stimulus can influence performance of behavior established by means of instrumental training. As described earlier, the first demonstration of PIT was reported by Estes and Skinner in 1941.[52] This consisted of showing that a tone that had been paired with shock could suppress lever-press responding that had previously been established using food pellets as instrumental reinforcement.

If anticipation of a shock can suppress lever-pressing for food, then perhaps anticipation of food can *increase* lever-pressing for food? (The term 'anticipation' was used by Estes despite his advisor, Skinner's, previous disapproval.[53]) This possibility was first explored in an experiment carried out at the University of Minnesota by Katherine Walker, who appears to have been a student working in Skinner's laboratory. There were three stages to her experiment. First, she trained 20 hungry rats to press a lever for food pellets on the same fixed-interval 4-minute schedule that Skinner and Estes had used in their experiments. Then, over a period of twenty-two days, the rats were trained to run down a 6-foot long elevated runway for several trials each day for food that was only available on half the trials; for the experimental group, a tone was sounded only on trials when food was to be found, while for the control group the tone was sounded on every trial. The final phase consisted of two sessions back in the Skinner box when lever-pressing was extinguished; over each 2-hour test, the tone was sounded for alternate 10-minute periods. Even though several rats in the experimental group showed no sign of learning about the significance of the tone during their runway training, the test results were encouraging, in that rats that had learned the tone discrimination increased their rate of lever-pressing when the tone was present, whereas the control rats did not show any change in their behavior.[54]

Soon after Walker's less than satisfactory experiment and in the same laboratory, Estes ran a similar, but simpler, experiment. Following the same stage of initial lever-press training, rats were given three 1-hour sessions of Pavlovian training in Skinner boxes in which the levers had been removed. Each session contained 10 trials in which a tone was sounded for a minute and followed immediately by delivery of a food pellet, independently of the rat's behavior. During two subsequent sessions, the levers were restored to the Skinner boxes and the test procedure used by Walker was implemented. During the two occasions in each test session when the tone was sounded for 10 minutes the experimental group increased their rate of lever pressing, while responding by a control group that was tested in silence simply declined in a steady manner.[55]

A possible objection to the conclusion that tone-evoked anticipation of food on test led to the increase in lever-pressing was based on what later became known as *superstitious conditioning*; given that the rats had previously been given lever-press training in the Skinner boxes, perhaps during the subsequent tone-food training the rats had made incipient lever presses to a now absent lever and these were reinforced by the arrival of a food pellet? Consequently, Estes repeated the experiment with one major change: The rats were now given tone-food training in a lever-less Skinner box *before* they were trained to press a lever for food. When the tones were presented during the subsequent extinction test, the results were essentially the same as those obtained in the previous experiment; the rats pressed the lever at a higher rate when the tone was present than when absent. Five years elapsed before a report of this second experiment was published.[56] In the intervening period, Estes had served in an army hospital in the

Figure 4.7 William ('Bill') K. Estes in 1976.
From Gluck and Roediger (2011). Reproduced with permission from the Association for Psychological Science.

Philippines and at the end of World War II returned to the USA to take up the position of an instructor in the Department of Psychology at Indiana University offered to him by his former advisor, Skinner, now chairman of the department; see Figure 4.7.

Given that both anticipation of a shock and anticipation of food can influence food-reinforced instrumental behavior, how might anticipation of a shock influence avoidance behavior? This was the question addressed by Richard Solomon and a research assistant, Lucille Turner. And how to test for such an effect in the absence of any behavior that could interfere with or even produce a transfer effect?

They based their approach on earlier work by another Harvard graduate student advised by Solomon, Abraham Black. This had included two new developments. One was to train avoidance behavior in a dog that was relatively restrained. As shown in Figure 4.8, Black's dogs were placed in a hammock that echoed the kind of harness that Pavlov had always used, even before he began to study conditioning. Shocks could be delivered to a dog's hind feet and, when a tone began to sound, the dog had five seconds in which to avoid a shock by turning its head to either left or right so as to operate the metal plates on either side. A "normal" procedure was implemented: An avoidance response terminated the WS, as well as preventing delivery of the shock, whereas an escape response made later than five seconds after onset of the tone terminated both the WS and the shock. Dogs reached the learning criterion of 10 trials in a row without a shock in anything from 13 to 134 trials.

The aim of Black's experiment was to assess the effect on extinction of exposing dogs to repeated presentations of the WS, the tone, in the absence of any further shocks, under conditions that prevented them from making any response when the tone was presented; see Figure 4.8. The second innovation was to use a recently refined form of curare to paralyze his dogs. The main point of using curare was to demonstrate learning under conditions in which no incipient habit could develop.

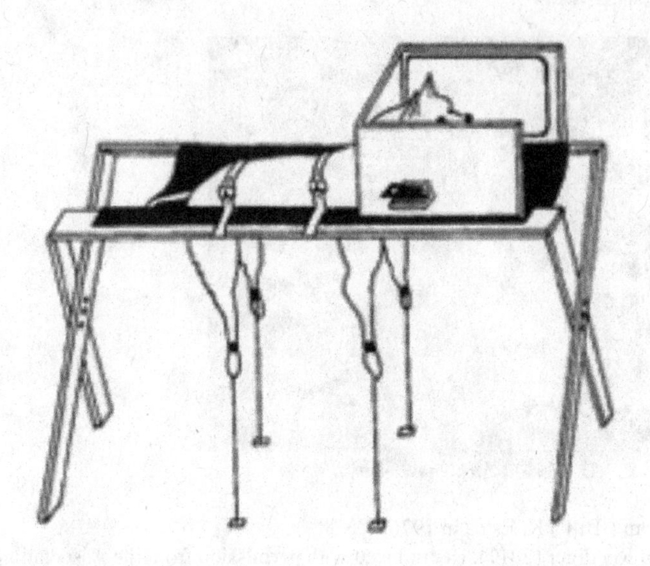

Figure 4.8 Dogs were first trained in this equipment to avoid shocks when a tone (warning signal) was presented by pushing against the side-panels with their heads. In a subsequent extinction stage, in which no more shocks were programed, a critical group was curarized while the tone was repeated 50 times.
From Black (1957). Public domain.

Over 20 years earlier Harlow had tried to obtain learning in curarized animals and thus show that not all learning was based on the Law of Effect.[57] However, there were problems with this and with later studies of a similar kind. The problems were mainly due to the poor quality of curare then available. By the time that Black commenced his Ph.D. project a form of curare had become available that completely paralyzed an animal's musculature – to the extent that artificial respiration was required – while leaving the animal fully conscious. This had been confirmed by a heroic researcher who could recall what had happened to him when completely paralyzed.[58]

Once Black's dogs had reached the criterion of 15 successive trials of successfully avoiding any shock, the shock generator was disconnected for the rest of the experiment. In the next stage, they were curarized and for the experimental group, the tone was sounded 50 times in the absence of shock. In the final extinction stage, these dogs rapidly stopped responding to this signal. In contrast, dogs in various control groups that had not been exposed to the tone when curarized continued to respond in extinction for many 100s of trials, many still responding when the experiment had to be terminated. Black concluded that no plausible form of S-R theory nor a Mowrer-type two-factor theory could explain his results. He did not offer his own alternative.[59]

The next step was taken by Solomon together with Lucille Turner. Their aim was to "show that a discriminative classical conditioning procedure carried out on curarized dogs can later control instrumental avoidance responses in the normal state."[60] Using the apparatus that Black had developed, they first trained dogs to avoid a shock when a WS was presented; this consisted of switching off for up to 10 seconds a light that otherwise shone directly at the dog's face. When the response of turning the head to one side or the other had been well established, the dogs were curarized and given Pavlovian discrimination training. For four dogs a high tone lasting 10 seconds was followed by shock, while no events followed a low tone; for the other four dogs the low tone served as the CS+ and the high tone as the CS-. In the final test stage, carried out when the dogs were in a normal state, presentation of the CS+ elicited a head-turning response at least as rapidly as did the WS, whereas the dogs responded more slowly, or not at all, to the CS-. The authors concluded that "transfer of training can occur without the benefit of mediation by peripheral skeletal responses or their associated feed-back mechanisms," noted that the possible role of conditioned autonomic responses of the kind central to Mowrer's two-factor theory "should be seriously explored in future experiments" but, like Black, were shy of suggesting their own theory as to the basis of such transfer effects.[61]

In 1957 Kamin moved to McMaster University where he refined the method for studying the impact of fear-eliciting stimuli on instrumental behavior. These changes included measuring such effects on a steady baseline of lever-pressing that was maintained by a variable-interval schedule; for example, his rats received a food pellet after pressing a lever at variable intervals that averaged 2.5 minutes. He also introduced what became the standard measure of what in 1961 he still referred to as the *conditioned emotional response* but by 1963 he adopted the name that has been used ever since, *conditioned suppression*. The measure consisted of a ratio of the number of responses that an animal made in the presence of the fear-eliciting stimulus to the total number of responses made both during this stimulus and during a pre-stimulus period of equivalent length; thus, 0 indicated complete suppression, while 0.5 indicated that the stimulus had no effect on the instrumental behavior.[62]

Black had joined Kamin at McMaster and together they used the improved conditioned suppression procedure in several studies. An important one had the aim of monitoring rats' fear during the course of avoidance training in a two-way shuttle box, in which a tone served as the 20-second warning signal. Fear was measured by the suppression ratio when on test trials the tone was presented in a Skinner box as the rats pressed a lever for food. A key finding was that fear increased during early stages of avoidance learning but then decreased as the avoidance response became well established. The authors concluded that their "results do not accord with the Solomon, Kamin and Wynne (1953) notion that fear of the CS is 'conserved' in the course of avoidance responding. The data on the whole reveal a considerable lack of parallelism between fear and instrumental behavior, and thus encourage speculation that variables other than fear of the CS are largely responsible for the maintenance of avoidance behavior."[63]

Meanwhile, Solomon had been recruited as a full professor by the University of Pennsylvania, where he again attracted particularly able graduate students to work with dogs in his laboratory. An important development was in parallel with Kamin's research, that of refining a procedure for studying – in this case – the effect of Pavlovian stimuli on an instrumental baseline maintained, not by food pellets, but by avoidance of shocks. How to achieve this had been first demonstrated by a Ph.D. student at Columbia University, Murray Sidman, one of very many students inspired by William Schoenfeld and Fred Keller with a productive enthusiasm for Skinner's ideas. Sidman managed to train rats to press a lever at a regular rate and maintain this behavior indefinitely under a schedule whereby, in the absence of any WS, a response prevented the arrival of a brief shock; if a rat stopped responding, then a regular train of shocks would be delivered.[64]

A similar schedule was also found to be effective in dogs given avoidance training in the two-way shuttle box that Solomon had developed when at Harvard (see Figure 4.7). If a dog sat still in one compartment, a 0.25-second shock would be delivered every 10 seconds, whereas jumping to the other side postponed the next shock for 30 seconds. This training was implemented by two of Solomon's students, Robert Rescorla and Vincent LoLordo. These researchers found that most dogs learned very rapidly to jump backwards and forwards in the shuttle box; within seven training days, dogs would average about eight jumps per minute, far above the minimum rate needed to avoid all shocks. To an observer who arrived at this point, it must have been very puzzling to see these dogs endlessly jumping back and forth for no apparent reason.

The point of establishing such baseline behavior was to investigate inhibitory learning, a topic that was central for Pavlov but one that had previously received little attention outside of Russia. Once their dogs were maintaining steady avoidance behavior, Rescorla and LoLordo began interspersed Pavlovian training using a high and a low tone, with one tone as a CS1 signaling the imminent arrival of an unavoidable shock and the other, CS2, signaling the start of a shock-free period. Over three experiments that varied the relationship between CS1 and CS2 the consistent finding for final test sessions was that the dogs jumped more rapidly when CS1 was presented – a result expected on the basis of the curare experiments by Solomon and Turner described earlier[65] – and the entirely novel finding that the dogs would slow down when CS2 was presented, showing signs of relaxation that contrasted with the agitation the dogs displayed when CS1 was presented.[66]

To a certain extent the ideas about avoidance learning and inhibition behind these important experiments by Rescorla and LoLordo had been anticipated by Konorski and Miller prior to the Second World War. By the 1960s Konorski had become a regular visitor to the University of Pennsylvania and a description in English of these very early avoidance experiments – usually with a single dog – was included within his *Integrative activity of the brain.*[67] However, it seems that neither Rescorla, Overmier, or LoLordo, as graduate students at Pennsylvania, were aware of this earlier work.

During his long career, Rescorla came to contribute more to the understanding of Pavlovian conditioning than anyone since Pavlov. This started with the study of inhibitory learning just described. This led to a concern with appropriate control procedures. In the penultimate paragraph of the report on their experiments on inhibition of avoidance behavior, the authors discussed possible shortcomings in the control conditions used in research on Pavlovian conditions. They went on to suggest that "possibly presentations of the CS and US which are truly random in time, as well as in order, would serve as a better control procedure for non-associative factors in Pavlovian conditioning."[68]

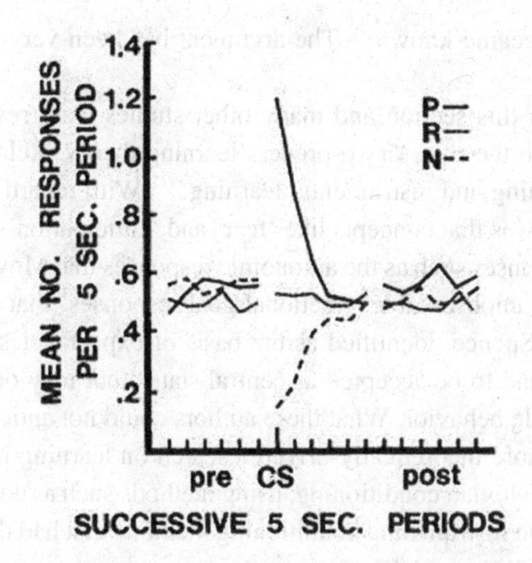

Figure 4.9 Test results indicating that a tone positively related to shocks increased avoidance responding (Group P), a tone negatively correlated with shock reduced avoidance responding (Group N) and a tone with a random relationship with shock (Group R) had no effect. From Rescorla (1966). Reproduced with permission from Springer Nature.

Such a control procedure was included for the first time in a follow up experiment that included important changes in how events were programmed during Pavlovian training. Dogs were first trained on the same Sidman schedule to jump to and fro in the shuttle box. Pavlovian training was then introduced while the dogs were confined to one compartment. On 24 occasions in each session and at variable times a 5-second tone was sounded. In Group P (for 'Positive') shocks could occur only during the 30-second periods that followed each tone. In Group N shocks could occur at any time, *except* during the 30-second periods that followed each tone. The most important group was a control group (Group R) for which both tones and shocks were programmed independently to occur at variable times; this arrangement meant that these dogs received just as many pairings of the tone and shock as the dogs in Group P. After five such Pavlovian sessions, a single test session was given in which on each of 24 test trials the tone was again presented for five seconds but now while the Sidman avoidance schedule ensured that the dogs continued to jump at a steady rate.

As shown in Figure 4.9, in Group P presentation of the tone initially doubled the rate of jumping, while the dogs in Group N almost stopped jumping immediately following the tone. Perhaps the most important result, however, was that in Group R the tone had no detectable effect. Given that these dogs had experienced as many pairings of the tone and shock as in Group P, Rescorla concluded that what is important in Pavlovian conditioning is the contingent relationship between the CS and US, that is, the degree to which the CS predicts the imminent arrival or non-arrival of the US.[69] As argued in an influential review paper, it follows that the proper control procedure for Pavlovian conditioning is one in which an animal experiences the same number of CSs and USs but in the absence any contingent relationship between the events, the *truly random control* as it became known.[70] The argument has been very generally accepted ever since.

The research described in this section and many other studies were reviewed in an important publication with the title, "Two-process learning theory: Relationships between Pavlovian conditioning and instrumental learning."[71] With regard to avoidance learning, a major point was that concepts like 'fear' and 'anticipation' could not be reduced to peripheral responses such as the autonomic responses that Mowrer identified as 'fear' or the usually unobservable 'fractional goal responses' that followers of Hull, such as Miller and Spence, identified as the basis of expectancies. Instead, emotions and expectancies had to be accepted as central states that may or may not be accompanied by observable behavior. What these authors could not anticipate was that for the next decade or more theoretically-driven research on learning in animals would be mainly based on Pavlovian conditioning, using methods such as conditioned suppression, rather than on the instrumental conditioning methods that had dominated previous learning research in Western laboratories.

Learned Helplessness

When planning studies of the kind described in the previous section, experiments that aim to test for transfer of Pavlovian conditioning to avoidance behavior, a question to resolve is: Should Pavlovian conditioning precede or follow avoidance training? A number of studies indicated that carrying out Pavlovian conditioning first can interfere with avoidance learning. Bruce Overmier and Martin Seligman, while students of Solomon at the University of Pennsylvania, decided that this *proactive interference* effect was interesting for its own sake; see Figure 4.10. Their experiments differed from previous transfer studies in that, during an initial stage, dogs were given a series of unsignaled shocks, that is, they omitted the conditioned stimulus from Pavlovian conditioning. During this initial stage, their dogs were placed in the kind of harness that Black had developed at Harvard (see Figure 4.8) and were subsequently given standard escape/avoidance training in the two-way shuttle box shown in Figure 4.5. One measure was how many dogs failed to escape the shock when given ten trials in the shuttle box. At most one in eight dogs that had never experienced shock prior to their avoidance training failed, whereas after experiencing unsignaled shocks 40–60% of these dogs never learned to escape the shock in the shuttle box.[72]

Previous researchers had speculated that the interference effect arising from giving Pavlovian conditioning prior to avoidance training could be based on response competition; any behavior that a dog acquired during Pavlovian conditioning might re-emerge in the shuttle box and get in the way of learning to jump to the other side. This possibility seemed unlikely after Overmier and Seligman found that dogs subjected to shocks when paralyzed by curare subsequently showed a similar level of interference with escape learning as non-curarized dogs. Furthermore, they found that the effect of inescapable shocks was transient; interference was pronounced when escape/avoidance training was given a day after a session with unsignaled and inescapable shocks but

Figure 4.10 Two of Solomon's graduate students from the University of Pennsylvania in the 1960s are shown in this group photo from 1976: Vin LoLordo is second from the left and Bruce Overmier on the far right. The other two researchers are Michael Rashotte on the far left and Jeff Bitterman, third from the left. All four were contributing to a NATO workshop on learning theory in Germany.
Reproduced with the permission of J. Bruce Overmier.

disappeared when two or more days had passed before escape/avoidance training was introduced. The two researchers concluded with the suggestion that "the source of the interference is a learned 'helplessness' (that) might well result from receiving aversive stimuli in a situation in which all instrumental responses or attempts to respond .... are no avail in eliminating or reducing the severity of the trauma. This interpretation .... suggests that the degree of control allowed to S over the shock exposure conditions is an important parameter for future investigation; high degrees of control over shock presentation allowed to S might 'immunize' S against proactive interference."[73]

Almost 20 years earlier Mowrer and Viek had introduced the term 'helplessness' in the context of an experiment that used a punishment procedure. They began their report by noting that "a painful stimulus never seems so objectionable if one knows how to terminate it as it does if one has no control over it."[74] Once again Mowrer's interest appears to have been triggered by Freud's treatment of anxiety. In this case, it was the link that Freud made between trauma and a child's feeling of helplessness.

Their experiment was based on the clever idea that, if a controllable shock is less painful than one over which an animal has no control, it should be less effective in punishing a response that precedes it than an uncontrollable shock. To test this idea Mowrer and Viek used an unusual procedure. On each daily trial, hungry rats were offered a bit of food and, if they ate the food, 10 seconds later a shock was delivered; if they failed to eat the food within 10 seconds, the food was withdrawn, and a shock was delivered. Rats in the *Controllable* condition could terminate the shock by jumping into the air. Paired with each rat in this group was a rat in the *Uncontrollable* condition.

As the name suggests, these rats had no control over the amount of shock they received but instead, a shock was delivered on each trial that lasted as long as the shock given to their partners. Within a few trials, a clear difference emerged between the groups: In the *Uncontrollable* condition, the shock was much more effective in punishing the rats for taking the food, whereas most rats in the *Controllable* situation continued to take the food on each trial. The paper is not an easy one to understand. Which may be one reason that over the next 70 years, it was cited only 216 times. In contrast, the paper reporting the next study to compare the effects of controllable and uncontrollable shock using a yoked design was cited 2,316 time over the 50 years since it was published.

Overmier was involved at the start of this study but then left to take up an academic position at the University of Minnesota. His place was taken by a new graduate student, Stephen Maier, who had started at the University of Pennsylvania in 1964, along with Seligman. The first experiment in this series used what became the classic *triadic* design for experiments on learned helplessness. A key group, *Escape*, were first given escape training in the harness that Black had devised and using the same kind of response panel on each side of a dog's head. During a single session, 64 unsignaled shocks were delivered to the dogs' feet; a dog in this group could terminate a shock by moving its head to one side or the other. All but two of the fifteen animals in this group soon learned to make this escape response almost as soon as the shock began. During this first stage of the experiment dogs in the second group, *Yoked*, received a pattern of shocks – ones that became progressively shorter – that was determined by the averaged behavior of the *Escape* group and over which they had no control. A final group, *Normal*, were not given any treatment on this first day. On the next day, all three groups were given 10 trials of standard escape/avoidance training in the two-way shuttle box.

The results completely confirmed the earlier prediction: The Escape group learned the avoidance response as well and as rapidly as the Normal group, while the Yoked group performed very poorly, in that they failed to escape the shock when in the shuttle box far more often than dogs in the other groups; of the fifteen dogs given the inescapable-shock treatment the previous day, eight failed to escape the shock on at least nine out of the 10 trials. These dogs seemingly had learned on their first day that their active responding was to no avail and transferred this learning from the hammock to the shuttle-box the following day. More generally the authors of the report drew the bold and broad conclusion that there are three kinds of learning: Together with learning that two events are positively related or that they are negatively related – as in inhibitory learning – they suggested that animals can learn that two events are unrelated.[75]

An alternative explanation as to why many dogs subjected to inescapable shock later failed to learn to escape in the shuttle box is that they had learned some kind of behavior that was incompatible with jumping in the shuttle box. This account predicts that dogs explicitly trained to keep still in order to escape shocks would later perform even more poorly in the shuttle box than dogs given equivalent inescapable shocks. As part of his doctoral research, Maier used the triadic design with dogs as subjects to test this prediction by giving one group initial training to keep still when placed in the harness. In the earlier experiments, dogs had to move their heads to operate a side panel to escape a shock. Now, over a series of sessions, they had to keep their heads still for increasingly longer times to escape a shock. Each of these dogs had a yoked partner that received exactly the same pattern of shocks independently of their behavior. The third, control, group were placed in the harness for the same amount of time but were never given any shocks during this initial stage. When given standard escape/avoidance training in the shuttle box dogs in the yoked group again failed to learn, whereas dogs that had had control over the shocks by remaining still learned to escape and avoid shocks, although not nearly as rapidly as the control group. Along with the previous finding based on curarized animals, this result appeared to rule out completely the possibility that animals given inescapable shock acquired some kind of behavior that was later incompatible with jumping the barrier in the shuttle box.[76]

To show that learned helplessness was important required demonstrations that it was a very general phenomenon. The obvious first step was to test for learned helplessness in a much less controversial, and also cheaper, animal than a dog. In the early 1970s, several researchers in different laboratories undertook many experiments with the aim of demonstrating learned helplessness in the rat. And many failed in this aim. At the University of Illinois, where Maier had obtained an academic appointment after leaving the University of Pennsylvania, a long series of experiments of this kind eventually proved successful. Maier and his colleagues found that prior treatment with inescapable shock interfered with learning to escape in a shuttle box or turn a wheel mounted on the side of the experimental chamber, only as long as the requirement for the escape response was made demanding. In the shuttle box the rats needed to run back and forth twice in order to escape the shock. In the wheel-turn chamber, the only effective responses were turns greater than a quarter of the circumference.[77]

Seligman was also concerned to show that rats could exhibit learned helplessness. He had remained at the University of Pennsylvania, obtaining an academic position there a few years after obtaining his Ph.D. His experiments used a new kind of experimental chamber that contained a retractable shelf mounted high on one wall and a lever mounted lower down on the opposite wall. In the first session rats in the *escapable* condition were given eighty trials in which unsignaled shocks of up to 10-seconds duration were delivered via a pin the rat's upper back. These were terminated when the rat pressed the lever, with in the required number of lever presses being gradually increased. Each rat in the *inescapable* condition was yoked to one in the escapable condition so that it received exactly the same pattern of shocks. The test session was given 24 hours later. This consisted of 20 trials, each of started with the simultaneous insertion of the high shelf and the onset of shock. These were maintained for one minute, unless the rat escaped by jumping up to the shelf. At the end of the one-minute trial, the shelf was retracted, thus "depositing the rat back on the floor." Rats in the *control* condition were simply given this 20-trial test session.

The first experiment of this kind yielded the opposite result to that expected on the basis of previous studies involving dogs. The test performance of the Inescapable group was poor, but no worse than that of the Control group. In contrast, the Escapable group rapidly learned to perform both the initial lever-pressing and the subsequent jump-up responses. Another three experiments were run before Seligman and his student, Beagley, obtained the pattern of results they wanted. The final experiment also used a quasi-triadic design. In the first stage, the Escapable group could escape by jumping up to the inserted ledge – some being helped early on by being pulled up by a researcher, while the same pattern of shocks was delivered to each Yoked partner. There was, however, no Control group. Instead, the researchers used data obtained under similar conditions from a control group in the preceding experiment. The abstract of the paper reporting these experiments concluded: "Rats, as well as dogs, fail to escape shock as a function of prior inescapability, exhibiting learned helplessness."[78]

The two studies described here in some details were widely accepted as demonstrating the generality of learned helplessness and inspired a large number of rat studies by a range of researchers. With the benefit of hindsight, one can see the limitations of these studies. For example, the experiments reported by Maier and his colleagues in 1973 did not include a group given escapable shocks and so the authors had no justification for the claim that lack of control was responsible for the effects they found. As for the study reported by Seligman and Beagley, apart from the limitations with their study that have just been noted, a crucial weakness was that both phases of the experiment were carried out within the same context. A key aspect of the original experiments was that the difference between dogs given escapable and inescapable shock in the initial phase was found in a quite different context.

By 1975 Seligman had begun to test for the generality of the helplessness effect in another species, namely humans. Instead of uncontrollable shocks his experiments used treatments such as insoluble anagrams. At the same time, he investigated the role of helplessness in depression.[79] Also, beyond the scope of this chapter is the very different direction that Maier took. He retrained himself as a behavioral neuroscientist and investigated the brain mechanisms involved when his rats were subjected to controllable or uncontrollable shock.

This later research by Maier led to a complete reversal of the claim that both he and Seligman had promoted for decades. Maier decided that learning that shocks were controllable was what transferred from one situation to another, not learning that shocks were uncontrollable. *Mastery* rather than *Helplessness*; *Hope* rather than *Hopelessness*. "The mechanism of learned helplessness is now very well-charted biologically, and the original theory got it backward. Passivity in response to shock is not learned … animals learn that they can control aversive events."[80] This was a return to an observation made at the very start of research that compared controllable and uncontrollable events: As previously quoted, in 1948 Mowrer and Viek commented: "A painful stimulus never seems so objectionable if one knows how to terminate it."[81]

Defense Reactions and Safety Signals

Like almost all other researchers who had studied avoidance learning, Solomon and his students showed little concern with the innate reactions animals make when frightened. It took until 1967 before anyone argued that these are not just important, but central to understanding avoidance learning.

Robert Bolles (1928–1994) was born in Sacramento, California. As an undergraduate at Stanford University, he majored in mathematics. This led to a Master's degree in this subject, followed by his first job. This was in the same US Naval Radiological Defense Laboratory in which Garcia and his colleagues carried out their pioneering studies of taste aversion learning; see Chapter 6. In 1954 Bolles enroled as a graduate student at the University of California at Berkeley to study psychology "under the tolerant umbrella of Edward Chace Tolman"[82] and with the more direct guidance of David Krech. There Bolles immediately began a series of rat experiments with a fellow student. These foreshadowed the interests that would shape his career and led to the idea he later brought to avoidance learning, namely that an animal's motivational state predisposes it towards certain patterns of behavior. For example, using an elevated T-maze, he and Petrinovich discovered that rats made hungry showed considerable variability in whether they entered the left or the right arm – and consequently rapidly learned a response-alternation problem – whereas other rats showed poor performance on this problem when made thirsty, a motivational state that induced greater behavioral stereotypy.[83]

After obtaining his Ph.D. in 1956, Bolles moved to the East Coast to take up positions, first at Princeton University and then at Rutgers University, prior to a more permanent appointment at Hollins College in Virginia in 1959. Seven years later he returned to the West Coast to take up a position at the University of Washington where he remained for the rest of his career.[84] In 1967 he published a magnificent book that, given the very large amount of research that he reviewed in great detail, must have taken him many years to write. A major theme in Bolles' *Theory of motivation*[85] was to demonstrate the inadequacies of drive theories of motivation. This criticism included theories of avoidance learning that focused on drive reduction, such as Mowrer's, and was based on reviewing what appears to be every experiment on avoidance learning that had ever been reported. The large number of avoidance studies reported in the late 1960s included a series he had run with students when at Hollins College, starting with experiments involving lever-pressing on a Sidman avoidance schedule.[86] Having listed eight major difficulties faced by current accounts of avoidance learning, Bolles went on to make the following suggestion. "Many of these difficulties … arise from studies of the acquisition of avoidance in different situations. It is becoming apparent that the acquisition of avoidance depends in a very important way upon what response the experimenter selects to be the criterion avoidance response." He went on to suggest that rapid avoidance learning occurs only when "the response is a natural defensive reaction when the animal is frightened …. in the rat fear is defined as the conditionable tendency to withdraw, freeze, or cower."[87]

Both in 1967 and in the highly influential paper he published in 1970, titled "Species-specific defense reactions and avoidance learning," Bolles distinguished between avoidance behavior that is acquired rapidly, such as one-way shuttle box avoidance, and behavior that is acquired more slowly, such as two-way shuttle box avoidance and – notably – lever-pressing on a Sidman schedule. An even greater challenge was that of training pigeons to peck at a response key whose illumination signaled an imminent shock if they failed to respond.[88]

While claiming that examples of rapid acquisition are based on little more than classical conditioning of defensive reactions, Bolles needed to appeal to a form of instrumental conditioning to explain the acquisition and maintenance of other kinds of avoidance behavior. His main solution to the longstanding challenge of explaining how the non-occurrence of an event (shock) could reinforce a response was to appeal to *safety signals.*

Following two earlier studies indicating that providing an animal with explicit feedback when it made an avoidance response led to more rapid acquisition, Bolles and one of his students tested what they termed the 'informational hypothesis.' The aim was to provide further evidence that "the introduction of a stimulus that is never paired with shock as an immediate consequence of responding leads to as good avoidance acquisition as the conventional CS-termination contingency."[89] Their experiments used various types of responding: Running in a one- or two-way shuttle-box, turning an activity wheel, and pressing a lever in a Skinner box. It turned out that providing feedback – in the form of turning the lights off for three seconds when a successful avoidance response was made – had its greatest effect on lever-pressing and two-way shuttling. Acquisition of these types of avoidance behavior was much more rapid when feedback was provided, but this informational effect was more limited in the wheel and hard to detect in one-way shuttle avoidance. The researchers explained these variations in the effectiveness of providing response feedback in terms of the degree to which intrinsic feedback is provided; notably, in one-way shuttle avoidance the response produces huge feedback in that animal finds itself in a new place in which it has never been shocked. They concluded that "the crucial aspect of all these stimulus changes is not the termination of a danger signal, i.e., a stimulus that has been paired with shock, but the onset of a safety signal, i.e. a stimulus that remains unpaired with shock in the situation."[90]

By 1970 Bolles had made the connection between his safety signals, Rescorla and LoLordo's experiments on conditioned inhibition of fear, and subsequent demonstrations that a fear inhibitor could function as a positive reinforcer. One such demonstration was reported by Rescorla from the laboratory he had set up at Yale University. His experiment used dogs in the harness arrangement devised by Black; see Figure 4.11. The dogs were initially trained under a Sidman avoidance procedure to operate both left and right head panels at a steady rate. The panels were removed for a second stage, in which the dogs were given Pavlovian training designed to establish a high tone, for example, as an inhibitor of fear. In a final test, the panels were returned and the avoidance schedule resumed; now pressing one panel both delayed the onset of a shock and produced the high tone, while pressing the other panel also delayed shock onset but produced the low tone that served as a control stimulus. The dogs showed a clear preference for pressing the panel that produced the inhibitory stimulus, the tone trained as a safety signal.[91]

Just five months after Rescorla submitted his report on the above experiment Ron Weisman and a student reported a similar experiment that they had run in Queens University in Ontario.[92] This used rats instead of dogs and turning a wall-mounted wheel as the avoidance response that was established and maintained by a Sidman schedule.

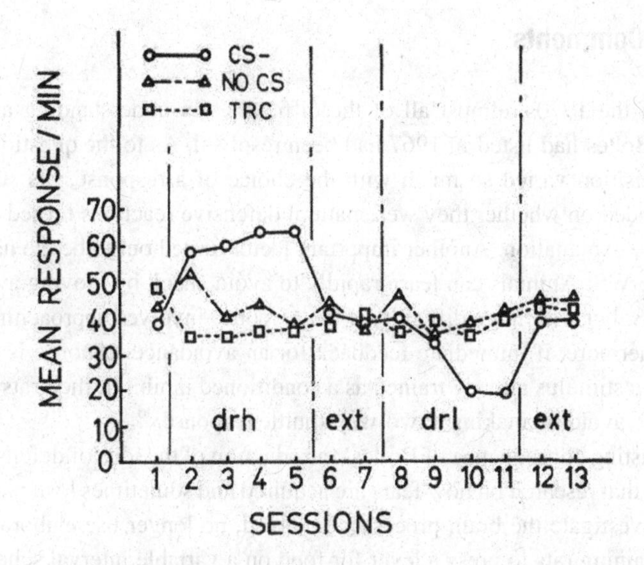

Figure 4.11 Rats trained to turn a wheel to avoid shocks delivered on a Sidman schedule *increased* their rate of responding when in an initial test phase such an increase (differential reinforcement of a high rate; *drh* schedule) was reinforced by presentation of a tone trained as a Pavlovian inhibitor. In a second test phase, the rats decreased their rate of responding when low rates were reinforced by this tone.
From Weisman and Litner (1969). Reproduced with permission from the American Psychological Association.

In a second stage, the rats were placed in a separate chamber and given three Pavlovian sessions of the kind pioneered by Rescorla and LoLordo: For the inhibitory group a 300 Hz tone was always followed by a variable period from 50 to 130 seconds that was free of shock, while otherwise 20 shocks were delivered in each one-hour session. The subsequent test sessions were conducted while baseline avoidance responding – turning the wheel – was maintained by the Sidman schedule. The tests used reinforcement schedules that had initially been developed for pigeons pecking a key to obtain grain. A *drh* schedule – differential reinforcement of a high rate – delivers a reinforcer only when an animal meets a requirement for responding rapidly, whereas a *drl* schedule – differential reinforcement of a low rate of responding – delivers a reinforcer only when an animal meets a requirement for responding at a low rate.[93] During four *drh* test sessions the rats in Weisman's experiment that were already turning their wheel at a steady rate, could produce the five-second tone by increasing their rate of wheel turning to more than 120 responses per minute for a brief period. In a subsequent series of four *drl* sessions rats needed to reduce their rate of responding to 12 or fewer wheel turns per minute in order to produce the tone. As shown in Figure 4.11, the results were clear. During the initial *drh* tests, rats in the inhibitory group (CS-) increased their rate of responding to more than 50% higher than the rat in the two control groups, while in the *drl* test that followed the response rate in the inhibitory group dropped to about half of that in the controls.[94]

Concluding Comments

By the end of the 1970s, almost all of the difficulties in understanding avoidance learning that Bolles had listed in 1967 had been resolved. As to the question of why speed of acquisition varied so much with the choice of a response, his suggestion that this depended on whether they were natural defensive reactions turned out to be only part of the explanation. Another important factor turned out to be the nature and location of the WS: Animals can learn rapidly to avoid shock by moving away from a localized WS, but with huge difficulty if the response involves approaching such a signal.[95] Furthermore, if immediate feedback for an avoidance response is provided in the form of a stimulus already trained as a conditioned inhibitor, then rats can very rapidly learn to avoid by making a two-way shuttle response.[96]

One long lasting consequence of Bolles' introduction of the idea of defensive reactions has been that research on how fears are acquired and sometimes lost, particularly studies that investigate the brain processes involved, no longer use elaborate procedure of first training rats to press a lever for food on a variable interval schedule and then measuring the fear evoked by a stimulus in terms of the degree to which it suppresses this instrumental behavior. Conditioned suppression has been largely replaced by simply measuring the amount of freezing a stimulus elicits when presented in a confined space.

The puzzle presented in 1953 by the dogs in Solomon and Wynne's experiment was: What kept them jumping backward and forwards in the absence of any perceptible consequence? A widely accepted solution was as follows. For an animal that has been successfully avoiding shocks for a long time the feedback from occurrence of the response becomes the safety signal that serves to reinforce the response. For reasons that had nothing to do with avoidance learning, various attempts were made in Rescorla's laboratory to extinguish a conditioned inhibitor; it turned out that simply presenting the stimulus on its own time after time made no detectable impact on its inhibitory properties.[97] Thus, an explicit safety signal or the feedback from making an avoidance response may continue to maintain the behavior long after fear of the WS or of the general situation has disappeared. Such fear and its termination can be very important very early in the acquisition of avoidance behavior, as Mowrer first suggested in 1927, but it seems to play an increasingly diminished role once the behavior is established.

5 Comparative Psychology

Species Differences in What Animals Can Learn?

According to the 17th-century Associationist philosopher, John Locke, at birth the human mind is a *tabula rasa,* a 'blank slate,' containing no constraints on the way that the mind will subsequently develop on the basis of sensory experience. It was a view with which most American behaviorists fully agreed. For Pavlov, one kind of stimulus was as good as any other, as expressed by his principle of *equipotentiality*, and what was true of how conditioned reflexes were acquired by dogs was presumed to be true of all other species; however, very late in life he may have modified this view; see Chapter 1. Similarly, according to Hull the principles of habit formation were the same no matter what response was involved and applied just as much to a human reacting to a signal that an electric shock was imminent as to a rat finding its way through a maze in order to obtain food. As for Skinner, critics pointed out that the title of his 1938 book, *The behavior of organisms*, was extraordinarily presumptuous in that the only organism mentioned was the laboratory rat and the only behavior analyzed was that of pressing a lever.

Among researchers concerned with learning processes active objections to the blank slate assumption took different routes. The one described in this chapter was to compare the learning and cognitive abilities of various species in the search for meaningful differences. The aims of this experimental tradition of comparative cognition echo those of the nineteenth-century anecdotalists, such as Darwin and Romanes. Some of this research examined the learning abilities of animals with large brains relative to their body mass, notably the chimpanzee and the dolphin, in the hope of demonstrating abilities that are not found in the rat. Other types of comparative research studied learning in animals with small brains, notably fish; in this case, it was expected that such species would show worse performance on tasks that rats find relatively easy.

Ape-Human Communication

Among the many challenges to his successors that Thorndike issued in his seminal paper of 1898 was to find evidence in support of the persistent and almost universally held view that the learning abilities of some species – especially those with large brains and evolutionary proximity to homo sapiens – are superior to those of 'lower' species.[1] As a Ph.D. student Thorndike had compared the rates with which cats, dogs, and chicks learned simple tasks and failed to find any differences. The superior performance of three Cebus monkeys he tested later did not make him change his view that all species learn by the same process, namely, one that involves the formation of habits, S-R connections, 'stamped in' by satisfying outcomes.

Among the first to meet Thorndike's challenge regarding species differences was Robert Yerkes. From when he was appointed in 1902 as an instructor in Harvard's Department of Psychology until cut short in 1917 by America's entry into the First World War, Yerkes, together with his students and associates, carried out a series of experiments on species that ranged from mice to pigs to macaque monkeys. His early studies tended to confirm Thorndike's failure to find species differences in the number of trials needed for an animal to learn a simple task. He then adopted the approach that was subsequently taken by several important successors. This was to devise more complex tasks that, for example, might be solved by a monkey but that would baffle a rat. Some of the new discrimination tasks were simple and involved just two choices, with a simple physical stimulus indicating the correct choice, while others presented multiple choices to an animal that was required to use the relationship between two stimuli in order to make the rewarded choice.

Inspired by early reports of Wolfgang Koehler's research on problem solving, Yerkes also tested an orang-utan – with some success – on the kind of problem that had convinced Koehler that chimpanzees were capable of *insight*.[2] For some years immediately after WW1 Yerkes' direct involvement in any kind of comparative research was limited. Following his appointment in 1924 as a professor at Yale University, he set up a laboratory there for studying the behavior of chimpanzees. More important in the long run, however, was Yerkes' increasing influence on various grant-giving agencies, such as the Rockefeller Foundation, that facilitated support for research enterprises that he considered important. One was to establish, starting in 1930, a research station for studying the learning and cognitive abilities of the great apes at Orange Park, Florida. When Yerkes retired in 1941, it was renamed the Yerkes National Primate Research Center and in 1965 relocated to the campus of Emory University where it has remained ever since.

Another important enterprise, to which Yerkes was able to influence funding, was the study of communication in great apes. An obvious challenge to the claim that the minds of both humans and of animals are blank slates at birth comes from evidence that, even under the most difficult circumstances, human children rapidly learn the language spoken by people in their environment, whereas no form of animal communication shows anything like the complexity and power of human language. Much of this evidence has come from studies of chimpanzees raised in a human household.

The initial aim of the important and fascinating study carried out by Winthrop Niles Kellogg (1898–1972) and his wife, Luella Kellogg, was, however, not primarily concerned with language. Rather the aim was to understand the role of the environment on early development in general. Kellogg's undergraduate career had been interrupted by World War I. As a mere 19-year-old, he flew in the UK and France for the US Army Air Service and was awarded a Croix de Guerre. After the war, he returned to study psychology and philosophy at Indiana University, where he married

Luella in 1920. On graduating he took up various jobs but was eventually persuaded to become a graduate student at Columbia University, where he obtained his Ph.D. in 1929. It seems that the idea for the study that made him famous came in 1927 from an article on 'wolf children,' two girls in India who had lived from a very early age with a pack of wolves and, when discovered, were found to behave in many ways as if they *were* wolves.[3]

Kellogg wanted to challenge what he saw as a widespread belief among psychologists that there is "a sharp qualitative demarkation between the behavior of man on the one hand and the behavior of infra-humans including the anthropoid apes."[4] This belief was misplaced, he argued, because the huge influence of environmental factors on early development was overlooked. Rather than study a human child raised in a nonhuman environment, Kellogg wanted to study the development of an ape – one as young as possible – raised in a human environment. "Suppose an anthropoid ape were taken into a typical human family at the day of birth and reared as a child. Suppose he were fed upon a bottle, clothed, washed, bathed, fondled, and given a characteristically human environment; that he were spoken to like the human infant from the moment of parturition; that he had an adopted human mother and adopted human father .... Under no circumstances should the subject of such an experiment be locked in a cage or led about on a leash..... The experimental situation *par excellence* should indeed be attained if this technique were refined one step further by adopting such a baby into a human family with one child of approximately the ape's age."[5]

The chance to undertake such a project arose in 1931 when Kellogg obtained a grant from the Social Sciences Research Council – possibly with help from Yerkes – that allowed him to take leave of absence from the academic position he had obtained at Indiana University and relocate, along with a reluctant Luella and their son, Donald, to Florida, where they lived close to the anthropoid research station at Orange Park. A female chimpanzee, Gua, soon joined the family. Gua was 7½ months old at the time and Donald, 10 months; see Figure 5.1. Over the next nine months, the Kelloggs maintained, with only an occasional respite, an intensive testing program that required them 12 hours each day for seven days a week. Donald was treated and tested in exactly the same way as Gua. The tests ranged from the purely physical – such as blood pressure and dexterity – to behavioral – such as locomotion and play – to cognitive – such as problem-solving and language comprehension. Figure 5.2 illustrates some of these tests. Thus, to compare Gua's speed of reaction to a startle stimulus with those of a group of children, several neighborhood children, who had been invited to a party, were lined up on a bench with Gua in the middle and filmed, when – without any warning – Kellogg fired a pistol behind them. Other photographs illustrate the equal interest in human faces shown by Donald and Gua, their equal interest and dexterity in playing with bricks, and Gua's superior ability to solve the problem of how to reach a cookie suspended from the ceiling by moving a chair directly underneath.

The Kelloggs collected a rich body of data, as reported in their book, *The ape and the child.*[6] Of particular interest to the Kelloggs were Gua's good understanding of spoken English and the much greater degree to which Donald imitated Gua rather than vice versa. "His capacities astonished even those closest to him when it

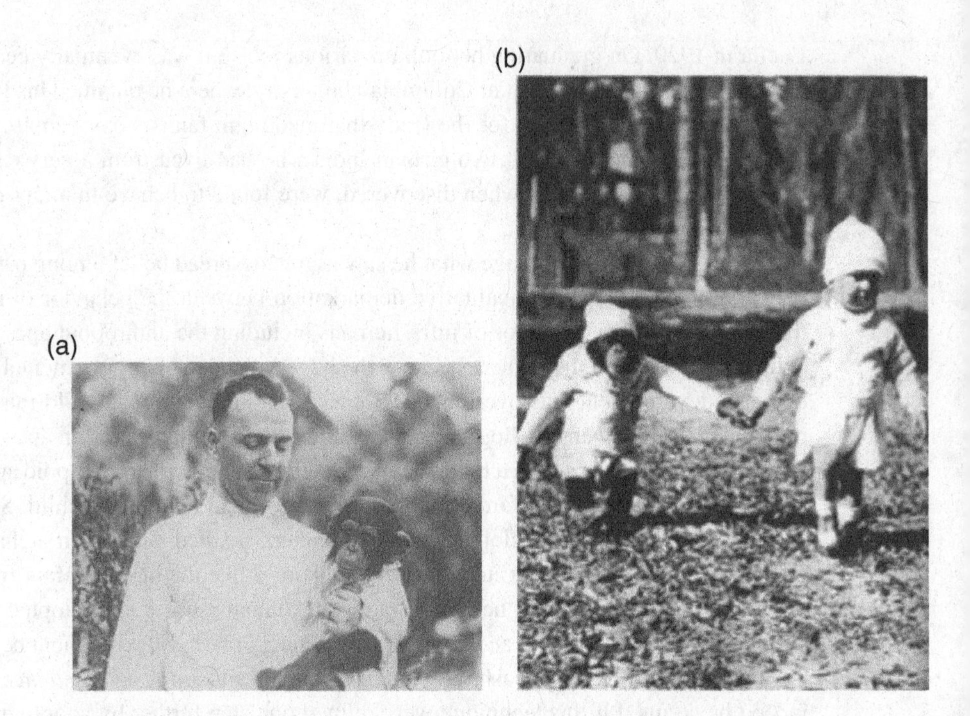

Figure 5.1 (a) Photo dates from 1932 and shows Winthrop Kellogg holding Gua. Reprinted by permission from Springer Nature, Copyright © 2017, Association of Behavior Analysis International.
(b) Photo shows Donald, the Kelloggs' son, with Gua; from Kellogg and Kellogg (1933). From Benjamin and Bruce (1982). Public domain.

became apparent that he was also vocally imitating his playmate." Such behavior was first observed during his fourteenth month in the reproduction of her 'food bark' or 'grunt.'[7] Later in their book the Kellogs noted: "The superiority of the child in vocal imitation stands also as a striking difference between the two subjects. His specific reproduction of certain noises made by the ape we have already considered. In view of the human's rapid adoption of such barks or grunts, and his subsequent employment of them in new and original situations, one can easily realize the full possibility of the so-called wild children taking over completely the growls and noises of the animal associates with which they are reported to have been found."[8]

The combination of constant testing and having an ape as his constant companion and closest friend seems to have delayed Donald's development, particularly in terms of language acquisition. Presumably for this reason the project was terminated earlier than planned, with Gua at 16 months of age returned to the primate colony, while the Kellogg family returned to Indiana.

On his return, Kellogg resumed a range of research projects, including conditioning experiments, many of which involved dogs as subjects. Another study examined maze learning in water snakes. With the outbreak of World War II, Kellogg, now 41 years old, again volunteered and this time flew in South America and Trinidad. After the war, Kellogg returned to the psychology department at Indiana where

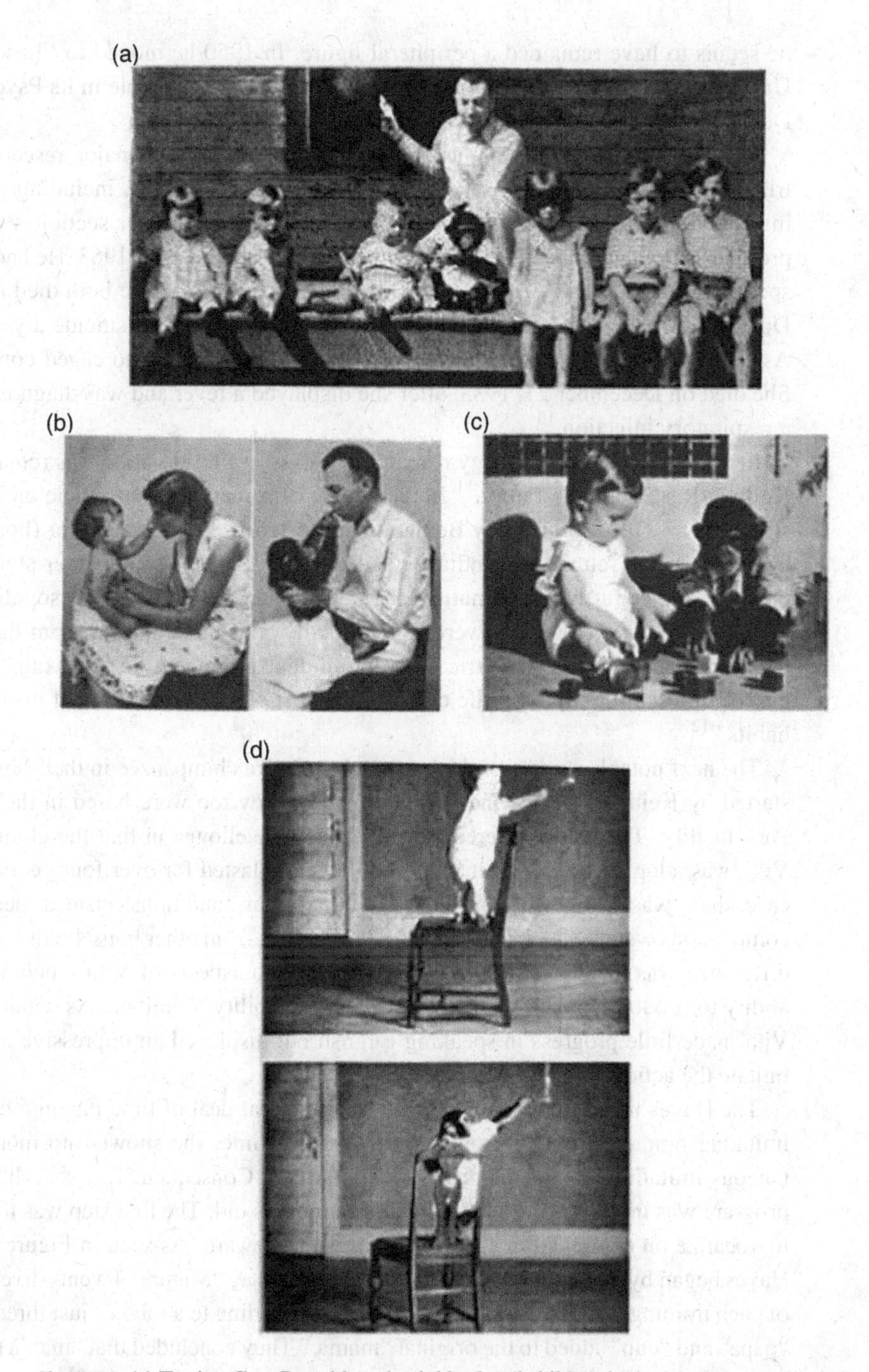

Figure 5.2 (a) Testing Gua, Donald, and neighborhood children invited to the party for their reaction times (startle response) by filming when Kellogg fired a starting pistol behind them without warning. (b) Testing Donald and Gua for their interest in human faces and in playing with toy bricks. (c) Testing their problem-solving ability, when they needed to move a chair to obtain a cookie suspended from the ceiling.
From Kellogg and Kellogg (1933). Public domain.

he seems to have remained a peripheral figure. In 1950 he moved to Florida State University, where he was less of a loner and played a central role in its Psychology Department.

Kellogg's move provided the opportunity for his second major research contribution. This was to carry out studies of bottlenose dolphins, including pioneering research on their use of sonar and – as described in a later section – on their problem-solving abilities.[9] Kellogg retired from Florida State in 1963. He and Luella spent a lot of their time traveling over the final five years before both died in 1972. Donald Kellogg, who had worked as a psychiatrist, committed suicide a year later. As for Gua, she did not survive for long following her return to caged conditions. She died on December 21, 1933, after she displayed a fever and was diagnosed with a respiratory infection.[10]

In 1982 the main psychology research building at Florida State was renamed the Kellogg Research Laboratory.[11] In the course of research for an article on Kellogg stimulated by this event, Ludy Benjamin contacted Kellogg's daughter (born when her parents had returned to Indiana) and over 30 of Kellogg's former students in order to gather further information about Kellogg, including his personality. The responses to the latter query were exceptionally consistent and far from flattering: "...more than a little egocentric ...overconfident might not be too exaggerated a description ...little time for idle chatter ...almost obsessive in some of his research habits."[12]

The next notable study, in which a couple raised a chimpanzee in their home, was started by Keith and Catherine Hayes in 1947. They too were based in the Orange Park facility. The study differed from that of the Kelloggs in that the chimpanzee, Viki, was adopted shortly after birth and the study lasted for over four years. In this case, there was no human child raised under similar conditions; instead, occasional comparisons were made with children of similar age from other households. A further difference was that the Hayes concentrated on two aspects of Viki's behavior, her ability to produce spoken English words and her ability to imitate. As detailed next, Viki made little progress in speaking English but displayed an impressive ability to imitate the actions of her human 'parents.'

The Hayes noted that, although Viki spent a great deal of time playing, of which imitating human activities occupied a lot of this time, she showed no more spontaneous imitation of human speech than Gua had. Consequently, a speech training program was introduced when Viki was five months old. The first step was to get her to vocalize on command in order to obtain some reward. As seen in Figure 5.3, the Hayes began by manipulating her lips to train her to say "Mama." Twenty-five months of such training resulted in Viki's vocabulary expanding to a total of just three words: "papa" and "cup" added to the original "mama." They concluded that "man's superior ability to use language may be his only important *genetic* advantage."[13]

In complete contrast, Viki's ability to imitate all sorts of actions strongly impressed the Hayes. Results from casual observations and from a series of tests continued to amaze – and amuse – Viki's 'parents.' One example of spontaneous imitation followed after Viki had on several occasions watched Catherine Hayes put on her make up: Viki

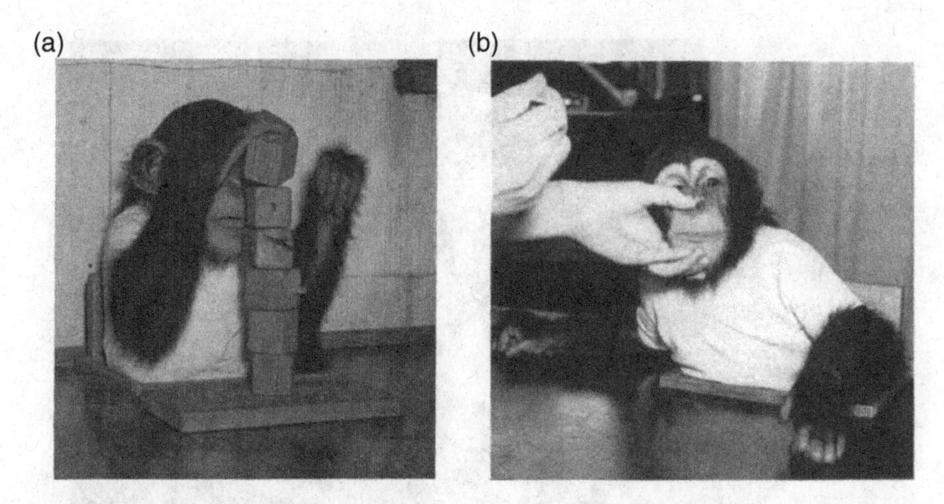

Figure 5.3 Photos illustrating Viki's manual dexterity (a) but problem with saying "Mama" (b). From Hayes and Hayes (1951). Public domain.

"appropriated a lipstick, stood on the washbasin, looked in the mirror, and applied the cosmetic – not at random, but to her mouth. She then pressed her lips together and smoother the color with her finger, just as she had seen the act performed."[14]

More formal tests included the *stick-and-tunnel* problem: An attractive lure was placed in the middle of a long tunnel and could be retrieved only by using a stick to poke it out. One demonstration of the solution by the experimenter was not enough. However, after the second such demonstration, Viki picked up the stick and pushed out the lure. Four children of around the same age needed either one or two demonstrations to solve the problem. On the other hand, a caged chimpanzee made no progress on this problem despite eleven demonstrations.

Viki also demonstrated delayed imitation. In one test she was allowed to watch the experimenter use a screwdriver to prize open the lid of several empty paint cans. She was then taken away for an hour and on return allowed to use the screwdriver. "She set to work immediately, and after some effort succeeded in opening two cans."[15]

As shown in Figure 5.4, Viki's tendency to imitate led her into what many years later would be regarded as bad habits. The Hayes even used the offer of a cigarette as a reward!

In deciding that Vicki's lack of progress in acquiring spoken English was of genetic origin, the Hayes implied that chimpanzees are born with a different kind of brain to that of humans. The primary aim of the next husband and wife team to raise a young chimpanzee was to train it to communicate. They also appealed to a genetic basis for previous failures to train chimps to speak English but one that gave rise to large differences in the vocal apparatus of the two species: Chimpanzees lack the fine control over tongue, lip, and mouth shapes that humans possess. In contrast, chimpanzees, especially home-raised individuals, display a great of spontaneous hand gestures and, as the Hayes, had found, readily learn to imitate gestures made by their human caretakers.

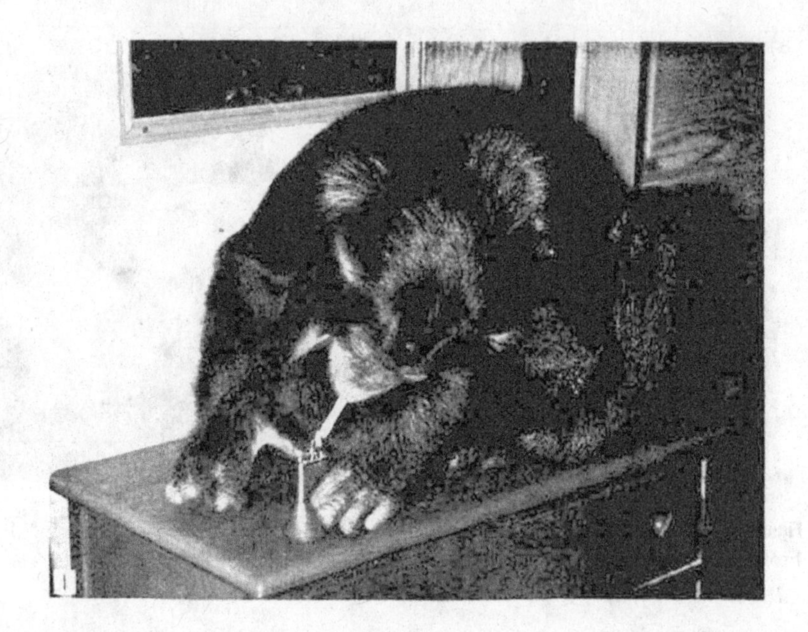

Figure 5.4 Viki at six years smoking a cigarette.
From Hayes and Hayes (1954). Public domain.

Many years earlier Yerkes had noted: "I am inclined to conclude from the various evidences that the great apes have plenty to talk about, but no gift for the use of sounds to represent individual feelings or ideas. Perhaps they can be taught to use their fingers, somewhat as does the deaf and dumb person, and thus helped to acquire a simple non-vocal 'sign language.'"[16] It took four decades before this suggestion was first implemented by Allen and Beatrix Gardner. "Psychologists who work extensively with the instrumental conditioning of animals become sensitive to the need to use responses that are suited to the species they wish to study. Lever-pressing in rats is not an arbitrary response invented by Skinner to confound the mentalists; it is a type of response commonly made by rats when they are first placed in a Skinner box.... We chose a language based on gestures because we reasoned that gestures for the chimpanzee should be analogous to bar-pressing for rats, key-pecking for pigeons, and babbling for humans."[17]

R. Allan Gardner obtained his Ph.D. at Northwestern University in Chicago and subsequently worked in a Defense Department research center in Natick, Massachusetts, where he continued his doctoral research on human learning. In 1960 he met the woman who was to become his wife and fellow researcher for the next 26 years. Beatrix Gardner (1933–1986) had been a refugee from Nazi-occupied Poland who had eventually reached the United States after spending some years in Brazil. What must have been an outstanding high school education enabled her to enter Radcliffe College – only later was this all-female college fully integrated into Harvard University – and, on graduating, to relocate to the UK and carry out ethological research – on the behavior of stickleback fish – at Oxford University, with Niko Tinbergen as her supervisor. After obtaining her Ph.D. in 1959, she returned to the

United States to take up a teaching position at Wellesley, another all-female college. She married Allen Gardner in 1961.[18]

In 1963 the two of them obtained academic positions in the psychology department at the University of Nevada. It seems that soon after their arrival they took the decision to embark on the project that was to make them famous; indeed, for a period in the early 1970s, they became celebrities. The female chimpanzee they named Washoe was wild-born and estimated to be from 8 to 14 months old when she arrived in their laboratory in June 1966. Rather than in the family home, Washoe's quarters were in a trailer home next to the house. "At the outset we were quite sure that Washoe could learn to make various signs in order to obtain food, drink and other things. For the project to be a success we felt that something more must be developed. We wanted Washoe not only to ask for objects but to answer questions about them and also to ask us questions. We wanted to develop behavior that could be described as conversation."[19] "A number of human companions have been enlisted to participate in the project and relieve each other at intervals, so that at least one person would be with Washoe during all her waking hours.…All of Washoe's companions have been required to master ASL (American Sign Language) and to use it extensively in her presence, in association with interesting activities and events and also in a general way, as one chatters at a human infant in the course of a day."[20]

Systematic training in ASL was at first based mainly on imitation; Washoe would frequently imitate some sign in spontaneous fashion but it took until 16 months into the project before the Gardners could get Washoe to imitate a sign on command (Figure 5.5). This was maintained by using tickling as a reinforcer. A later addition to the training program was to use a guidance procedure: "Washoe could acquire signs that were first elicited by our holding her hands, forming then into the desired configuration, and then putting them through the desired movement."[21]

Three features of Washoe's signing were of particular interest. First, whether she would produce signs spontaneously "for no obvious motive other than communication"; they reported the example of the first use of the sign for 'toothbrush' as one of the first examples.[22] The second was whether, having learned the sign for a particular referent, Washoe would show transfer of such learning to other members of the same class; again, many examples of such transfer were reported, of which the sign for 'flower' was given as an example. The third, and most controversial aspect of Washoe's use of ASL, was the question of whether she ever combined signs in a new and meaningful way. The Gardners were convinced that Washoe had this ability. The most famous example occurred when Washoe was taken on an excursion and saw for the first time a duck fly off from a pond; "Water bird," she signed.

Very late in his life, Pavlov had mused about the *second signal system*, one unique to humans.[23] He does not seem to have developed the idea, but later Soviet followers "treated the second signal system as a major part of his thought.-- probably because it made it easier to portray him as a non-reductionist who (as per dialectical materialism) recognized a qualitative difference between lower animals and humans."[24] Most behaviorists, including Hull and Spence, had generally avoided discussing the human capacity for language. The notable exception was B. F. Skinner who published in

Figure 5.5 Washoe and Beatrice Gardner signing "drink."
From Gardner and Gardner (1978; p. 143). Public domain.

1957 an analysis of human language within the framework of his concept of operant conditioning. As described in Chapter 8, his book, *Verbal behavior,* was savagely criticized by a graduate student in linguistics.[25]

Chomsky's review and his theories of grammar stimulated a revival of interest in language among psychologists. Thus, the Gardners' first report on their Washoe project was published at a time of passionate debate between those who held that language should be regarded as a human-specific 'instinct,' with a divide separating this from forms of communication in other species, and behaviorists who, like Skinner, claimed that 'verbal behavior' was as much subject to the laws of learning as any other kind of behavior. The Gardners were firmly in the latter camp. "From time to time we have been asked questions such as, 'Do you think that Washoe has language?' or 'At what point will you be able to say that Washoe has language?' We find it very difficult to respond to these questions because they are altogether foreign to the spirit of our research. They imply a distinction between one kind of communicative behavior that can be called language and another class that cannot. This in turn implies a well-established theory that could provide the distinction. If our objectives had required such a theory, we would certainly not have been able to begin this project as early as we did."[26]

By 1966 word of the Washoe project was creating great interest well before any report was published. Among those excited by the prospect that communication between apes and humans might, at last, be established was a psychologist at

Columbia University in New York. Herbert Terrace had obtained an academic appointment there in 1961, directly after completing a Harvard Ph.D. based on the experiments on discrimination learning by pigeons that are described in Chapter 7. In 1967 Terrace visited the Gardners and Washoe and was greatly impressed. The following year he collected his first chimpanzee from a primate center in Oklahoma and flew back to New York with 5-week old, home-raised Bruno. Bruno then lived for fourteen months as part of the family of a former student of Terrace's, before being returned to Oklahoma.

Since this trial run had gone extremely smoothly, Terrace went ahead with plans to set up his own project for teaching sign language to a chimpanzee. Preparations included time spent back at Harvard discussing the nature of language with Skinner and the psychologist who pioneered modern studies of language development in children, Roger Brown. They also included collaboration with a colleague at Columbia, the MIT-trained linguist and student of Noam Chomsky, Tom Bever, and the search for funding to support the project.[27]

Project Nim started in November, 1973 with the arrival in New York of a 5-day old male chimpanzee from Oklahoma that Terrace and Bever decided to name, 'Nim Chimpsky.' "The goals of Project Nim are easily stated. I wanted to socialize a chimpanzee so that he would be just as concerned about his status in the eyes of his caretakers as he would about the food and drink they had the power to dispense. By making our feelings and reactions a source of concern to Nim, I felt we could motivate him to use sign language, not just to demand things, but also to describe his feelings and tell us about his views of people and objects. I wanted to see what combinations of signs Nim would produce without special training, that is, with no more encouragement than the praise that a child receives from its parents (see Figure 5.6). I especially wanted to find out whether these combinations would be similar to human sentences in the sense that they were generated by some grammatical rule."[28]

Compared to Washoe's home life and training, Nim's were far more variable. Serious problems beset the project. These included repeated changes in his caretakers and teachers and in his living quarters. A major headache for Terrace was that of obtaining adequate funding. Nevertheless, Nim's use of sign language continued to develop. By the end of the project, at the age of four years, he had apparently acquired around 120 signs. More importantly, he regularly produced combinations of signs in what appeared to be systematic and meaningful ways. "When we finally completed them during the summer of 1977, the tedious statistical analyses of Nim's signing behavior gave us just what I had hoped for – a solid basis for demonstrating that a chimpanzee can create a sentence."[29] In other words, Terrace had hard evidence for the claims that the Gardners had made for Washoe. Lack of funding and the difficulty in finding appropriate people who were sufficiently proficient in sign language led to the decision to end the project and return Nim to Oklahoma in September 1977.

Terrace's project did not end with Nim's departure. Further analysis of his 2-sign, 3-sign, or longer strings was needed. Terrace noted two features. First, Nim did not show any tendency to produce longer strings as he became older; this contrasted with data from even deaf children whose signing sequences rapidly became longer and

Figure 5.6 Herb Terrace communicating with Nim by signing "mine," while Nim then replies with "me; hat."
From Terrace (1979). "Mean Length of Utterance vs Age in Months" from NIM by Anna Michel, Introduction by Herbert S. Terrace, Photographs by Susan Kuklin and Herbert S. Terrace, copyright © 1980 by Anna Michel. Introduction copyright © 1980 by Herbert S. Terrace, Photographs copyright © 1979, 1980 by Herbert S. Terrace. Used by permission of Alfred A. Knopf, an imprint of the Knopf Doubleday Publishing Group, a division of Penguin Random House LLC. All rights reserved.

longer; see Figure 5.7. Second, the longer strings did not contain any more information than the shorter strings; most were long because one or two signs were repeated.

There were also videotapes to be analyzed. Up to this point, the only data were from the reports of Nim's teachers. What analysis of a proportion of the videos revealed was that a large proportion of Nim's signs were repeats of signs that the teachers had just made; only around 10% were novel, in that, there was no overlap between the teacher's and Nim's signs. A similar analysis of the limited film footage of Washoe's signing that was publicly available suggested the same problem with the Gardner's data. Although Terrace does not use the term in his book about the Nim project, he implied that the apparently striking performances of both Nim and Washoe were largely based on unconscious cueing by the teachers, an effect named after the horse, Clever Hans, who appeared to be able to count. "I must therefore conclude – though reluctantly – that until it is possible to defeat all plausible explanations short of the intellectual capacity to arrange words according to a grammatical rule, it would be

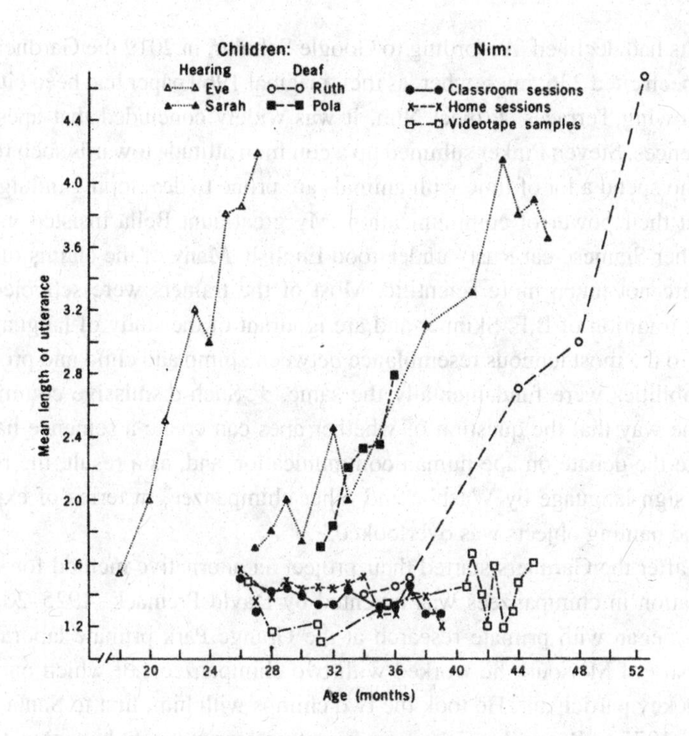

Figure 5.7 This graph shows the increasing length of children's utterances as they become older compared to the lack of increase in the length of Nim's utterances.
From Terrace, Petitto, Sanders & Bever (1979). Reproduced with permission from the American Association for the Advancement of Science.

premature to conclude that a chimpanzee's combinations show the same structure evident in the sentence of a child."[30]

The Gardners were not deterred by this and other highly negative accounts of their research.[31] However, they do seem to have taken on board the criticism that their early reports on Washoe were heavy on anecdotes and light on reporting the basic data. Washoe was followed by three further home-raised chimpanzees who were also taught American Sign Language and also subjected to a double-blind vocabulary test. The fully reported results are convincing that all four chimpanzees were able to produce from 16 to 35 correct signs when shown pictures of a range of objects. The authors note that "controls for Clever Hans errors have been standard practice in comparative psychology. To date, most, if not all, research on human children has been carried out without any such controls. It is as if students of child development believed that, whereas horses and chimpanzees may be sensitive to subtle nonverbal communication, it is safe to assume that human children are totally unaffected.… A striking exception to this rule has been the work of Terrace (1979) with the chimpanzee Nim which included no controls for Clever Hans errors, whatsoever."[32]

Research on signing in chimpanzees, including possible communication between chimpanzees, was also carried on by the Gardners' Ph.D. student, Roger Fouts.[33] However, by the early 1980s interest in signing by chimpanzees among psychologists and linguists had declined. According to Google Scholar, in 2019 the Gardners' 1984 paper had been cited 236 times, whereas their original 1969 paper had been cited 1978 times. Following Terraces' Project Nim, it was widely concluded that apes cannot create sentences. Steven Pinker summed up a common attitude towards such research: "People who spend a lot of time with animals are prone to developing indulgent attitudes about their power of communication. My great-aunt Bella insisted in all sincerity that her Siamese cat Rusty understood English. Many of the claims of the ape trainers were not much more scientific. Most of the trainers were schooled in the behaviorist tradition of B.F. Skinner and are ignorant of the study of language; they latched on to the most tenuous resemblance between chimp and child and proclaimed that their abilities were fundamentally the same."[34] Such dismissive commentaries reflected the way that the question of whether apes can create a sentence had come to dominate the debate on ape-human communication and, as a result, the remarkable use of sign language by Washoe and other chimpanzee, in terms of expressing requests and naming objects was overlooked.

Shortly after the Gardners started their project an alternative method for studying communication in chimpanzees was invented by David Premack (1925–2015). His career also began with primate research at the Orange Park primate laboratory. At the University of Missouri, he worked with two chimpanzees, of which one, Sarah, became his key participant. He took the two chimps with him, first to Santa Barbara and then in 1975 to Pennsylvania, where he set up a primate study center, linked to the University of Pennsylvania. The center held nine chimpanzees. Notably, from the start all of his subjects remained caged animals.

The communication task for the individuals he studied, notably Sarah, was to learn the 'meaning' of a large number of plastic objects varying in shape, color, and texture. Sarah became proficient in the use of around 100 of such symbols, learning to both produce 'sentences,' that is, a number of symbols arranged in a correct sequence, and to respond appropriately to such a sequence, for example, "Sarah give Mary apple."[35] A variation of this approach was developed by Duane Rumbaugh. In this case, the symbols were different shapes – *lexigrams* – on a special computer keyboard. Several chimpanzees were trained on this system with varying success. The approach was further developed by Sue Savage-Rumbaugh so that pressing a key produced a spoken English word. Bonobos appeared to be exceptionally good at learning to use this system. A criticism of earlier research was that "early ape language studies were predestined to fail because … they concentrated on production to the virtual exclusion of comprehension."[36] In the context of research showing that in babies understanding speech precedes talking, this group implemented a training program with a bonobo, Kanzi, starting when he was only six months old. When he was eight years old, blind tests using novel spoken English sentences of increasing complexity revealed that his level of comprehension was similar to that of a two-year-old child given identical tests.[37] In terms of producing of sequences of lexigrams, like Nim, Kanzi failed to produce longer and longer strings as he grew older. Since then, there appear to have been no more substantial projects on the controversial topic of ape language.

In the context of the present chapter what this series of studies indicated was that there appear to be profound differences in cognitive abilities between humans and apes, and between apes and other primates, that are not easily explained in terms of motor control, sensory ability, motivation, or environmental opportunities. A major achievement was to show that communication between humans and other species could be established to an extent that was previously unimagined. Arguably the most influential finding from this body of research was obtained from some tests given to Premack's Sarah. The results suggested that, on the basis of a series of pictures, she could understand what goal some human might have in mind.[38] The suggestion inspired Josef Perner to develop tests that would reveal the stage at which young humans develop a *theory of mind* and thus launch what became a massive research enterprise in developmental psychology.[39]

Harry Harlow, Learning Sets, and Maternal Attachment

There have been only a few systematic studies of the great apes' behavioral and cognitive abilities and these have involved at most a handful of individuals. A major reason for this has been the high costs involved. Cost was also an important reason why studies of the learning and cognitive abilities of monkeys were also limited to a few individuals until the middle of the twentieth-century, when Harry Harlow (1905–1981) set up a breeding colony of rhesus monkeys at the University of Wisconsin.

Harry was born as the third of four sons to the impoverished Isaac family in a small town in Iowa. He was soon recognized as particularly argumentative. The family gave high priority to education and somehow supported Harry for the year he attended Reid College in Oregon and then at Stanford University, where he chose to major in psychology and then continue as a graduate student. An important intellectual influence was Lewis Terman, whose major work was the development of the Stanford-Binet IQ test, an endeavor that assumed the importance of genetic differences in intelligence, in contrast to Watson's and other behaviorists' claim that such influences were probably negligible. On obtaining his Ph.D. in 1930 Harry obtained a junior teaching position in the small and poorly equipped Department of Psychology at the University of Wisconsin; see Figure 5.8. Following Terman's advice, before taking up the position he changed his last name from Isaac to Harlow, even though he was only 1/64th Jewish.[40]

On arriving in Wisconsin Harlow carried out some behavioral tests of primates on visits to the local zoo. The results inspired him to set up the first primate facility entirely dedicated to behavioral research. He commandeered an abandoned building near to the university and, with some help from volunteers, converted it into a laboratory.[38] Harlow's earlier experiments with rats when a Ph.D. student at Stanford University had already made him highly critical of Hull's theories. A major aim of the Wisconsin research program was to demonstrate the inadequacy of Hullian theory to account for the learning abilities of monkeys.

One line of experiments attacked Hull's claim that reinforcing events were confined to those that reduced drives based on a few biological needs. Harlow demonstrated

Figure 5.8 Harlow in 1931, with Clara Mears, whom he married a year later. From Blum (2002). Reproduced with permission of their son, Bob Israel.

the power of intrinsic motivation by showing that his monkeys would work repeatedly to solve mechanical puzzles in the absence of any explicit reward.[41] His student, Robert Butler, placed his monkeys in a chamber containing small windows that were normally closed and showed that the mere opportunity to see what was going on in the world outside, whether just people passing by or a model train going around its circular track, was sufficient to motivate the monkeys to solve discrimination tasks when the only reward was a brief opening of a window;[42] see Figure 5.9. And if monkeys could be motivated by curiosity alone, then clearly this was true of humans. "Can anyone seriously believe that the insatiable curiosity-investigatory motivation of the child is a second-order or derived drive conditioned upon hunger or any other internal drive?"[43] As a jibe at Skinner, Harlow referred to the apparatus that Butler had built as the 'Butler Box.'[44] Ever fond of puns and wordplay, Harlow commented: "I do not want to discourage anyone from the pursuit of the psychological Holy Grail by the use of the Skinner box, but as far as I am concerned, there will be no moaning of farewell when we have passed the pressing of the bar."[45] Tennyson's poem, *Crossing the bar,* was possibly well known in Harlow's day.

An even more influential series of experiments involved the use of a more complicated piece of equipment that Harlow developed. The Wisconsin General Test

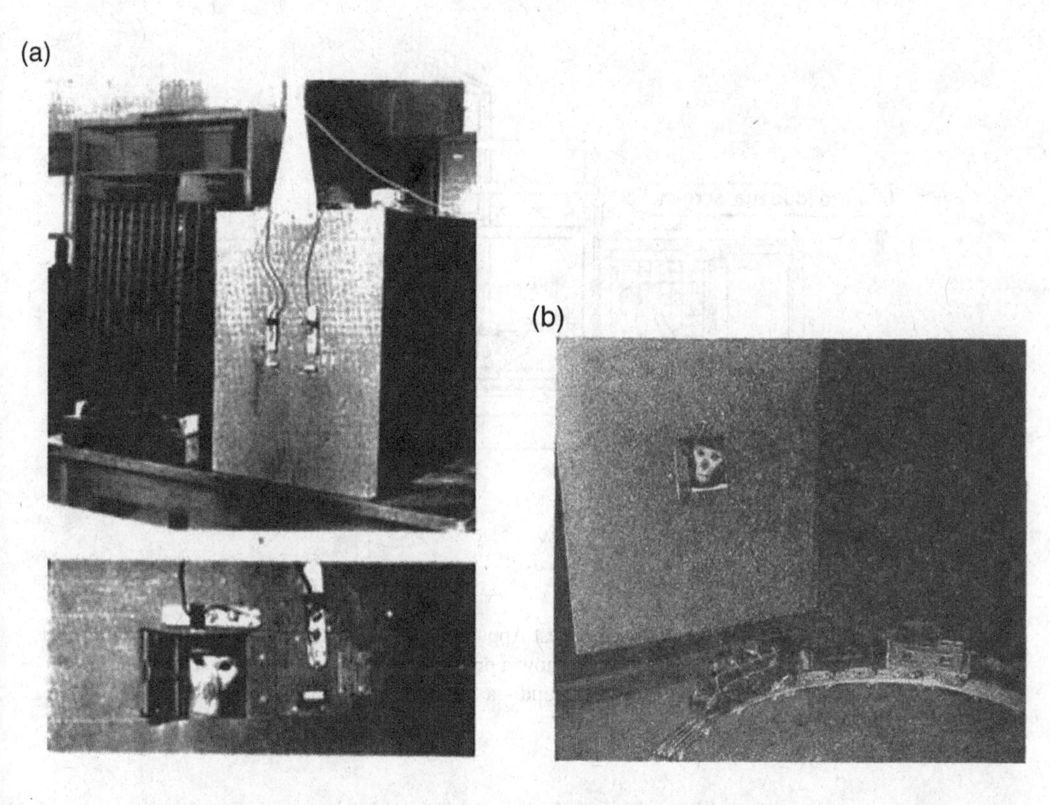

Figure 5.9 If a monkey pressed a lever, then a window would open that allowed the monkey a view into the next room for a limited time. The sight of a moving toy train sustained lever-pressing. The sight of a stationary banana did not.
(a) Photos are from Butler and Harlow (1954); (b) Photo from Butler (1954). Public domain.

Apparatus, or *WGTA*, is shown in Figure 5.10. It allows a monkey to reach out through the bars of its cage and choose one out of an array of objects that cover food wells. Choice of the object designated correct on a given trial results in discovery of a food item in this well, while other wells, covered by other objects, are revealed to be empty. Pulley systems allow the experimenter to raise or lower screens so that prior to a trial the monkey is unable to see which well is baited and during the trial is unable to see the experimenter.

The training programs for chimpanzees – for example, those used by the Hayes and the Gardners, as described in the previous section – were greatly influenced by the results Harlow obtained using the WGTA. Whereas almost all previous animal experiments on discrimination learning had given the subjects a single problem to solve, using the WGTA Harlow gave his monkeys problem after problem of the same kind. Thus, in a series of object discriminations initially, a monkey might find food under object A but not under B, in the next under C but not under D, and so on. This became known as a *learning set* procedure. What Harlow found was that the speed with which a monkey solved a given problem increased with

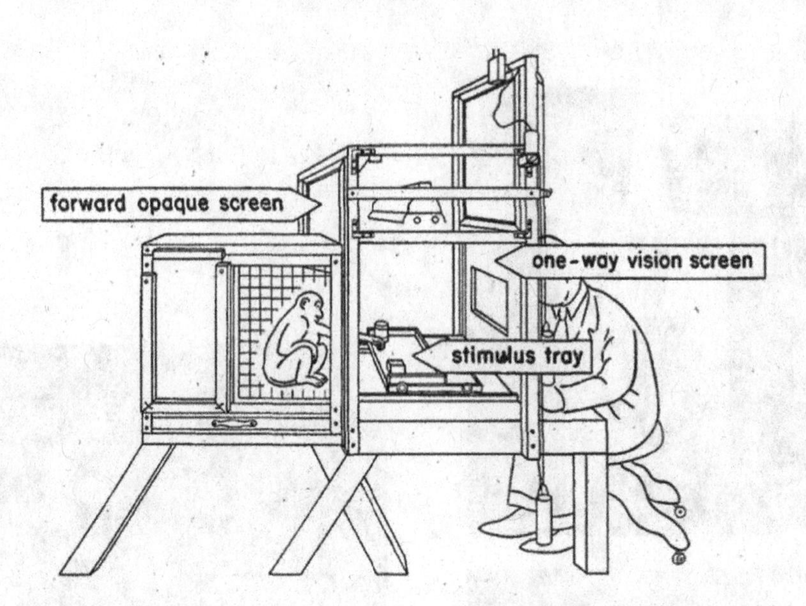

Figure 5.10 This Wisconsin General Test Apparatus (WGTA) was used in Harlow's and other labs to test whether monkeys would show a progressive improvement in learning a series of discrimination problems of the same kind – a "learning set." Other tasks using the WGTA included the oddity problem.
From Harlow (1949). Public domain.

the number of problems it was given until it could solve a new problem in a single trial; see Figure 5.11. For Harlow, this demonstrated that the monkey had shown *learning-to-learn*.[46]

As the following quote makes clear, Harlow believed that theories based on rat experiments had failed to provide any insight into human behavior and that this was because almost all such studies were based on a 'single learning situation.' 'The obligation of the theoretical psychologist is to discover general laws of behavior applicable to mice, monkeys, and men. In this obligation, the theoretical psychologist has often failed. His deductions frequently have no generality beyond the species he has studied, and his laws have been so limited that attempts to apply them to man have resulted in confusion rather than clarification. One limitation of many experiments on subhuman animals is the brief period of time the subjects have been studied.'

"The behavior of the human being is not to be understood in terms of the results of single learning situations but rather in terms of the changes which are affected through multiple, though comparable, learning problems.... This learning to learn transforms the organism from a creature that adapts to a changing environment by trial and error to one that adapts by seeming hypothesis and insight.... If learning sets are the mechanisms that, in part, transform the organism from a conditioned response robot to a reasonably rational creature, it may be thought that the mechanisms are too intangible for proper quantification. Any such presupposition is false."[47]

In his major paper reviewing the results from learning set experiments, Harlow also cited data from young children who were given up to 84 object discrimination problems

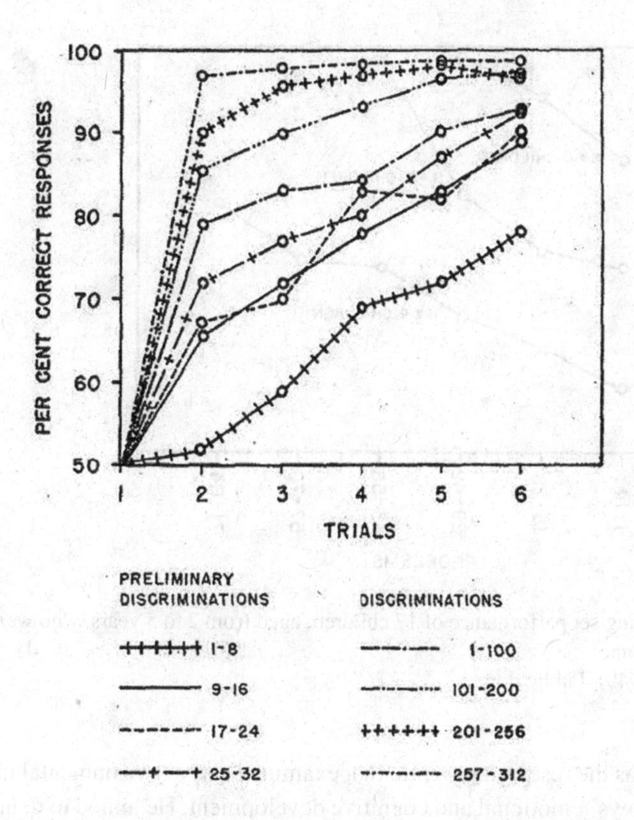

Figure 5.11 The average performance of eight rhesus monkeys when given a total of 312 6-trial discrimination problems, each involving two objects different from the objects used in any previous problem (*learning set* task). The most important data are those from Trial 2, where after 100 or so such problems the monkeys were at least 85% correct and after 257–312 trials almost 100% correct, thus showing strong learning based on Trial 1 alone.
From Harlow (1949). Public domain.

to learn. As shown in Figure 5.12, these revealed a learning set effect similar to those obtained with monkeys. These data were collected by Margaret Kuenne (1918–1971). Two years after arriving in Wisconsin Harlow had married a graduate student in the psychology department, Clara Mears. They were divorced in 1946. This was the year in which Margaret Kuenne arrived as an associate professor in the same department as Harlow; two years later she became Margaret Kuenne Harlow. She and Harry worked closely together until her death from breast cancer in 1971. Intriguingly, Kuenne had obtained her Ph.D. in 1944 from the University of Iowa following research on children supervised by the leading behaviorist, Kenneth Spence. One wonders how often intra-marital debates took place on the merits or otherwise of the Hull-Spence version of behaviorism. Less than a year after her death Harlow remarried Clara Mears.[48]

Harlow became a highly influential figure within academic psychology. In 1958–1959 he served as President of the American Psychological Association. Rather than his research on learning sets, what gave him lasting fame – and notoriety – among a

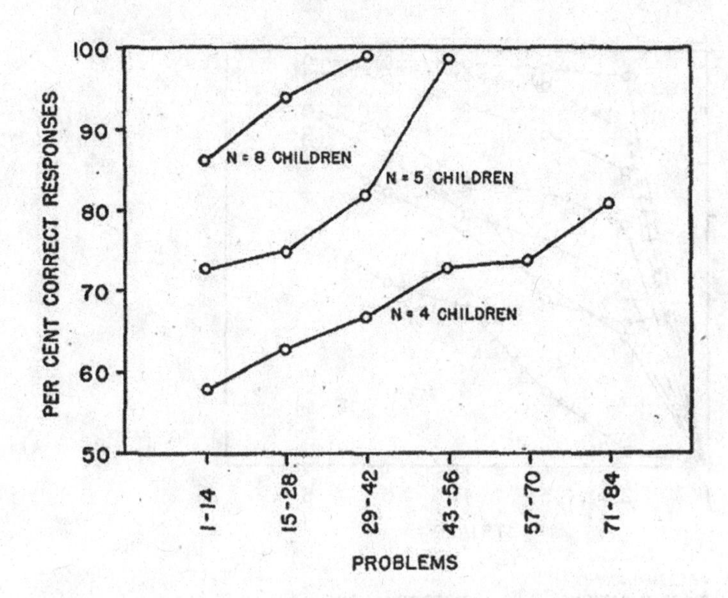

Figure 5.12 Learning set performance of 17 children, aged from 2 to 5 years, who were tested by Margaret Kuenne.
From Harlow (1949). Public domain.

wider public was the research program that examined early environmental influences on infant monkeys' emotional and cognitive development. He aimed to demolish the behaviorist prescription for child raising dating back to John Watson: "When you are tempted to pet your child remember that mother love is a dangerous instrument."[49]

Most of these studies entailed separating an infant from its mother shortly after birth. In a particularly notorious, study, an infant was shown to prefer a surrogate 'mother,' made out of a metal frame, but covered in cloth, to one consisting of only a frame and a bottle from which the infant could obtain milk; see Figure 13. This research was inspired, once again, by the aim of showing the inadequacy of behaviorists' analysis of 'mother love.'

"Love is a wondrous state, deep, tender, and rewarding. Because of its intimate and personal nature, it is regarded by some as an improper topic for experimental research. But whatever our personal feelings may be, our assigned mission as psychologists is to analyze all facets of human and animal behavior into their component variables.... From the developmental point of view, the general plan is quite clear: The initial love responses of the human being are those made by the infant to the mother or mother surrogate. From this intimate attachment of the child to the mother, multiple learned and generalized affectional responses are formed.... The position commonly held by psychologists and sociologists is quite clear: The basic motives are, for the most part, the primary drives – particularly hunger, thirst, elimination, pain, and sex – and all other motives, including love or affection, are derived or secondary drives. The mother is associated with the reduction of the primary drives – particularly hunger, thirst, and pain – and through learning, affection or love is derived."[50] Perhaps this 'commonly held position' became less common as a result of Harlow's research on infant monkeys?

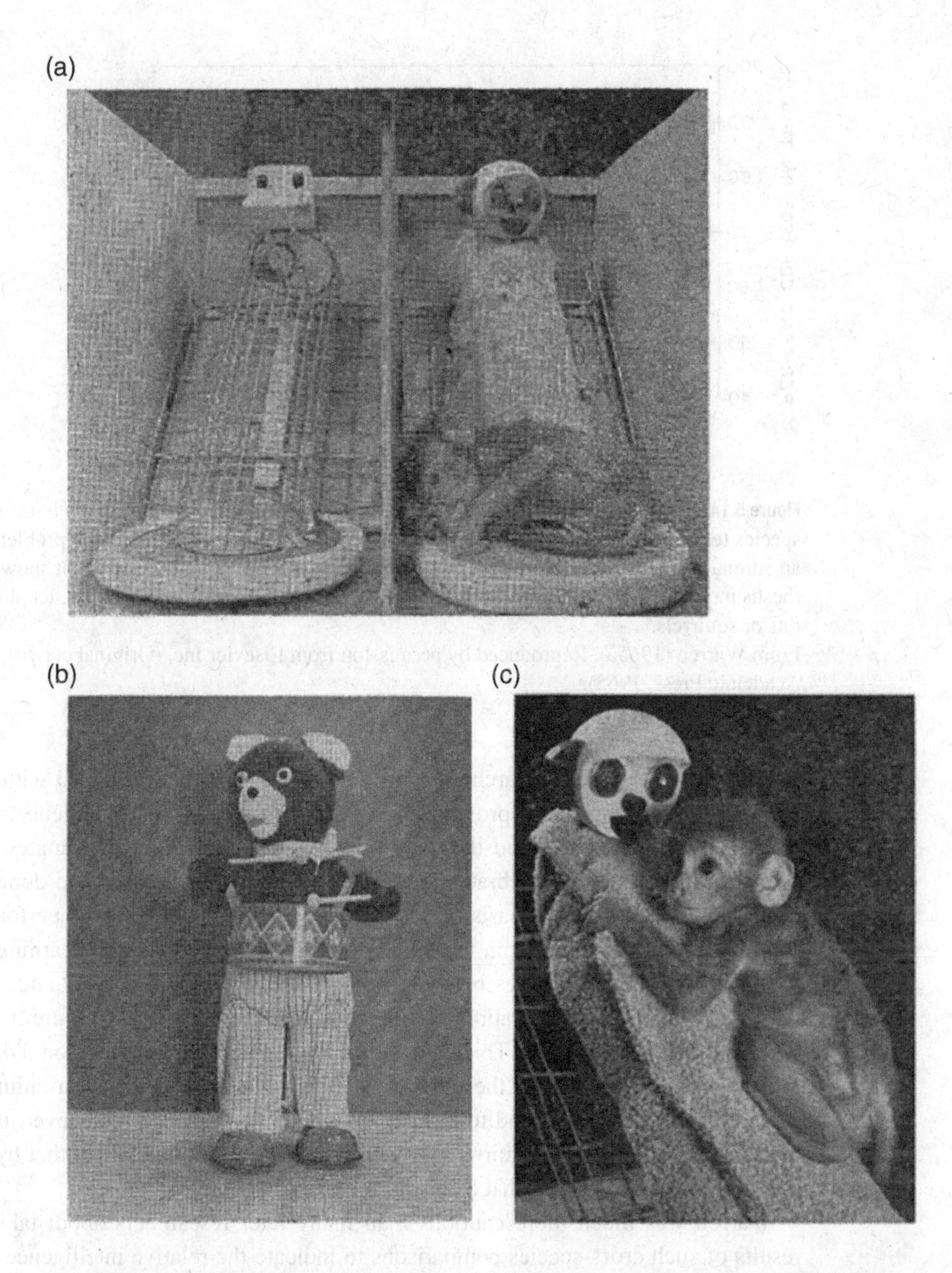

Figure 5.13 (a) Infant monkeys preferred the cloth-covered mother surrogate even when milk was available only from the wire "mother." (b) Toy figure used in fear test. (c) Reaction of infant monkey in fear test.
From Harlow (1958). Public domain.

A graduate student of Harlow's, John Warren, obtained his Ph.D. in 1953 after completing several experiments in which he tested monkeys on learning sets. On leaving Wisconsin he was one of several researchers who followed Harlow by giving learning sets to other species. In Warren's case, he first studied cats on a modified

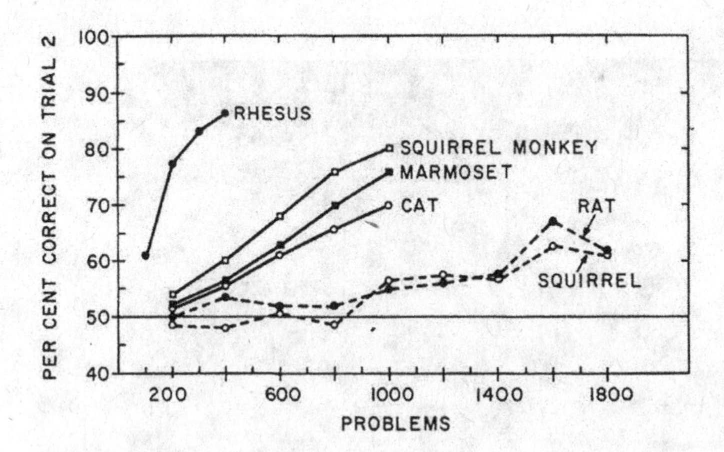

Figure 5.14 Performance on Trial 2 of a series of discrimination problems given to different species tested on a learning-set procedure, as a function of the number of such problems that an animal has previously been given. It may be seen that the rapid improvement shown by rhesus monkeys was not, for example, shown by the South American monkeys, let alone by rats or squirrels.
From Warren (1965a). Reproduced by permission from Elsevier Inc. (Original copyright: Academic Press, 1965).

WGTA. Much of such research was inspired by the idea that the speed with which learning-to-learn took place provided a cross-species measure of intelligence.

In 1965 Warren addressed the question of whether learning by primates differs from learning in other vertebrates. First, he concluded – as Yerkes had done many decades earlier but on the basis of less substantial evidence – that the performance of primates did not differ from that of any other vertebrate on simple learning tasks. However, species differences began to appear on more complex tasks and, in particular, in learning set formation. He included the graph shown in Figure 5.14 that was frequently reproduced. This shows that, in terms of performance on Trial 2 of a problem as a function of the number of discrimination problems an animal had solved, no species was found to rival Harlow's rhesus monkeys. However, the rate of *learning-to-learn* by a squirrel monkey was found to be faster than that by a cat, that in turn was faster than that by a squirrel or a rat.[51]

Warren was much more cautious than many later researchers about taking the results of such cross-species comparisons to indicate the relative intelligence of different species. "These conclusions must be regarded as tentative for two reasons. Generalizations of negative results are very hazardous; it is quite possible that dogs or dolphins would succeed where cats have failed, or that cats tested under more appropriate conditions would perform more adequately than the groups discussed in this chapter. In either event, the suggested qualitative difference between primates and other mammals would vanish. In the second place, the very similar learning set performance of cats and marmosets raises considerable doubt about differences between orders in learning capacity."[52]

Dolphin Intelligence and Communication

In the 1960s there was growing interest in marine mammals and, in particular, the intelligence of dolphins and their apparent ability to communicate with each other. There were several reasons for this interest. Studies of the dolphin's senses and of its large brain raised questions for psychologists and other scientists about the possible functions of all that gray matter. A second reason was the US Navy's investment in the use of marine mammals for military purposes; see Figure 5.15. For many decades dolphins and other sea mammals were trained at the US Naval Undersea Warfare Center (Wood, 1973) Furthermore, research grants became available from the Office of Naval Research for university-based research on dolphin behavior.

One of the first studies of learning by a dolphin – referred to as a 'porpoise' in the initial research paper – was carried out by Winthrop Kellogg and his student, Charles

Figure 5.15 The US Navy started training dolphins and other marine mammals in 1949. They have been used ever since by the US and other countries for a variety of purposes. This photo shows a dolphin used for mine clearing during the Iraq War.
U.S. Navy photo by Photographer's Mate 1st Class Brien Aho, Public Domain, Wikipedia.

Rice. Intriguingly, at the start of his research career Kellogg had studied the home-raised chimpanzee, Gua, and now towards the end of his career, Kellogg worked with a dolphin that was as close to home-raised as such an animal could be. "At the time of the experiment (Paddy) was approximately 8 years old and had lived in captivity for 7 years. As a kind of compensation for being without other dolphins, Paddy has had throughout his captivity, a nearly 24-hour contact with human beings. Mr. John H. Hurlbut, owner of the display, constructed his aquarium so that the tank is, in effect, a large room in Mr. Hurlbut's home and directly adjoins his apartment. He or others are with the porpoise almost constantly. This 7-year intimacy with human beings has given the animal a sophistication and familiarity with man which has probably never been duplicated with any other dolphin."[53]

What prompted this study was the apparent contradiction between claims that the dolphin's visual abilities were very poor – especially compared to the superb sonar-based ability to perceive its underwater world – and informal observations of its behavior when jumping out of the water. Commenting on the performance of dolphins trained to perform in various commercial aquaria, Kellog and his student, Rice, wrote: "Leaping from the water for vertical distances up to 16 feet, they seize a fish held as bait in the air. How would such tricks be possible, it is argued, if the animals possessed poor eyesight?"[54]

The study started in the Spring of 1962. Using a special piece of equipment that Kellogg had designed and built, Paddy was trained to discriminate between successively presented pairs of shapes, such as a triangle, a heart, or star; see Figure 5.16. These were cut out of thin sheets of brass and painted black and were presented against a white background both below and above the water line. Reward for making a correct choice was a small piece of herring. An errorless training procedure was employed, of the kind that Herb Terrace had recently developed to train pigeons,[55] as described in Chapter 7. It turned out that Paddy was long-sighted, at least when the shapes were presented in the air: "It appeared as though a starting distance of 8 feet or more was necessary for visual perception from water to air and that when Paddy approached from points closer than 8 feet, his perception was poor."[56]

Following training on many simple shape discriminations, Paddy was given more complex tasks. For example, he needed to learn that "when paired with a circle, four triangles are all positive shapes and to be approached. Yet when paired with an apex-upward triangle, the other three triangles are negative shapes and are to be avoided.... In spite of such difficulties the dolphin had little or no trouble responding to, and retaining, these combinations and could respond to one after another in a single experimental session without making a mistake. It made two discriminations without error after an interval of 7 months."[57]

Popular interest in dolphins was prompted partly by displays of dolphin behavior in commercial aquaria. By the 1950s there were three such aquaria in Florida and one in California. The displays became ever more impressive as training procedures improved. This partly reflected the adoption of operant conditioning techniques developed for the laboratory rat or pigeon. Keller Breland, an early student of B.F.

Figure 5.16 The 'home-raised' bottlenose dolphin, Paddy, discriminating between two shapes presented underwater.
From Kellogg and Rice (1963). Public domain.

Skinner (see Chapter 8), was particularly important in this development; he and his wife, Marion Breland, trained a variety of species for commercial purposes.[58] A particularly important innovation, especially when training dolphins, was first to establish some auditory stimulus, such as a whistle or a clicker, as a conditioned reinforcer by pairing it intermittently with food.[59]

Another influence was a very popular book, titled *Man and dolphin.*[60] It included an exciting speculation; the first two sentences read: "Eventually it may be possible or humans to speak with another species. I have come to this conclusion after careful consideration of evidence gained through my research experiments with dolphins."[61]The author was John Lilly (1915–2001). He was born to a wealthy family in Saint Paul, Minnesota, trained as a physician, and then became involved in research projects, eventually ones involving dolphins. A personal account of this research was included in his 1961 book, along with a great deal of factual information about dolphins, about the ways he recorded the sounds made by dolphins, and about the trials and tribulations of setting up his own dolphin research station on the Caribbean island of Saint Thomas in the Virgin Islands; see Figure 5.17.

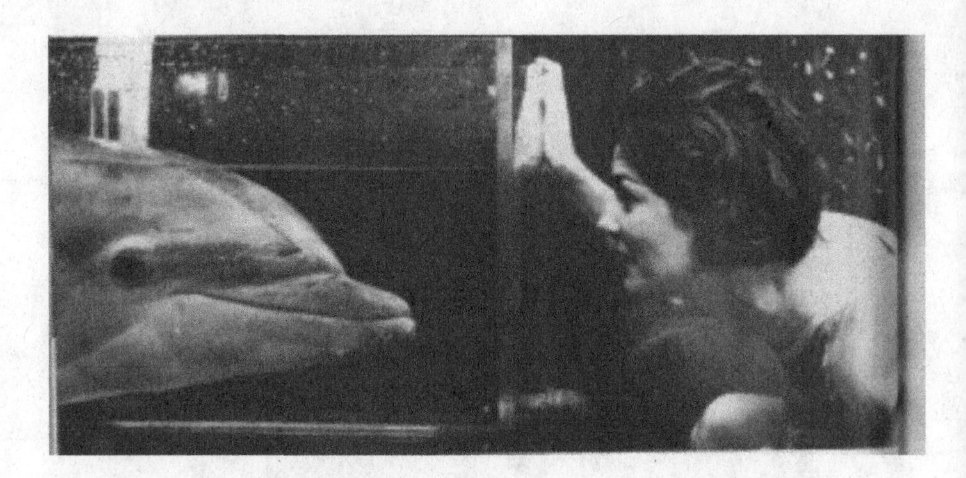

Figure 5.17 Elvar, one of the first dolphins to be tested in John Lilly's research institute, interacting with Lilly's wife and coresearcher, Elizabeth Lilly.
Photographs from MAN AND DOLPHIN: ADVENTURES OF A NEW SCIENTIFIC FRONTIER by John C. Lilly, copyright © 1961 by John C. Lilly. Used by permission of Doubleday, an imprint of the Knopf Doubleday Publishing Group, a division of Penguin Random House LLC. All rights reserved.

Man and dolphin contained some very prescient speculations. Well before the launch of US Navy program described above, Lilly wrote: "Cetaceans might be helpful in hunting and retrieving nose cones, satellites, missiles and similar things that men insist on dropping in the ocean. They might be willing to hunt for mines, torpedoes, submarines, and other artefacts connected with our naval operations."[62] Lilly did not actually claim that dolphins possessed a language, something far more 'intellectual' (his term) than the communication systems identified in other species, such as bees. But he suggested that it was a distinct possibility. Press coverage of his research portrayed him as a "man that talks with dolphins."

Lilly's subsequent fame as a member of the 1960s counter-culture – he was a friend of Allen Ginsburg and Timothy Leary – and his involvement in the search for extra-terrestrial intelligence further undermined respect on the part of his scientific peers. Meanwhile, research on dolphins was continued at his private institute, which survived his death.[63]

An early test of whether a dolphin could transmit information to another dolphin was carried out in the late 1960s. First, the two dolphins, Buzz and Doris, were given some pre-training when they could see both each other and visual stimuli anywhere in the circular tank in which they lived. For the important test, the tank was divided in half by a canvas curtain that prevented the dolphins from seeing, but not from hearing, each other. Two paddles were suspended side-by-side in Buzz's half of the tank. His task, as *receiver*, was to push the right-hand paddle whenever – unseen to him – a light was flashing in Doris's half of the tank but push the left-hand paddle if a steady light came on in her half of the tank. If Buzz made the correct choice, then fish rewards were delivered both to Buzz and to Doris as the *sender*; if he was wrong, then the trial ended without any such reward. Thus, the animals could obtain the fish reward on a high percentage of trials only if Doris could somehow tell Buzz whether the light was steady or flashing; otherwise, performance would remain at chance, fish delivered on just 50% of trials. Performance of the two dolphins gradually improved, reaching a level of more than 90% correct after around one thousand such trials.

Throughout the experiment the researchers, William Evans from the US Naval Warfare Center in Pasadena and Jarvis Bastian from the Department of Psychology at the University of California, Davis, recorded the sounds made by Buzz and Doris and then spent a very large number of hours analyzing these records, hoping to identify the acoustic signals that carried the message from Doris to Buzz as to what she could see. Eventually, they discovered that Doris responded to the steady light by emitting a short burst of sonar pulses but did not respond in this way to the flashing light. Furthermore, Doris behaved in the same way in the absence of Buzz; this indicated that the information she had previously communicated to Buzz was not intentional. Evans and Bastian concluded that "the basis of this rather complex performance is very much what might have been expected had the female been trained by herself to emit signals differentially in the presence of the different visual signals and if the male had been separately trained to press his paddles differentially with respect to these acoustical signals. This is what actually happened, except that, in effect, they trained themselves."[64] The experiment did not disprove the claim that dolphins possess some form of language but it seems that it is not one capable of expressing whether a light is flashing or not.

How Buzz and Doris trained themselves was suggested by an experiment I modeled on this dolphin study, using pairs of pigeons. The sender pigeon was presented with either a red or a green light; and these lights were out of sight of the receiver that was faced with a choice between two white response keys. The rule for some pairs of pigeons was 'if green, choose left; if red, choose right'; this was reversed for other pairs. Several pairs solved the problem in around the same number of trials that Buzz and Doris had needed. How the pigeons 'trained themselves' turned out to be based on a combination of Pavlovian and instrumental conditioning.[65]

The person who made the greatest contribution to the study of dolphin intelligence and communication was Louis Herman (1930–2016). Herman was born and raised in the Queens borough of New York where his parents ran a small store. He studied psychology at City College, New York, where he obtained a bachelor's and then a master's degree. After serving in the US Air Force, he enrolled in Pennsylvania State University where he obtained his Ph.D. in 1961. This was based on research into human performance related to the attentional theory of Donald Broadbent, a theory described in Chapter 7. For a few years, Herman held a teaching position at Queens College, part of the City University of New York, where his research on reaction times using undergraduates was very productive. Probably on this basis, he obtained a more secure academic appointment in a very different environment, that of the University of Hawaii in Honolulu. Although during his first year or so there he continued his reaction time research, at some point he was persuaded – allegedly by a student – to make the most of the new research opportunities his environment offered and switch from humans to dolphins.[66]

Figure 5.18 Louis Herman with three of his graduate students late in 1979. Reproduced with permission from Gordon Bauer, who is on the right of the group of students.

Many years later Herman wrote: "In my earliest work on dolphin cognition, I sought to make contact with comparative animal studies seeking measures of 'intelligence' that might order species in a way consistent with their brain development. The most popular model for such study in the 1950s through to the early 1970s was the learning set task devised by Harry Harlow."[67]

The dolphin that served in his team's first study was named Wela. She had been in captivity for three years at the Sea Life Park in Hawaii, living in a large oval-shaped tank where the experiment took place, and was over 10 years old when the experiment began. The discrimination procedure was similar to one that had been used by Kellogg and Rice: Wela was trained to discriminate between pairs of shapes presented underwater, where the shapes were attached to the end of long levers, and on pressing the correct shape a tone signaled that she would be given a fish on returning to the opposite end of the tank. Twelve trials were given on each of the first 40 problems and 6–8 trials on the remaining 40. Wela's performance was disappointing, in that neither the average number of correct trials per problem nor her Trial 2 performance improved very much above chance and she developed a very strong position preference.[68] It seems that Herman had never before worked with any species other than humans and his lack of such experience may have contributed to the disappointing outcome. Clearly, the researchers needed to adopt more effective procedures in future; see Figure 5.18.

It took another few years to develop a training procedure that worked well. One major change was to use auditory stimuli, tones of various frequencies that were modulated at various frequencies. Each problem consisted of a discrimination between two such signals presented in succession from separate loud speakers. The single dolphin that served in discrimination learning set experiment that used this procedure was a female named Keakiko. She needed to be trained first to position herself in the optimal place for listening to the signals and to remain there until both signals had been presented. The discrimination task was then to approach the speaker emitting what had been designated the positive stimulus for that problem and press the paddle next to this speaker, whereupon a fish reward was delivered. The researchers also used a *fading* procedure so that for early experiments in the series the sound serving as a negative stimulus was presented very briefly, just 0.5 seconds, and then increasingly longer until it reached 2.5 seconds, the same duration as that of the positive stimulus.[69] All in all, the study must have taken several months, with many hours of testing carried out six days each week; the procedures were much more complicated and time-consuming than those used by Harlow to test his monkeys in the WGTA.

The huge effort paid off in that Keakiko eventually performed at a level equal to that of Harlow's monkeys. "Runs of errorless problems in some cases rivalled those reported for individual rhesus monkey subjects (Harlow, 1949). This suggests a revision to Warren's (1965) statement that 'no non-primate mammal yet tested has approached the level of one-trial learning (in discrimination learning set tasks) observed in primate species.'"[70]

Herman and his students next tested Keakiko's memory capacity, including her ability to retain 'lists' of sounds. In general, the results were remarkably similar to those obtained from human participants given words to remember.[71] Then the research program changed direction. "Our learning-set results, together with diverse findings on memory, began to establish that dolphins were flexible learners with apparent requisite skills necessary for managing even more complex cognitive tasks. We thus begun a study of Keakiko's ability to learn to understand instructions given within an arbitrary acoustic language we created."[72]

In this language, three arbitrary sounds served as the names for a ball, a ring, or a cylinder, while three further sounds represented the actions *touch*, *fetch* or *mouth*. "Keakiko was able to carry out two-word instructions ordered as *object-name* + *action-name*, so that, for example, the sequence glossed as *ball* + *touch* resulted in her touching the ball, and not the other objects."[73]

This program and the training program for a second dolphin, Puka, came to an end in May, 1977, when two disgruntled ex-employees decided one night to 'liberate' the two dolphins and abandoned them in an area of the sea known to have a large shark population. "They were never recovered and almost certainly died not long after their abandonment," Herman reported.[74]

Just over a year later Herman's research resumed after two young female dolphins arrived at the University of Hawaii; see Figure 5.19. The new program concentrated on their ability to understand a new artificial language, one based on visual signals. These were gestures involving a trainer's arms and hands that could be either directly observed or seen on a television screen. The achievements of this program go beyond the scope of the present book; they are summarized in Herman's overview of his research.[75]

Figure 5.19 Louis Herman and a new dolphin in 1980.
© Walter Sullivan/The New York Times/Redux/Headpress.

The Comparative Psychology of Jeff Bitterman: Serial Reversal and Probability Learning

An arresting title of an article or book can have a major impact on the attention that it attracts. One of the most extreme examples of a title that remained eye-catching for decades was published in 1950 by Frank Beach, whose main interest was in hormonal influences on behavior. The title of his paper was *The Snark is a Boojum*. Beach based his argument around Lewis Carroll's nonsense poem about a group of characters who went hunting for a creature, the Snark, and then discovered that it was a Boojum; as a result, the Baker who saw it first "softly and suddenly vanishes."[76]

Beach's paper was based on his presidential address to the experimental division of the American Psychological Association. His argument was that the research tradition of comparative psychology, one that had started with so much promise, had been hijacked by learning theorists who had no interest in any species but that of the laboratory rat. As a result, 'true' comparative psychology was in danger of vanishing completely. He bolstered his argument with graphs illustrating the steady decline in the range of species reported in key psychological journals, one that was accompanied by

Figure 5.20 The white rat bewitches experimental psychologists and leads them to their doom. From Beach (1950). Public domain.

an equally steady rise in studies of rat behavior. Another graph showed the steady rise of studies investigating learning processes and the relative decline of interest in other aspects of behavior. Beach illustrated his argument with a cartoon based on another literary classic, the Pied Piper of Hamelin in this case; as shown in Figure 5.20.

Judging by a comprehensive review by Warren of "the comparative psychology of learning" published 15 years later, Beach was wrong to predict that interest in the behavior of nonhuman species other than the rat would disappear but correct in predicting that comparative psychologists would continue to concentrate on learning. Warren's review covered a very large number of different species, focusing on attempts to answer three basic questions: "At what phyletic level do we first encounter animals capable of learning? How do learning performances differ among the higher animals? What is the phylogenetic significance of the differences in learning that have been observed?"[77]

As for the first question, Warren was skeptical of claims that conditioning could be obtained in protozoa and, although accepting that habituation can be demonstrated in flat worms (planaria), was not convinced by the then extensive – and highly publicized – research with such animals that they could show both Pavlovian and instrumental conditioning. As for the second question, regarding differences in learning "among the higher animals," Warren found very little advance in finding answers because, in general, differences in rates of simple forms of learning or in performance on complex tasks were much more variable *within* a species than across species. However, as mentioned earlier in this chapter, Warren believed that performance on learning set tasks could well reveal important differences between species: "…there is reason to believe that the ordering of avian and mammalian species in respect to learning-set capacity is meaningful."[78]

Warren and his colleagues had also carried out cross-species studies of performance on a simpler task than learning sets. The procedure involved repeated reversals of simple discriminations. For example, a rat or a chicken might first be rewarded for turning left in a T-maze and, when its performance reached a criterion, the rule was reversed in that it now needed to turn right. Once the criterion was reached for this second stage, the rule was reversed once more so that the left arm was now correct again. A similar experiment might involve a visual discrimination. For example, in 1962 Bullock and Bitterman trained four pigeons on reversals of a position preference, finding a rate of improvement comparable to that reported for rats, and another four on reversal of a simultaneous color discrimination – a red response key vs. a yellow response key – and found a somewhat slower rate of improvement over the series of reversals. On the other hand, when a final group of four pigeons was given reversals of a successive color discrimination – for example, both keys red, peck the left key; both keys yellow, peck the right key – only one pigeon showed substantial improvement. Since they found that, even with spatial reversal task, improvement was found only under specialized conditions, they advised caution when making cross-species comparisons: "… the negative results obtained with simpler animals may be due only to the inadequacy of the experimental conditions employed."[79]

In his review of research on serial reversal learning Warren was less cautious. His summary of experiments in his own laboratory, in that of Bitterman, and in a few other laboratories concluded that rats and birds showed "a progressive reduction in the number of trials taken to learn successive reversals, often approaching or attaining a level of one-trial learning…. Fish and invertebrates either make no improvement in reversal learning or make an improvement which is independent of reversal training and presumed to be the result of general adaptation to the conditions of the experiment."[80] On the basis of such evidence, Warren added *repeated reversal learning* to his very short list of learning tasks that "suggest an orderly improvement in the efficiency of learning within the vertebrate series."[81]

The researcher that most energetically pursued this line of research was Martin Edward Bitterman (1921–2011), who was always known to friends and colleagues as 'Jeff.' Like Louis Herman, Bitterman was raised in New York, went to college there, went on to graduate school in psychology, and then held various academic appointments before relocating in 1971 to the University of Hawaii where he spent the rest of his career. There the similarities end. As an undergraduate at New York University the influence of a radical behaviorist, a follower of Watson, led Bitterman to abandon the plan of becoming a lawyer and to take up psychology instead. As a graduate student, he studied comparative psychology, first at Columbia University and then at Cornell University. He obtained his Ph.D. in 1945 following work on Pavlovian conditioning at the 'behavior farm' under the supervision of Liddell;[82] see Chapter 1.

For the next 12 years Bitterman held a number of appointments at various institutions until ending up at Bryn Mawr College in Philadelphia in 1957. During these years – as subsequently – he was enormously productive, publishing on a variety of research topics that involved both human and rat participants. He also gained a reputation for what the understatement in his obituary described as "the sharpness of his mind and tongue."[83] Bitterman crossed swords with a number of other psychologists, notably with Spence, as described in Chapter 7. In 1957 his review of Spence's latest book ended with the conclusion: "Despite certain novel features and some interesting implications for experimental design, it is a thing, not of tomorrow, but of yesterday. It will intrigue specialists in learning, not in its own right, but as the work of its author – loyal and dauntless servant of a once-powerful, now tired tradition, painfully passing to its reward."[84] A large man, Bitterman could be as combative in person as in print.

The year Bitterman arrived in Bryn Mawr also marked the start of his subsequent concentration on comparative psychology. This began in collaboration with a former student, Jerome Wodinsky, who was now based at the American Museum of Natural History. Their subjects were fish; more specifically, 10 African mouthbreeders, *Tilapia macrocephala*, each about two inches long. The task was a discrimination reversal: A fish was required to push against one of two circular panels, one painted with black-and-white vertical stripes and the other with horizontal stripes. A correct choice was followed by delivery of a small food pellet into the tank. Performance on successive reversals did not improve in the way typical of when rats were tested on this task. "Analysis of initial errors showed … a pattern of recovery from negative transfer qualitatively distinct from that previously obtained for the rat, a finding who significance for the problem of phylogenetic differences in learning ability remains to be assessed."[85]

This initial study was followed a series involving fish that lasted many years. However, instead of focusing on serial reversal learning, for the next few years, most of the experiments with fish examined the effects of extinction. In the late 1950s research on the *partial reinforcement extinction effect* – the PREE – was one of learning theorists' most popular topics. As confirmed in dozens of experiments using both rats and humans, responses that were reinforced only on some occasions – usually a 50% schedule was used – persisted much longer when reward was no longer available than responses that had been consistently – 100% – reinforced; see Chapter 2.

Whether such an effect is also shown by other species was the question first addressed in 1959 by Wodinsky and Bitterman.[86] The answer was 'sort of.' Following preliminary training, the African mouthbreeders used in their experiment were first given 20 trials a day in which pressing the target was always followed by a pellet in the Consistent group and followed by a pellet on only half the trials in the Partial group. Following 10 such sessions, extinction sessions were introduced; these ended when a fish failed to respond in less than 30 seconds for five trials in succession.

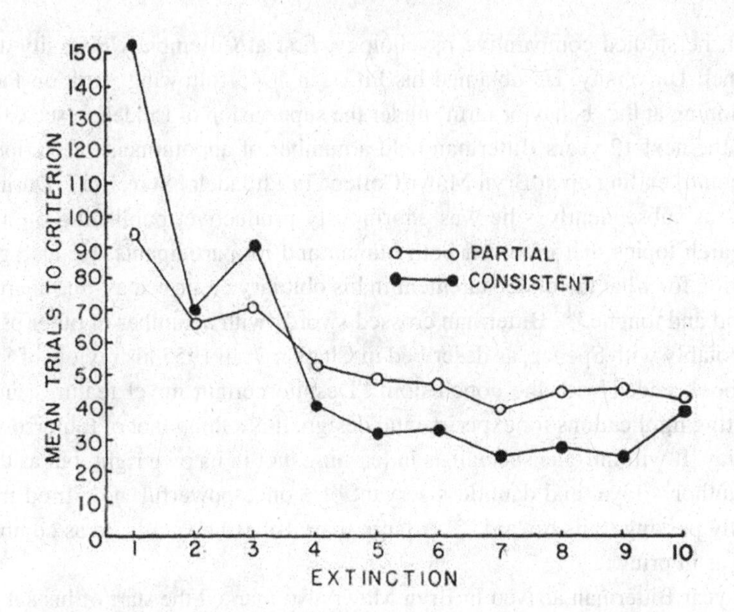

Figure 5.21 During each training phase of this experiment the fish were given either Partial reinforcement (50% of responses reinforced) or Consistent reinforcement (100%). During the following extinction phases, responses were not reinforced and trials continued until a fish had stopped responding. A further cycle of training followed by extinction then followed. The figure shows that over the first 10 such cycles, these fish failed to show the standard Partial Reinforcement Extinction Effect displayed by rats and humans; thus, there was no significant difference between the two groups in terms of number of trials to extinction.
From Wodinsky and Bitterman (1959). Public domain.

A series of training and then extinction cycles was given. In the early cycles the Partial group was just as quick to extinguish as the Consistent group; see Figure 5.21. This was a result quite at odds with what was consistently found with rats. On the other hand, after many such training and extinction cycles, greater resistance to extinction emerged in the Partial group.

Following this initial study and for at least eight years Bitterman, his students, and colleagues ran dozens of experiments that examined resistance to extinction following partial reinforcement. Across experiments, they varied such factors as the inter-trial interval, the time between trials, and the magnitude of reward. The general conclusion that eventually emerged was that fish were not so different from rats after all. Given favorable conditions such as a large reward and a short inter-trial interval, they show a similar, though smaller, PREE, to that shown by the rat.[87]

Another kind of task used by Bitterman and his colleagues was probability learning. For example, an animal might be rewarded 80% of the time for choosing the response on the left, or the blue stimulus, and 20% for choosing the alternative. They reported that for both spatial and visual versions of a probability learning task both rats and monkeys 'maximized'; that is, they always chose the side or stimulus associated with higher probability of reward, whereas fish always showed 'matching' by, for example, responding 80% of the time to the stimulus rewarded 80% of the time.

TABLE 2

BEHAVIOR OF A VARIETY OF ANIMALS IN FOUR CLASSES OF PROBLEM WHICH DIFFERENTIATE RAT AND FISH EXPRESSED IN TERMS OF SIMILARITY TO THE BEHAVIOR OF ONE OR THE OTHER OF THESE TWO REFERENCE ANIMALS

Animal	Spatial problems		Visual problems	
	Reversal	Probability	Reversal	Probability
Monkey	R	R	R	R
Rat	R	R	R	R
Pigeon	R	R	R	F
Turtle	R	R	F	F
Decorticated rat	R	R	F	F
Fish	F	F	F	F
Cockroach	F	F	—	—
Earthworm	F	—	—	—

Note.—F means behavior like that of the fish (random probability matching and failure of progressive improvement in habit reversal). R means behavior like that of the rat (maximizing or nonrandom probability matching and progressive improvement in habit reversal). Transitional regions are connected by the stepped line. The brackets group animals which have not yet been differentiated by these problems.

Figure 5.22 Categorization of animals as to whether they show rat-like or fish-like patterns of behavior when given habit reversal or probability learning tasks, based on either spatial or visual discriminations.
From Bitterman (1965b). Reproduced with permission from the American Psychological Association.

Bitterman's research become very widely known in the mid 1960s, partly because of his article in the *Scientific American*, ambitiously titled *The evolution of intelligence*.[88] Figure 5.22 shows a figure from another of his papers that compares animals from five species on habit reversal and probability learning. It can be seen that a line is drawn between rat-like patterns of behavior in these two tasks and fish-like behavior, with, the pigeon and the turtle falling in between rats and fish.

An interesting exchange between Bitterman and a major critic, Nick Mackintosh, took place in 1969. It started with Mackintosh suggesting a reason for comparing the performance of two species on some task, justification that was very different from traditional appeals to understanding the 'evolution of intelligence.' "For suppose there exists a theory that purports to explain the behavior of rats in a given experimental situation, and that when goldfish are trained in this situation their performance differs markedly from that of the rat. The theorist, if sufficiently versatile, ought to be able to account for this difference in the behavior of rat and fish by postulating,

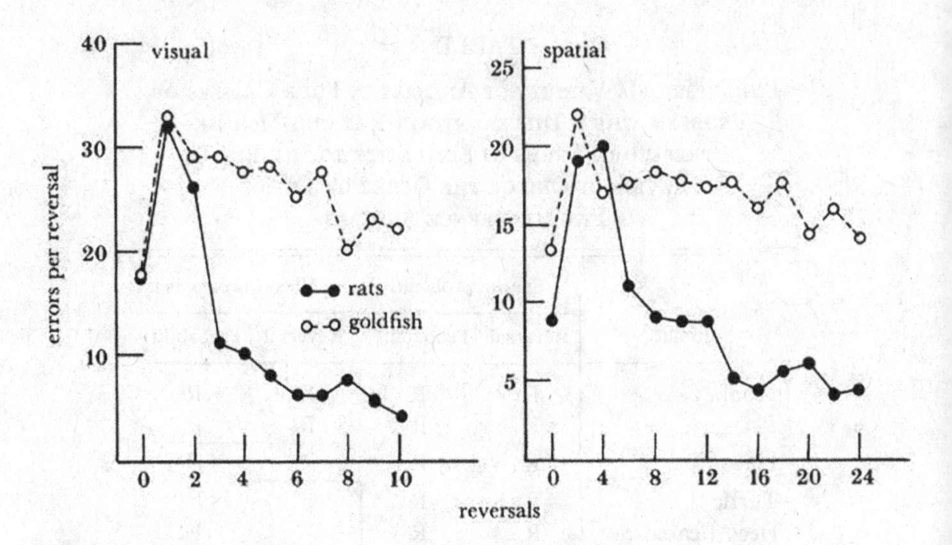

Figure 5.23 These graphs show that, whether given a series of red-green or spatial discrimination reversals, the number of errors made by rats declined rapidly with each reversal, whereas the improvement shown by goldfish was only gradual.
From Mackintosh, Wilson, Boakes and Barlow (1985). Reproduced with permission from the Royal Society (UK).

say, some differences between the parameters used to explain the behavior of rats and those used for fish; and he ought then, if the model is of any generality, to be able to predict further differences in the behavior of rat and fish in some different experimental situation. Confirmation of such predictions would provide and interesting and rather powerful line of support for the appropriateness of the model……. The argument is not entirely simple, but before it is developed it is necessary to show that these behavioural differences are real ones (a matter disputed by some), but that Bitterman's analysis is nonetheless misleading."[89]

The first step for Mackintosh was to confirm that rats differ from birds (pigeons, doves, and chicks but not mynahs or magpies) and birds differ from fish in terms of performance on serial reversal and probability learning tasks. His review of evidence from other laboratories and from his own experiments that compared rats and chicks suggested that such differences were quantitative and not qualitative, as Bitterman had maintained; see Figure 23. Mackintosh went on to suggest that an attentional theory of discrimination learning could provide a partial account for these species differences: Performance on both tasks depends on how well an animal can maintain attention to a relevant cue.[90]

In reply to Mackintosh, Bitterman marshaled evidence to dismiss Mackintosh's claim that species differences were quantitative and at least partly explicable in terms of difference in maintaining attention to relevant cues. He doubted, for example, that there is anything like attention in the fish.[91] Mackintosh then came back with a detailed analysis of various experiments with the aim of demonstrating that Bitterman's interpretation of their results were misleading. The debate illustrates the difficulty of obtaining agreement on what to conclude from such comparative research; however, its details are beyond the scope of the present chapter.

The first time I met Bitterman was in 1971, when he presented a paper at a conference on *Inhibition and learning* that Sebastian Halliday and I had organized at the University of Sussex in the UK. Many months earlier we had drawn up a list of potential participants, all of them researchers who had published on the topic of inhibition and had sent them letters of invitation to the Sussex meeting. Bitterman was not on the list, since we were not aware that he had any active interest in the topic. We had almost completed the program for the meeting, when a letter arrived from Bitterman. He explained that his invitation seemed to have gone astray but that we would be pleased to know that nonetheless he was able to contribute to the meeting.

The series of experiments on reversal learning in the goldfish and the pigeon he reported were important for two main reasons. The first was that the experimental procedures involved *successive* discriminations (or as he called them, 'unitary conditions') instead of the simultaneous discrimination procedures that had been used almost exclusively in over a decade of research on serial reversal learning. For example, the first discrimination given to pigeons was one in which pecking 20 times at the single response key when it was red was followed by access to grain for a few seconds, whereas 20 pecks at the key when it was green ended that trial without any other consequence. Two daily sessions on this initial discrimination were followed by two daily sessions with the discrimination reversed. The important aspects were, first, that learning the reverse of the previous discrimination improved and, second, that most of this improvement resulted from a pigeon's increased ability to inhibit pecking at the new negative color. When simultaneous discrimination procedures are used to study serial reversal learning, it is normally difficult to tell whether any improvement results from faster learning to respond to the new positive stimulus or faster learning *not* to respond to the new negative stimulus.

Bitterman's second important result was obtained from experiments with fish and involved improved spatial contiguity between correct responses and the reward that followed. When using a reward to train an animal to acquire some new response or to ensure that some already learned response is performed promptly, it is important to ensure that there is as small a delay as possible between performance of the response and delivery of the reward. Less well known than this principle of temporal contiguity is the principle of spatial contiguity: The closer the reward is to the target response, the more effective is its reinforcement of the response. More often than not, lack of spatial contiguity is confounded with lack of temporal contiguity. This was true of the fish experiments described earlier in this section; when a fish made a correct response to the target, a food pellet was dropped into the tank at some distance from the response paddle, thus ensuring a delay between the response and consuming the pellet.

Together with his long-term colleague and collaborator at Bryn Mawr College, Dick Gonzalez, Bitterman had earlier demonstrated the importance of spatial contiguity in an experiment on color discrimination learning with pigeons: They learned the discrimination more rapidly when the colors were presented within magazine apertures in which grain could be delivered than when – conventionally – the colors were presented on response keys mounted a few centimeters from the magazine.[92] He now explained how such spatial contiguity between response and reward could be obtained when working with fish: "Instead of striking at an illuminated target in one location and being rewarded with a worm discharged from a feeder at another location, the animals struck at an illuminated Plexiglas nipple through which liquid food could be delivered with a Davis pump."[93]

The new arrangement revealed that goldfish could now show improvement in serial reversal learning. This improvement was mainly based on more rapid inhibition of responding to the new negative stimulus, as was the case with pigeons. However, differences between the two species had not completely disappeared, in that the fish, but not the pigeons, also showed increased reluctance to respond to the positive stimulus. The final experiment that Bitterman reported found, as expected, that with the new way of delivering a reward to a fish serial reversal performance also improved when a simultaneous procedure was used. He entertained the idea that the previous poor performance of fish in over a decade of experiments on reversal learning might be due "to the inadequacy of the techniques employed in the previous work."[94]

By the time of the Sussex meeting Bitterman had left Bryn Mawr and was about to take up a full professorship at the University of Hawaii. In Hawaii he continued to publish several research papers each year. At first these were mostly on discrimination learning in the goldfish. Then in 1976 he published his first paper on a species that became his animal of choice for the next three decades, the honeybee. One of his Ph.D. students at Hawaii, Patricia Couvillon, became his major collaborator in this research and continued to study learning in honeybees after Bitterman's death in 2011. However, the fascinating work of Bitterman and Couvillon on honeybees is beyond the time span of the present book.

Matching-to-Sample and Short-Term Memory

One of the last studies Harlow published using the WGTA employed a procedure known as *oddity learning*.[95] In this experiment, sixteen rhesus monkeys were trained on a form of learning set consisting of repeated 6-trial problems, in which new objects were used for each problem. On each trial, a monkey was faced with a line of three objects, of which two were identical, while the different, *odd*, object was either on the left or right. After training on about 200 such problems, the monkeys averaged nearly 90% correct on the first trial of a new problem, thus, demonstrating that they had – in some sense – learned the rule, 'choose the object that is different from the other two.'

The oddity task is one of two approaches that have been taken towards studying whether animals can learn on the basis of same/difference relationships. The complementary task, *matching-to-sample,* dates back to early in the twentieth-century when a Russian psychologist trained a female chimpanzee to select from an array of objects the one that matched the object she was shown by the experimenter.[96]

To further illustrate his argument that productive research is more likely to take place in the absence of theories, in 1950 Skinner sketchily reported an experiment in which pigeons were given an array of three illuminated response keys – for example, two red and one green – and pecking at the side key with the color that matched the color of the center, *sample*, key was reinforced. Skinner's implied message was that, anything a chimpanzee can learn, a pigeon can learn too.[97] Subsequently, one question driving such research has been whether pigeons can respond in primate-like fashion on the basis of a same or different relationship, as opposed to merely learning to respond appropriately to specific configurations of cues.

The first study to address this question was carried out in the early 1960s at Columbia University. This had become a center for research on operant conditioning that rivalled the pigeon laboratory set up by Skinner at Harvard University; see Chapter 8. Cummings and Berryman trained their pigeons on a three-color matching task, using red, blue, and green, and afterwards they were tested when yellow was substituted for blue; whereupon the birds' performance dropped immediately to chance.[98]

Evidence that, nonetheless, pigeons could display generalized matching was later reported by Zentall and Hogan in experiments that used a savings measure.[99] Subsequently, after replicating one of their key findings, Wilson found that this resulted from the negative effects of initial training on an oddity task rather than positive transfer from training on matching.[100] In line with this conclusion, since the mid-1980s the most common finding has been that pigeons trained on matching-to-sample with a certain set of stimuli fail to respond above chance when tested on new stimuli. However, there is some evidence that with special intensive training, pigeons learn to respond on the basis of relational cues.[101]

Involvement with Bundy Wilson, later to become Bundy Mackintosh, in a project largely designed by Nick Mackintosh, brought home to me the difficulties of working with species other than rats or pigeons. The aim of this project was to compare pigeons with various species of European corvids on matching-type tasks, using exactly the same apparatus and experimental parameters. For every hour spent testing corvids, Wilson needed to spend dozens of hours hand-rearing and taming her subjects, many of which were first housed in her bathroom. To help her obtain some rooks, my children and I would accompany Wilson to local rookeries to find nestlings lying on the ground that had been pushed out of their nest by older siblings; we needed to get to them before the local foxes and stoats did; see Figure 5.24.

The major effort involved in this project at least yielded some interesting results. For example, one of a series of experiments comparing pigeons with jackdaws, jays and rooks involved an initial stage in which birds were either trained on a matching-to-sample task or on an equivalent conditional – or non-matching – task; if A, choose C; if B, choose D. In this stage, various colors were used as the stimuli. In the tests that followed all birds were given a matching task using other colors as stimuli. When four jackdaws were used as subjects, in the test stage the pair trained first on a matching task showed good transfer to the new matching task and over six sessions sustained their superior performance to that of their fellow jackdaws that had first been trained on the conditional task. In comparison, eight pigeons learned more

Figure 5.24 Bundy Mackintosh (née Wilson) and one of the jackdaws that served in her experiments designed to compare the performance of this species with that of pigeons in a matching-to-sample task.
Reproduced with permission from Bundy Mackintosh.

rapidly and reached a higher level of performance than any of the jackdaws but there was no difference in the test stage was between the four pigeons previously trained on the first matching task and the four trained on the conditional task. Thus, the corvids trained on matching in the initial stage appeared to have learned the rule, choose whichever comparison stimulus is the same as the sample stimulus, under the same conditions in which pigeons failed to learn the rule.[102]

The small number of birds from each corvid species meant that only tentative conclusions could be drawn as to their learning abilities when compared with those of pigeons. A review of these and related experiments ended as follows: "The real interest of the comparison between pigeons and corvids, for example, is not to prove that pigeons are stupid, but rather to take advantage of the pigeon's apparent failure to transfer, say, the matching rule to new stimuli, to establish the point that such transfer is not an automatic consequence of teaching an organism a matching discrimination. When transfer does occur, therefore, this is precisely something that demands

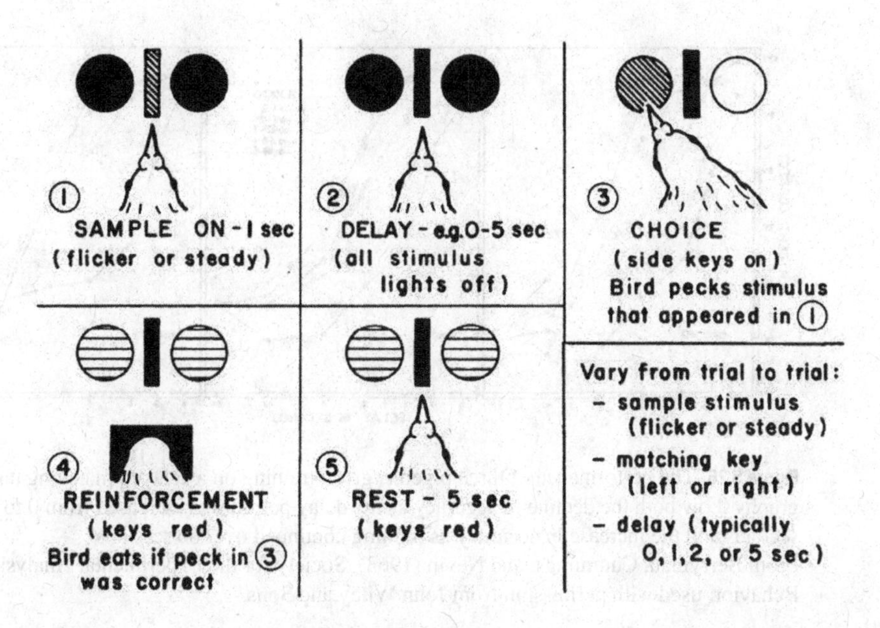

Figure 5.25 In the first experiment to use a delayed matching procedure a trial began with either a steady or flickering white light (sample) displayed on a vertical center panel (1). There was then a delay of 0, 1,2, or 5 seconds with all lights off (2). The circular side keys were then illuminated, one with the flickering and the other with the steady light (3). If the pigeon pecked at the key with the matching light, reinforcement was delivered, that is, grain in the hopper (4). If it pecked at the wrong key, a five-second rest period followed (5); an interval then occurred before the next trial began.
From Blough (1959). Public domain.

explanation; it requires the postulation of a process or set of processes that probably lies outside the scope of our simple theories of conditioning or associative learning."[103]

An important variant of the matching-to-sample task described above was first introduced by Donald Blough in 1959, when he was employed at the National Institute of Mental Health. His name for the procedure, *delayed matching,* has persisted ever since. Instead of color lights – avoided since they might produce aftereffects – he trained his four pigeons to match either a 10 Hz flickering or a steady white light. Unlike the *simultaneous* condition used in the experiments described earlier in this section, the condition for Blough's birds was that, on pecking the central panel when either a flickering or steady light appeared, this sample stimulus would disappear and be followed by the display of the two comparison stimuli on the side keys after a delay that ranged from 0 to 5 seconds. The procedure is illustrated in Figure 5.25.

On average the birds' performance was better when the delay was short or zero. However, as Blough emphasized, averaging over the four birds was misleading in that two of them showed no decrement at all as the delay was increased to five seconds. These two birds were those that showed the most pronounced forms of stereotyped behavior. For example, on trials when the flicker stimulus had been displayed Bird 1 continued to peck in a rapid and hard manner at the top edge of the sample panel,

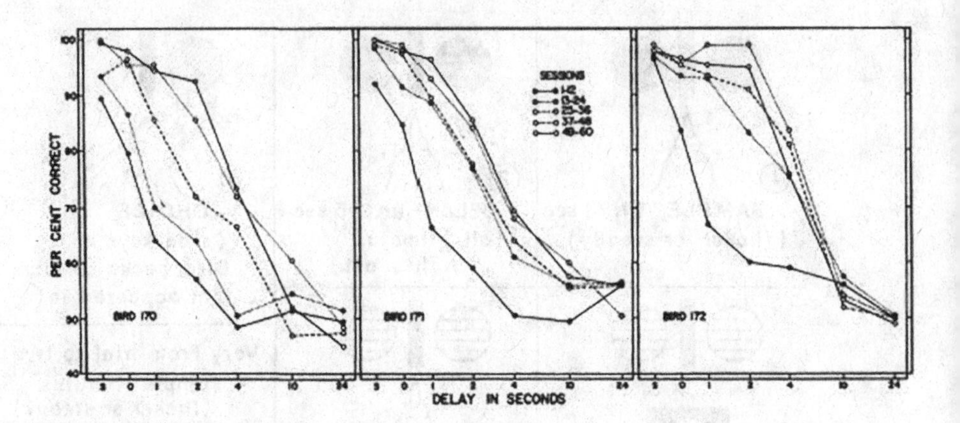

Figure 5.26 The performance of three pigeons given training on a delayed matching task. The graphs show both the decline in accuracy as the delay period was increased from 0 to 24 seconds and the increase in accuracy as training continued over 60 sessions.
From Berryman, Cummings, and Nevin (1963), Society for the Experimental Analysis of Behavior, used with permission from John Wiley and Sons.

whereas when the steady light had been displayed it pecked in a slower and more gentle manner at the center of the sample panel.[104]

Blough discussed the behavior of these last two birds in terms of Skinner's concept of superstitious behavior. As for the other two birds, he might have been tempted to describe their systematic decline in accuracy as the delay interval was increased as reflecting decay of short-term memory for the sample. Instead, Blough avoided such 'cognitive' language and, as befitted the Skinnerian journal in which his report was published, he suggested that their data "might represent a 'decay time' of stimulus control that is a basic behavioral fact in its own right."[105]

Blough's experiment was followed up at Columbia University by Berryman, Cummings and Nevin in 1963. Using seven pigeons with a previous history of simultaneous matching-to-sample, this study became a model for the very many experiments on delayed matching in pigeons that followed. Three colors – red, green and blue – were used as the stimulus set. The range of delays was expanded from that used by Blough to 0–24 seconds and on each trial a bird was required to peck five times (FR5) at the sample stimulus for it to disappear at the start of the delay period.[106]

The delay gradients that Berryman and his colleagues obtained from three of their pigeons are shown in Figure 5.26. Four other pigeons had been excluded on the basis of displaying aberrant simultaneous matching prior to training on delayed matching. In contrast to the individual variation in delayed matching performance that Blough's four pigeons had displayed, the terminal performances of the three pigeons shown in Figure 5.24 were very similar. The experimenters looked for, but failed to find any, "repetitive chains of different topography" of the kind that Blough had seen in two of his birds. They pointed out that various aspects of their procedure would work against the development of mediating behavior. In particular, "in our experiment, employing three standard stimuli, would require three differentiated chains, while two chains sufficed for Blough's procedure."[107]

In many subsequent studies that used a delayed matching procedure to study short-term memory, the procedure was modified to make it unlikely that an animal's performance was based on acquiring different behaviors to each sample stimulus. One example of this – and of how delayed matching was used to measure short-term memory in other species – is provided by one of the last experiments that Herman and his colleagues performed on their dolphin, Keakiko, before she was abducted.

The aim was to test the dolphin's short-term memory for auditory stimuli. The delayed matching procedure was as follows. Each trial started with a sample sound – one from the large set of complex tones used previously – being produced via both of the speakers and then the dolphin needed to wait in the 'listening' location until two more sounds were presented in succession, one from the left and one from the right speaker. One of these was identical to the sample and the other novel for that trial. Keakiko was rewarded with a fish if she then pressed the paddle next to the speaker that had emitted the sound that matched the sample. Over a large number of trials that always used a different sample the delay between the sample and the two comparison sounds increased from 15 to 120 seconds.

Once she had been trained on 150 or so such problems, Keakiko rarely made an error on the first trial with a new problem even with the 120-second delay. No systematic tests with longer delays were attempted as she displayed "emotional responding" when required to wait in the listening area for more than a couple of minutes. Overall, her performance was similar to that displayed by capuchin monkeys – and far superior to that of pigeons – given delayed matching tests using visual stimuli. As with her performance on the prior discrimination learning set task she was set, "the evidence remains consistent with expectations of learning capacity of the bottle-nosed dolphin based on the advanced development of its brain."[108]

Delayed matching procedures of the kind first introduced in 1959 by Blough have been used extensively ever since then to study memory in various species.[109]

From the *Comparative Psychology of Intelligence* *to* Comparative Cognition

The comparative psychology of intelligence was the title of a highly controversial and deliberately provocative article published in 1987 by Euan Macphail.[110] His argument was built on the comprehensive review of research in comparative psychology that he had published five years earlier.[111] He pointed out that the original impetus for the discipline of comparative psychology was Darwin's vision of tracing the evolution of human intelligence. Macphail's definition of intelligence was based on "two salient characteristics of human intelligence … The first is that human intellectual capacity appears to be very general in its range of applications"; the second, he suggested "is the evident reliance of intellectual activity on previous experience, learning and memory."[112]

He repeated his argument in the earlier review that none of the various claims to have demonstrated that one species was more intelligent than some other had ruled out the possible role of *contextual variables*. "Suppose two species are tested in identical versions of a given problem and that one species solves it and the other fails. Such a result *could* be due to some intellectual advantage of the one species over the other, but other less interesting possibilities clearly exist. One species might find the reward less motivating than the other, or there might be sensory or motor demands in the problem that were satisfied more easily by one species because of a more developed visual system or the possession of hands rather than wings."[113]

Following his conclusion that no study had conclusively ruled out the possible role of contextual variables, Macphail proposed his *null hypothesis*: "That there are no differences, either quantitative or qualitative, among the intellects of nonhuman vertebrates." As for humans, "they possess the same 'basic' general intelligence as non-humans, plus a species-specific language acquisition device which, when it becomes operational, also has very general application and so raises the level of general intellectual capacity far above that of non-humans."[114] On the much more positive side, he suggested that research in comparative psychology had very clearly demonstrated "that so many learning phenomena appear consistently in groups of widely differing ecology and phylogeny." He proposed, a little tentatively, that such commonality reflected the general challenge for any species "to detect the causal structure of its world and thus mechanisms of learning of general applicability, notably the acquisition of associations as studied by means of various conditioning paradigms."[115]

Whether Macphail's highly-cited article of 1987 was influential or not is hard to tell. Nevertheless, its publication coincided with the near disappearance of studies that attempted to compare the 'general' intelligence of different species and an increase in studies that were grouped under the title *comparative cognition*. The basic assumption of this field of research is that the cognitive processes of animals are of interest for their own sake and how they might relate to human cognition is of limited or no concern. On the other hand, how such processes might relate to the way that a particular species survives or flourishes in its particular environment is important to many researchers in this field.

The modern birth of this branch of psychology is generally seen to have occurred at a conference held at Dalhousie University in 1976. The proceedings were published two years later. While some topics were central to the study of associative learning, the titles of other chapters provide a useful guide to what kinds of process were of interest at the birth of comparative cognition. The titles covered topics that included short-term and working memory, timing, and spatial memory.[116] A second meeting of researchers involved in comparative cognition took place in 1982 at Columbia University. The proceedings indicate the degree to which the field had broadened in the previous six years. In addition to those listed above, the topics now covered the processing of pitch and rhythm structures by birds, temporal pattern recognition, concept formation, form recognition, geometric factors in spatial representations, and counting.[117]

The growth of this field is beyond the time frame of the present book. The reader can, however, find out about such research from two excellent books both first published in 1998, one by Sarah Shettleworth[118] and the other by William Roberts.[119]

6 Imprinting and Constraints on Learning

The view that laws of learning of considerable generality and precision could be found was so firmly established that those experimental psychologists who questioned it did it so in the teeth of their own training. For biologists, however, the reverse has been the case. Until recently, at any rate, all biology students were subjected to a review of the animal kingdom. Although such courses often required the memorization of a vast assemblage of facts, they had one outstanding advantage – they brought the student immediately face to face with diversity. This included not only the diversity of organisms, but also the diversity of mechanisms within organisms…. Of course, biologists inevitably assimilate some envy for the physicists who can say $e = mc^2$ in every contingency, and they have indeed had two good breaks themselves with the theory of evolution and the unraveling of the genetic code. But by and large, they are still sufficiently impressed by diversity to be suspicious of such assumptions as 'all behavior of the individuals of a given species and that of all species of mammals, including man, occur to the same set of primary laws.' (Hull, 1945)[1]

The author of this 1973 claim about an important difference between experimental psychologists and biologists was Robert Hinde, who – as suggested by the quote – was trained as a biologist but was also very familiar with research on learning carried out by psychologists. He belonged to a research tradition that began early in the twentieth century and that became known as *classical ethology*. The central concerns of these researchers were how a particular species behaves in its natural setting and the differences between species. Although they concentrated on instinctive behavior, two important examples of learning were studied within this tradition. One was the phenomenon of imprinting that was systematically documented in Austria by Konrad Lorenz.[2] The other was the acquisition of songs by certain bird species, such as chaffinches, as studied in the UK by William Thorpe.[3] Both kinds of learning are found in only a limited number of species. While few psychologists took an active interest in birdsong, many carried out important studies of imprinting and, more generally, of types of learning that occur only in the very young of many species.

Konrad Lorenz on Imprinting

"It is a fact most surprising to the layman as well as to the zoologist that most birds do not recognize their own species 'instinctively,' but that by far the greater part of their reactions, whose normal object is represented by a fellow member of the species, must be conditioned to this object during the individual life of every bird. If we raise a young bird in strict isolation from its kind, we find very often that some or all of its social reactions can be released by other than their normal object. Young birds of most species, when reared in isolation, react to their human keeper in exactly the same way, in which under natural conditions they would respond to fellow members of their species."[4]

The phenomenon of imprinting in various species of birds had been described for centuries before it was first studied experimentally in the 1870s by Douglas Spalding. His results were discussed at some length in 1890 by William James in the latter's influential book, *Principles of psychology*.[5] Spalding covered the eyes of chicks with a small hood shortly before they were due to hatch and later removed the hood at different times. He reported what later became known as the *critical period*: If the hood was removed 2–3 days after hatching, then the chick would persist in following whatever large moving object it first saw, whereas, if removal was delayed for five or more days, the chick would run away from the same object.[6]

Early in the twentieth century reports of imprinting in swans, geese, and ducks by the German zoologist, Oskar Heinroth, revived interest in the phenomenon; the results he obtained with hand-raised goslings were particularly striking.[7] Heinroth inspired his student, Konrad Lorenz, to continue this research. Lorenz first reported his studies in a highly influential paper published in German in 1935, titled *Der Kumpan in der Umwelt des Vogels* (*The Companion in the Bird's world: the fellow-member of the species as a releasing factor of social behavior*). Later, he complied with a request to publish a shorter version in an English journal.[8]

Lorenz was born in Vienna, Austria in 1903. Following his father's wishes, Lorenz studied medicine, first at Columbia University in New York and then at the University of Vienna where he received his medical degree in 1928. He must have been an outstanding student, since on graduating he was appointed to a teaching position in the university's anatomy department. His childhood had been filled with pets of various kinds and, even as a student living at home with his tolerant parents, he filled the apartment with animals that ranged from various fish to a monkey; see Figure 6.1. His love of animals led him to take a second degree in 1933, this time in zoology. His 1935 paper was based largely on his work for this degree, studies that were almost entirely based on birds, such as jackdaws, crows, and various species of ducks, that he raised in isolation from their conspecifics. Most of the work was carried out in and around the Lorenz family home in Altenberg, a small town on the Danube upstream from Vienna. The species that provided the most material for his theories was the grey lag goose. The way that goslings he hand-raised imprinted on him – or, maybe, on his rubber boots – gave rise to a number of famous images, such as the photo shown in Figure 6.2.

Lorenz's research was interrupted by the German invasion of Austria in 1938. Soon afterward he joined the Nazi Party and wrote about the dangers of domestication "couched in the worst of Nazi terminology," activities he later described as "ill-advised."[9] Curiously, in 1940 he was appointed to the position of professor of

Figure 6.1 Konrad Lorenz as a young man with two of his hand-raised crows. The photograph was taken by Lorenz's best friend from childhood, Bernhard Wolfgang Hellmann, who later died in a Nazi concentration camp.
Reproduced with permission from Paul Hellmann, son of Bernhard Hellmann.

psychology at the University of Koenigsberg. However, only a year later, he was drafted into the German army as a medical officer. After being sent to the Eastern Front, he was captured by the Russian army and served four years in a prison camp before being released in 1948. On return first to Germany and then to the family home in Altenberg, Lorenz resumed his research.[10]

Lorenz's 1935 paper stimulated active interest in imprinting among zoologists and psychologists in Europe and in North America; from the late 1940s, it became a very lively topic of research.[11] One reason was the list of clearly specified but contentious claims that Lorenz made. "…I wish to call the reader's attention more especially to the points in which this process differs from what we call associative learning. (1) The process is confined to a very definite period of individual life, a period in which in many cases is of extremely short duration; …. (2) The process, once accomplished, is totally irreversible, so that from then on, the reaction behaves exactly like an 'unconditioned' or purely instinctive response. This absolute rigidity is something we never find in behavior acquired by associative learning, which can be unlearned or changed, at least to a certain extent…. (3) The process of acquiring

Figure 6.2 Konrad Lorenz followed by goslings that had imprinted on his boots. Reproduced with permission from the Science Library, UK.

the object of a reaction is in very many cases completed long before the reaction itself has become established ...... This offers some difficulties to the assumption that the acquiring process in imprinting is essentially the same as in other cases of 'conditioning,' especially in associative learning. To explain the process in question as one of associative learning, one would have to assume that the reaction is, in some rudimentary stage, already present at the time when its object is irreversibly determined, an assumption which psychoanalysts would doubtless welcome, but about which I have doubts. (4) In the process of imprinting, the individual from whom the stimuli which influence the conditioning of the reaction are issuing, does not necessarily function as an object of this reaction. In many cases, it is the object of the young bird's begging-reactions, or the following reactions, in short, the object of the reactions directed towards the parent-companion, that irreversibly determines the conditions which, more than a year later, will release the copulating reactions in the mature bird. This is what we might call a super-individual conditioning to the species and certainly, it is the chief biological task of imprinting to establish a sort of consciousness of species in the young bird, if we may use the term 'consciousness' in so broad a sense."[12]

In what could be seen as a challenge to learning theorists, Lorenz went on to write: "It is a purely conceptual dispute whether imprinting is to be regarded as a special sort of learning, or as something different. The decision of this question depends entirely upon the content we see fit to assign to the conception of 'learning.' Imprinting is a process with a very limited scope, representing an acquiring process occurring only

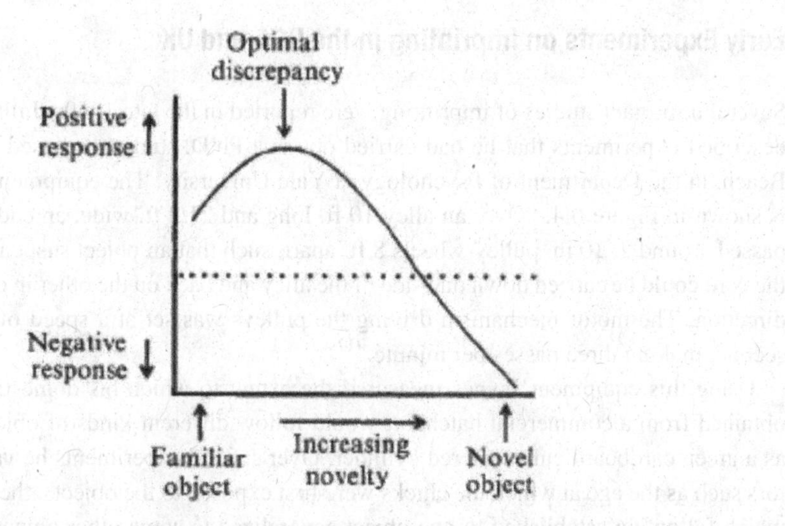

Figure 6.3 This graph illustrates Bateson's claim that sexual imprinting serves to optimize mate choice: Not too familiar and not too novel.
From Bateson (1978). Reproduced with permission from Springer Nature.

in birds and determining but one object of certain social reactions…. I should think it unwise to widen the conception of learning by making it include imprinting."[13]

Beginning in the late 1940s Lorenz acquired worldwide fame on the basis of English translations of a series of popular books, such as *King Solomon's Ring*[14] in 1949 and *On aggression*[15] in 1966. For his contribution to ethology, he was awarded the Nobel Prize in 1973, along with his friend and collaborator from the 1930s, Niko Tinbergen, and Karl von Frisch, who was the first to describe the 'dance of the bees.'

By this time, the large number of studies of imprinting that Lorenz had inspired showed that many of his claims were either misleading or wrong. For example, what became known as filial imprinting was recognized as the process by which a young bird developed attachment to a mother-figure and not, as Lorenz had claimed, the process by which the young came to recognize members of their species.

Lorenz's claims about sexual imprinting were also widely rejected in the long run. In discussing these claims, in 1978 the biologist, Patrick Bateson, wrote: "Subsequent experimental work has, however, suggested that …. a bird may also show a preference for its own species in the absence of experience with any of them but itself … sexual imprinting merely refines this bias in natural conditions. If birds are able to identify and mate with their own species without prior experience … what is the function of sexual imprinting?" On the basis of experiments using quail that exhibited differences in plumage, Bateson's answer was that sexual imprinting functions to promote optimal outbreeding. The choices of his male quail supported the prediction that "the strongest mating preference of a bird should be for something a little different (but not too different) from the object with which it had been imprinted." see Figure 6.3.[16]

Early Experiments on Imprinting in the USA and UK

Several important studies of imprinting were reported in the late 1950s. Julian Jaynes described experiments that he had carried out as a Ph.D. student, advised by Frank Beach, in the Department of Psychology at Yale University. The equipment he used is shown in Figure 6.4. "Over an alley 10 ft. long and 21/2 ft. wide, an endless cord passed around 2, 10 in. pulley-wheels 8 ft. apart, such that an object suspended from the cord could be carried down one side of the alley and back on the other in clockwise direction. The motor mechanism driving the pulleys was set at a speed of 1 ft. per second, making three passes per minute."[17]

Using this equipment Jaynes measured the extent to which his domestic chicks, obtained from a commercial hatchery, would follow different kinds of objects, such as a green cardboard cube or a red cylinder. Over several experiments he varied factors such as the age at which the chicks were first exposed to the objects, the extent to which following established to one object generalized to some other object – sometimes finding considerable generalization to objects of a quite different shape, and the amount of exposure they were given to their object. In broad terms, he replicated Spalding's original results with this species, such as the critical period in which the following response was likely to develop and the later development of fleeing from the same object. What Jaynes added was a large amount of empirical data, for example, on the conditions that gave rise to following and the role of practice in the long-term retention of imprinting.[18]

There is a telling contrast between Jaynes' approach and that taken by the other major study that was reported in 1956, the year of Jaynes' first paper. In the USA Jaynes' experiment were carried out in a laboratory setting, using a single, easily obtainable domesticated species tested in semi-automated equipment. In the UK Robert Hinde, William Thorpe and Margaret Vince tested two related species of wild birds, the moorhen, and the coot, in an outdoor setting using a similar set up to Jayne's but one that was manually operated.[19]

No one was more responsible for the growth of research on animal behavior in the UK than William Homan Thorpe (1902–1986); see Figure 6.5. His undergraduate degree in 1923 from Cambridge University was in agriculture and in 1929 he obtained his Ph.D. there for research on various insects, especially pests. In the 1930s he studied the behavior of various bird species, concentrating on the chaffinch and the way that individuals from this species learn to sing. These studies were the first in which a sound spectrograph was used to provide a detailed analysis of samples of birdsong. Also in the 1930s, Thorpe made visits to the Netherlands and Austria and became closely acquainted with both Tinbergen and Lorenz. He was highly effective in lobbying Cambridge University to establish in 1950 the Cambridge Ornithological Station in Madingley, a small village on the outskirts of Cambridge. Thorpe became its first director and one of his first major decisions was to appoint Robert Hinde as 'curator' of the station.[20]

After serving in the Royal Air Force during the Second World War, Robert Aubrey Hinde (1923–2016) studied natural sciences as an undergraduate, while serving

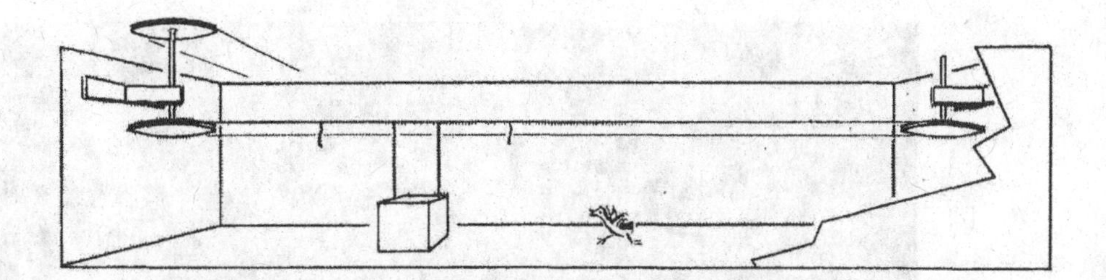

Figure 6.4 This apparatus was constructed for Julian Jaynes. A motor-driven pulley system could move a suspended block backwards and forwards in the long alley. Providing it was young enough, a chick placed in the alley would start to follow the block.
From Jaynes (1956). Public domain.

Figure 6.5 William Homan Thorpe in his late 40s.
From Hinde (1987). Reproduced with permission from the Royal Society (UK).

as secretary of the Cambridge Bird Club and developing a friendship with David Attenborough, who later became world-renowned for his documentary films about animals. On graduating in 1948, Hinde enrolled as a Ph.D. student at Oxford University to study the behavior of the great tit. He was heavily influenced by Niko Tinbergen who arrived in Oxford in 1950. After obtaining his Ph.D. that year and immediately joining Thorpe in Madingley, the two of them became involved several projects on various aspects of bird behavior. One of them was the imprinting study on coots and moorhens referred to above.[21]

Figure 6.6 Experimental pen used to study the following response in coots and moorhens. Unlike in Jaynes' alleyway, shown in Figure 6.4, the various suspended objects were moved by hand.
From Hinde, Thorpe, and Vince (1956). Reproduced with permission from Brill Publishers.

The experiments run by Thorpe and his colleagues were far less systematic than those reported by Jaynes. They involved a large range of seemingly arbitrary objects that were moved by hand – at speeds judged to be suitable, given the range of ages of the birds they tested – when suspended from the line shown in Figure 6.6. The objects ranged from the simple black box shown in this figure to a wooden model of a moorhen to an inflated orange football bladder. Not all objects were suspended; the 'Man' was an experimenter walking backwards and the 'Hooded Man' was an experimenter wearing a sack and a raincoat to conceal his head and arms. These researchers took less care than Jaynes to restrict their birds' early visual experience; after hatching and being hand-fed twice each hour during the day, most of the birds "were allowed a full view of the person who was feeding them, and could see people around throughout the daylight hours."[22]

The main aim of this study was to test two of the claims made by Lorenz: that the following response develops only during a critical period in a bird's early life, and that, once established, it is irreversible. In summarizing the results from what must have been a very lengthy study, the authors concluded that "following can be evoked by many objects quite different from each other in size and shape, but having in common the property of being in motion.... birds trained on one model would generalize to other throughout practically the whole period in which they would follow at all. In this sense, the learning is not irreversible." They were agreement with Lorenz about the critical period for following to develop, at least in their moorhens: These birds "were more likely to follow if tested from the first day after hatching than if several days were first allowed to elapse. Fleeing increases during the first few days of life, and this may be an important factor limiting the 'sensitive period' for the establishment of the following response.... There is no evidence that 'imprinting' is fundamentally different from other types of learning."[23]

The study seems to have marked a turning point in the research careers of two of the authors. The third author, Margaret Vince, had previously been based in the Applied Psychology Unit at Cambridge where she had carried out experiments on human reaction times; subsequently, she studied the behavior of domestic chicks and problem-solving by great tits.[24] The first author, Robert Hinde, developed a close friendship with the developmental psychologist, John Bowlby, who had recently become well-known for his theories on the importance for a child's future of early bonding with his or her mother. Their collaboration led to the introduction of rhesus monkeys into the Madingley Field Station, where Hinde studied interactions among the infants and between infants and their mother.[25] Harlow's reports on the effects of separating infant monkeys from their mothers, as well as human studies indicating that disruptions of a child's contact with mother can have long-term effects, provided the context for Hinde and his collaborator, Yvette Spencer-Booth, to carry out one of their rare – or maybe, only – experimental studies. Effects of removing the rhesus mother even for just a few days from the cage containing her infants was found to affect their behavior up to two years later.[26]

Following the 1956 report of his study of imprinting with Thorpe and Vince, Hinde's only published research involving birds was a study of nest-building and other behavior in canaries. In 1960 he handed over research on imprinting at Madingley to a new Ph.D. student, Patrick Bateson.[27]

As for Thorpe, his main research concerned the development of song by chaffinches and, more generally, on vocalizations in a range of bird species. In this work, he was joined by a number of Ph.D. students, notably two from the USA who were to become major contributors to this field, Peter Marler and Fernando Nottebohm.[28] Thorpe also had a great influence on subsequent imprinting research via the many lectures he gave on the subject.

Among those inspired by Thorpe to carry out experiments on imprinting were psychologists who became influential heads of some of the new psychology departments that were established in the late 1940s and 1950s in both British and Canadian universities. One such academic was Wladislaw Sluckin, who carried out a series of experiments on the following response in domestic chicks at Leicester University in the UK. Together with his friend and collaborator, Eric Salzen from Liverpool University, Sluckin analyzed the relationship between perceptual learning, associative learning, and imprinting.[29] In Canada, a long-term head of the psychology department at Dalhousie University, who came from the UK to establish its strong psychobiological tradition, was Henry James. His major contribution was to show that chicks could be imprinted to a flickering light and that the effectiveness of moving objects was largely based on the equivalent of a flickering visual stimulus that such stimuli produced in young birds.[30] This explained why the following response that a young bird developed to the first moving object it was exposed to would subsequently be elicited by moving objects of quite different shapes and sizes, as both Jaynes and Hinde and his colleagues had discovered.

Imprinting and Associative Learning

The earliest attempt to relate imprinting to more general learning processes was the suggestion by Sluckin and Salzen that imprinting can be understood as a form of perceptual learning.[31] The idea was that animals gradually come to discriminate between familiar and unfamiliar objects or environments, and this provides the basis for attachment to what is familiar. One problem with this approach was that perceptual learning itself was poorly understood. Another was that it gave no special status to stimuli such as movement and flashing lights that clearly have a particularly powerful effect on recently hatched chicks and ducklings. The importance of such stimuli was central to a later theory that more explicitly tied imprinting to contemporary theories of conditioning. This was developed in the USA by Howard Hoffman and his collaborators. It appealed to what later became known as *simultaneous associations* or *within-compound conditioning*, concepts that were not clearly identified and named until a key paper on the topic was published by Rescorla in 1980.[32]

Howard S. Hoffman (1925–2006) was born in New York City. His father was a salesman and both parents were grand master bridge players. On graduating from high school Howard enlisted in the army in 1943 and nine months later was sent to Europe; see Figure 6.7. His experience of a considerable amount of serious combat is described in a notable exercise in oral history carried out by his historian wife, Alice. On return to the USA in 1945 the GI Bill made it possible for him to enrol at the University of Chicago and to start studying physics. Once there, however, he discovered a love of painting, a love that endured throughout his life, and he left to study art in New York. While training to become an art teacher, his experience in a nursery school aroused his interest in child behavior and inspired him to study psychology at the New School of Social Research in New York. Graduation in 1952 was followed by a Master's degree in psychology in 1953 from Brooklyn College and a Ph.D. in 1957 from the University of Connecticut. Also in 1957, he obtained his first academic position, an Associate Professorship at Pennsylvania State University. In 1970 he was invited to become a Professor of Psychology at Bryn Mawr College where he remained of for the rest of his career.[33]

Hoffman's doctoral research was on speech perception. However, he was unable to continue this line of research at Penn State because no anechoic chamber was available there. On the other hand, the Department of Psychology had good facilities for behavioral research involving animals. Working with both pigeons and rats, Hoffman soon became well known within this area for his studies of conditioned suppression and avoidance learning, notably for demonstrations that barely diminished effects of such learning could be detected many years later. What made him more widely known

Figure 6.7 Howard Hoffman when newly enlisted into the US Army.
From Hoffman and Hoffman (1990). Public domain.

was his rediscovery in the early 1960s of a phenomenon that became known as *prepulse inhibition* of the startle reflex; this refers to the decrease in the intensity of a reflex response to, say, a sudden gunshot, if this is shortly preceded by some kind of mild stimulus.[34] Hoffman's experiments on this effect started with rats and continued, now mainly with human infants, well into the 1980s.

In parallel with his other lines of research, in the mid 1960s experiments on imprinting in ducklings were begun in Hoffman's laboratory. A notable aspect of many of these experiments – one that possibly reflected his early training in speech perception – was that a duckling's distress calls were recorded. The suppression of such calls by a moving object or a flashing light was seen as just as important a feature of what had become known as the filial response as whether following occurred.

Hoffman's theory of imprinting in precocial birds was based on five premises. First, that certain kinds of stimulation are innately capable of evoking filial behavior such as the suppression of distress calling. Second, such stimuli function as reinforcers in a learning-theory sense. The third premise was the assumption of simultaneous conditioning, not that this term was used: "As a result of their consistent pairing with reinforcing stimulation, the initially neutral features of the stimulus gradually acquire the status of a conditioned stimulus in that they come to elicit a conditioned filial response and as such become sources of conditioned reinforcement." A nod towards earlier perceptual learning accounts noted that "as this process transpires, the imprinting stimulus becomes increasingly familiar." The fourth premise followed from several previous studies by other researchers and was directed at explaining evidence for a critical period: "There is a maturationally-based increase in the tendency for an immature precocial bird to respond fearfully to imprinting stimuli that are unfamiliar." The strength of such a fear response depends on how incongruous a novel stimulus is. The fifth and final premise was that "the behavior an immature precocial bird displays to a given imprinting stimulus reflects a resolution of the competing tendencies aroused by the stimulus to respond filially and fearfully."[35]

Although the theory was based on evidence mainly from chicks and ducklings, it was intended as a general developmental theory. Thus, in immature primates certain forms of tactile stimulation – such as the soft cloths that Harlow's monkeys clung to when frightened – play a similar role to that of movement, flashing lights, or certain calls for a precocial bird; or so it was argued.[36]

One implication of the fifth premise, that of competition between filial and fear responses, is the need to modify Lorenz's concept of an 'all-or-nothing' critical period, that of a window of opportunity that is firmly shut when a bird reaches a certain age. Instead, the premise predicts that, if habituation of a fear response occurs, filial behavior will emerge at an age well beyond that at which normally imprinting no longer occurs. The idea of a fixed critical period was perhaps based on Lorenz's use of observations in a natural setting in which birds could flee from a strange object. Thus, in an artificial setting where a bird was unable to flee from an imprinting object fear would subside as the object became familiar and a filial response emerge.

The prediction that the opportunity to escape is an important factor was confirmed in an experiment with 8-day old ducklings that were placed in the chamber shown in Figure 6.8. This contained a foam block that could move backwards and forwards behind a window that allowed the block to be seen by a duckling in the front compartment. The floor of the compartment was marked into four quadrants, with quadrant #2 closest to the moving block. The main outcome measure was the time that a duckling spent in quadrant #2.

The condition for each of the four *Escape* ducklings was that, after being placed in quadrant #2 at the start of each session, if they moved to another quadrant when the block was moved, the light illuminating the block was switched off and its movement terminated. The behavior of the four *Control* ducklings had no effect, so that they were exposed to the moving block throughout each of the 30-minutes sessions that were given twice each day for three days. By the sixth session, the *Control* ducklings were spending all their time close to the stimulus, while the *Escape* ducklings continued to escape from quadrant #2 as soon as the illuminated block moved.[37]

It seems unlikely that the imprinting effect obtained in the *Control* condition from the above experiment would have been obtained if the ducklings had been very much older. Just as fear of the new increases as precocial birds get older, the effectiveness

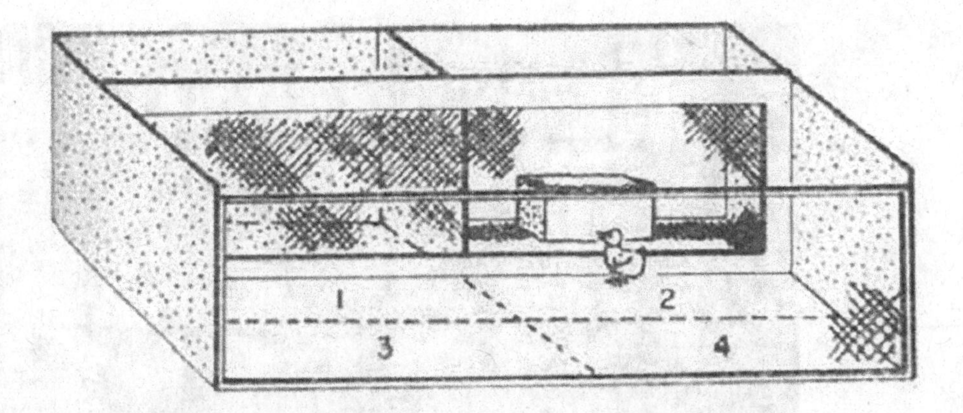

Figure 6.8 This apparatus was used to test whether the opportunity to escape was important in the development of filial imprinting. While the block was moved backwards and forwards, the experimenters recorded which of the four quadrants in the viewing chamber a duckling occupied. Remaining in Quadrant 2 indicated strong attachment. Remaining in the other quadrants, particularly #3 indicated fear of the moving block.
From Ratner and Hoffman (1974). Copyright © 1974 Published by Elsevier Ltd.

of movement or a flashing light as a reinforcing stimulus appears to wane with age. Does their effectiveness also decrease once imprinting has occurred to some object? This raises the issue of whether imprinting is self-terminating or - more broadly – how *secondary imprinting* is affected by some initial imprinting episode.

A general belief that an initial imprinting experience decreases an animal's 'imprintability' was partly based on the well-documented finding that it is difficult to imprint precocial birds that have been housed with their peers. In contrast, according to the theory in terms of associative learning proposed by Hoffman and his colleagues, imprinting a bird on a second object should not be at all difficult.

One test of this prediction was an experiment in which four ducklings were first imprinted on a rotating lamp 12–16 hours after they had hatched. Successful imprinting was indicated by suppression of distress calling when the lamp was present. Then on Day 6 of their life they were subjected to a second imprinting procedure, one involving the moving foam block shown in Figure 6.8. Further test sessions revealed that the block could suppress distress calling when it was stationary and did so without any decline in effectiveness in tests repeated throughout Days 7–16 of the ducklings' life. The researchers concluded from this and similar experiments that "a prior history of imprinting does not prevent a second imprinting object from exhibiting these capacities."

In the early 1980s, I gave a lecture that mentioned secondary imprinting in the context of Hoffman's theory. Afterwards, an undergraduate approached me to express his belief that secondary imprinting would be difficult following primary imprinting to a stimulus that was much more natural than, for example, a rotating lamp. David Panter needed a research project for his final year and as a result, I became involved for the first – and last – time in research on imprinting.

Figure 6.9 Chicks exposed to this domestic hen during primary imprinting later showed weak or no secondary imprinting to an otherwise effect stimulus object.
From Boakes and Panter (1985). Reproduced with permission from David Panter.

We decided on an experiment with three groups that would differ only in their primary imprinting experience. Subsequently, all were given a secondary imprinting condition similar to that shown in Figure 6.8; in our case, an upturned coffee cup moved backwards and forwards behind a window. To obtain ducklings would have been difficult and expensive and so we hatched chicks instead. The most 'natural' stimulus we could think of was a domestic hen and, fortunately, a neighbor was willing to lend hers for the experiment; see Figure 6.9.

The primary imprinting procedure took place during three brief sessions on Days 1 and 2 after the chicks had hatched. While one group of chicks could watch the hen walking around in its hold cage, a second group were able to view an 'artificial' stimulus. This consisted of a plastic windmill consisting of six shapes of various colors that would spin when a fan – out of sight of the chicks – was directed at the windmill. The third group served as a control; these chicks had an equally clear view of the windmill but its colored shapes remained stationary. These chicks showed little interest in the windmill and remained inactive, while the other chicks became increasingly active and stayed as close to their stimulus – the hen or the windmill – as their holding cage would allow.

The results from the crucial secondary imprinting stage were very clear. The *Control* chicks that had never seen anything moving before showed strong imprinting, both in terms of following the cup and in ceasing to make distress calls even when the cup was stationary. Imprinting to the cup was also displayed by the *Windmill* chicks, although not as strongly as displayed by the controls. As for the *Hen* group, the results were just as Panter – but not the learning theorist – had predicted: These chicks followed the cup only some of the time and their distress calling was unaffected by the cup's presence, whether it was moving or stationary.[38]

These results were consistent with a capacity theory based on the idea of a limited number of receptors that are active only during the first few days of a precocial bird's life and the assumption that some particularly effective stimuli – such as our hen – use up more receptors than a weaker stimulus – such as our windmill.[39] An alternative that remains close to Hoffman's associative theory is our suggestion in 1985 that the reinforcing power of some stimulus element such as movement becomes weaker as it is repeatedly experienced; that is, it habituates, just as the effectiveness of a loud noise in conditioning fear towards some stimulus diminishes as it is repeated. Following several further experiments in the 1980s and 1990s on secondary imprinting and related issues, no single theory has been found to provide a satisfactory explanation of filial imprinting.[40]

Experimental Psychology Meets Ethology in Cambridge, UK

In the first few decades of the twentieth century, the mutual disdain that existed between European-based ethologists and many American behaviorists was maintained by geographical separation. However, in the UK, the appointment in 1949 of the Dutch ethologist, Niko Tinbergen, to a lectureship at Oxford University and the establishment by Thorpe in 1950 of the Ornithological Field Station at Cambridge University marked the start of an era in which ethologists and the few British psychologists interested in learning processes took considerable interest in each others' ideas. Indeed, Thorpe's vision for research carried out in the new field station was that it would concentrate on the 'key problem' of the relationship between instinctive and learned behavior. Tinbergen's classic book, *The study of instinct*, of 1951 made the work of ethologists easily accessible to English-speakers and helped towards his appointment 1968 as Professor of Animal Behavior at Oxford.

In the early 1950s, Cambridge was still one of the few universities in the UK where a student could study psychology. The head of the Department of Experimental Psychology was Oliver Zangwill, a neuropsychologist. He and some colleagues had a strong interest in learning theory and its relationship to ethological research on animal behavior. In 1961 Zangwill and Thorpe jointly edited a book, *Current problems in animal behavior*; undergraduates like myself who were studying psychology in Cambridge in the 1960s were encouraged to read this book, along with Tinbergen's *A study of instinct*.

It was then no accident that Cambridge was the setting for an important conference held in 1972 that was organized by Robert Hinde and his wife, Joan Stevenson-Hinde. The following year they published a book titled *Constraints on learning* based on contributions to the conference. The book included chapters by locals like Patrick Bateson, who described constraints on what stimuli were effective in filial imprinting. For example, a flashing red light was much more effective than a green one; also, a chick was more likely to imprint onto a large object than a small one – something matchbox-size was more likely to be treated as a potential food object than as a mother figure.[41]

A temporary local was Jerry Hogan, an experimental psychologist from the University of Toronto who was spending a sabbatical year at the Madingley Field Station. His research on newly hatched chicks followed the assumption that they need as much to learn what to eat as to what to adopt as a mother figure. Hogan reported that up to 48 hours of age chicks were as likely to peck at sand as at commercial chick food but by 72 hours of age they ignored sand and chose the food; it seemed that they had learned to associate the food with its nutritious consequences, despite a delay of 80 minutes between ingestion and any nutritional benefit.[42] That learning could proceed despite such a long delay was still not widely accepted, despite research a few years earlier indicating that rats can associate a taste with unpleasant consequences occurring hours later, as described later in this chapter.

Three more psychologists based in Canada made important contributions to the conference. Sarah Shettleworth, also from the University of Toronto, seems to have been the first person to use the phrase 'constraints on learning' in print.[43] In her contribution to the Cambridge meeting, Shettleworth reported experiments in which she attempted to use delivery of food to reinforce a range of different behaviors that her golden hamsters displayed in a large arena; see Figure 6.10. As expected, all animals pressed a lever with increasing frequency when food was made contingent on this response, while other behaviors such as rearing, digging, and scrabbling against a wall also increased when reinforced. In contrast, various types of grooming behavior, such as face washing and scratching with a hind leg, together with scent marking, proved essentially insensitive to food reinforcement. "Not everything an animal does can be modified equally readily by reinforcement" was her conclusion.[44]

Shettleworth's research revived interest in scattered reports of similar phenomena that had previously received little attention. The earliest example was in 1911 when Thorndike reported his failure to increase the frequency of grooming behaviors of chicks or cats when such responses were rewarded.[45] A much later„ and especially interesting, example was Konorski's success in use food reinforcement to increase the frequency with which a cat scratched its ear as long as a wad of cotton had been placed in the ear.[46] A similar effect was later demonstrated in rats by Pearce, Colwill, and Hall. They found that using food to reinforce scratching behavior had only a limited effect, except when a Velcro collar was fitted to an animal's neck. In rejecting the idea of a 'biological constraint,' whereby animals cannot *learn* to associate grooming behaviors with an outcome such a food reward, these researchers suggested that *performance* of scratching depends on the presence of "an unobservable eliciting stimulus, an itch."[47]

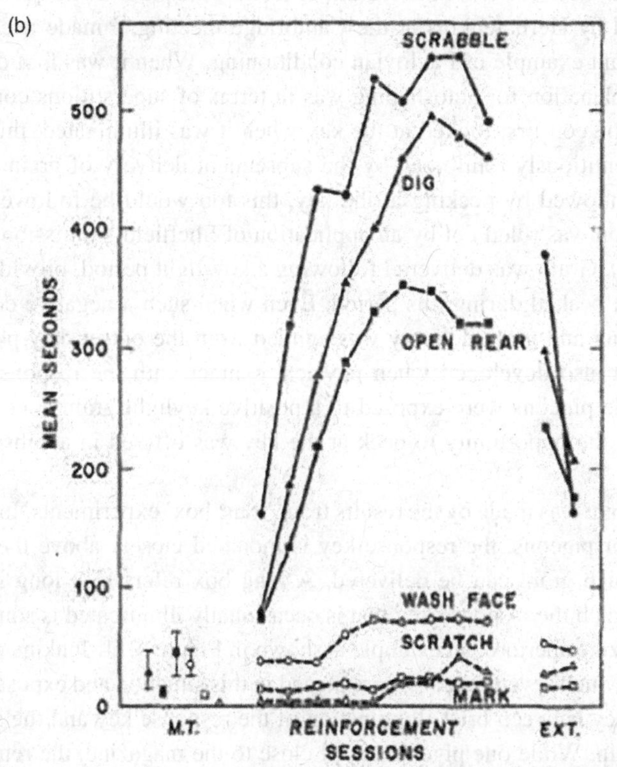

Figure 6.10 Panel (a) shows various action patterns displayed by golden hamsters. As shown by the graph in Panel (b), some actions, for example, "scrabble," increased when reinforced with food, while others, for example, "wash face," did not.
From Shettleworth (1975). Used with permission from the American Psychological Association.

Autoshaping, Sign- and Goal-Tracking

In terms of its subsequent impact by far the most important body of research reported at the Cambridge conference was on *autoshaping*. For some decades researchers working with pigeons had used a *discriminated operant* procedure in which, for example, only when a light came on to illuminate a response key was a pigeon rewarded for pecking at the key. A seemingly minor modification of this procedure was introduced in experiments carried out in Herb Jenkins' laboratory at McMaster University. Hungry pigeons always received access to grain following a period in which the key was lit, *whether or not* they had pecked at the key during this period. Almost all pigeons exposed to this procedure came to peck at the key at high rate.[48] From then on no one studying key-pecking in pigeons needed to manually shape each individual bird; researchers could now use the Brown-Jenkins procedure to shape the response automatically; hence the title 'autoshaping.'

It soon became clear that the discovery of autoshaping was of far greater significance than that of release from the tedium of manual shaping. Subsequent research, as summarized by Herb Jenkins at the Cambridge meeting,[49] made it clear that it was an important example of Pavlovian conditioning. When it was first discovered, a plausible explanation for autoshaping was in terms of superstitious conditioning: Thus, when a pigeon first looked at the key when it was illuminated, this response would be adventitiously reinforced by the subsequent delivery of grain and, when looking was followed by pecking at the key, this too would be followed by food. This explanation was ruled out by an application of Sheffield's omission procedure (see Chapter 2). Grain was delivered following a key-light period, provided that the pigeon had not pecked during this period. Even when such a negative contingency between pecking and grain delivery was applied from the outset, key-pecking still developed.[50] It also developed when physical contact with the response key was prevented while pigeons were exposed to a positive keylight-grain contingency, as revealed when the opportunity to peck at the key was offered in a subsequent test in extinction.

The same point was made by the results from 'long-box' experiments. In a standard Skinner box for pigeons, the response key is mounted closely above the magazine aperture in which grain can be delivered. A long box refers to a long rectangular enclosure in which the response key that is occasionally illuminated is some distance from the magazine aperture; an example is shown in Figure 6.11. Jenkins reported an experiment in which seven pigeons were placed in this situation and exposed to a positive contingency between brief illumination of the response key and the subsequent delivery of grain. While one pigeon stayed close to the magazine, the remaining six approached and pecked at the key for as long as it remained lit. "It is of interest that because the key-light was located almost three feet from the grain dispenser, which remained up for only four seconds, the behavior of pecking at the light until it went off caused a substantial loss in the amount of grain received."[51] Watching a pigeon rush away from the place in which food will shortly arrive can convey the power of sign-tracking!

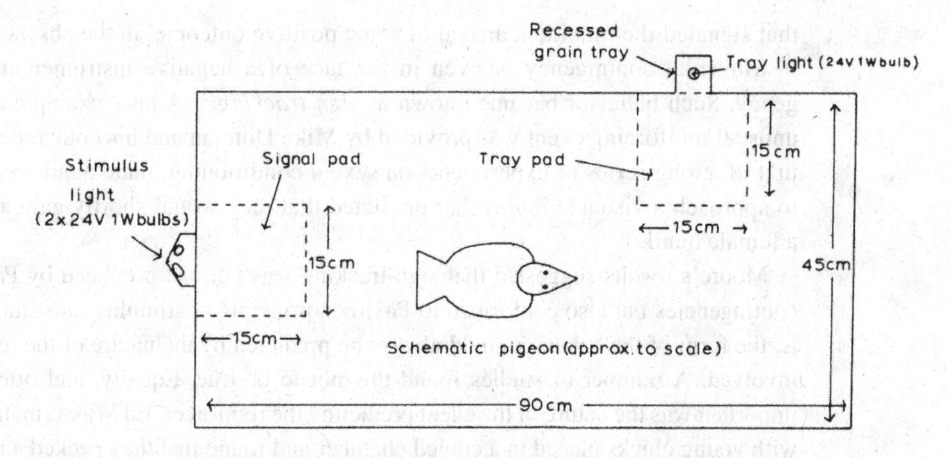

Figure 6.11 In an experiment using this 'long box' occasional trials would start with the stimulus light, shown on the left, coming on for a few seconds. As this was switched off, the grain tray was presented for a few seconds. Most pigeons would remain close to the stimulus light when it was lit (*sign-tracking*) and dash to the grain tray when it was operated. Only a few moved towards the grain tray when the stimulus light was lit (*goal-tracking*). From Boakes (1979). Reproduced with permission from John Wiley and Sons.

Furthermore, as detailed by the third contributor from Canada, Bruce Moore, the form of a pigeon's key-peck differed according to whether the key-light was paired with food or with water, just as the type of saliva produced by a dog in response to a signal for meat powder differs from that produced when the signal indicates that dilute acid is about to be injected into its mouth.[52]

These results came as a great surprise to a generation of learning theorists that had accepted Skinner's claim that Pavlovian conditioning only affects autonomic responses such as salivation and changes in skin conductance or those that had embraced Mowrer's theory that this type of conditioning affects only emotional states. Thirty years earlier Karl Zener had filmed the behavior of dogs undergoing traditional Pavlovian salivary conditioning, with a bell to signal that food was about to be delivered. He described the kind of preparatory behavior they displayed when the bell was sounded.[53] His conclusion that Pavlovian training can produce changes in external behavior was generally disregarded on the grounds that the behavior he described "was almost certainly instrumentally, rather than classically conditioned."[54] For researchers in the Skinnerian tradition that relied on rate of pecking by pigeons as their main dependent variable the finding that autoshaped pecking was produced by Pavlovian contingencies raised the disturbing possibility that what they were studying was respondent, rather than operant, behavior; perhaps Skinner's switch from studying lever-pressing by rats to key-pecking by pigeons (see Chapter 8) had not been such a good move.

It soon became clear that autoshaped pecking was not a special phenomenon but, instead, an example of a very general type of behavior: A range of species, including human children, were found to approach and interact with localized events that signaled the imminent arrival of some positive outcome, in the absence of any instrumental contingency or even in the face of a negative instrumental contingency. Such behavior became known as *sign tracking.*[55] A later example using an unusual reinforcing event was provided by Mike Domjan and his colleagues; in the first of a long series of experiments on sexual conditioning, male quail were found to approach a visual stimulus that predicted that they would shortly gain access to a female quail.[56]

Moore's results suggested that sign-tracking was not just produced by Pavlovian contingencies but also conformed to Pavlov's principle of stimulus substitution; that is, the form of the behavior could always be predicted by the nature of the reinforcer involved. A number of studies found this not to be true. Equally, and often more, important was the nature of the event predicting the reinforcer. Ed Wasserman worked with young chicks placed in a cooled chamber and found that they pecked a response key when its illumination predicted the onset of a heat lamp, even when such pecking prevented the onset of the lamp; this behavior was quite different from that evoked by the lamp itself.[57] And Peter Holland found that a diffuse light paired with food would come to evoke frequent rearing, while a tone evoked a head jerking response that was rarely observed when a visual stimulus was predicted food delivery.[58]

The importance of the kind of predictive event used in such experiments was also shown in dramatic fashion in Bill Timberlake's laboratory at Indiana University. When the arrival of another rat acted as a signal that food would be delivered, the first rat, although hungry, treated the newcomer as a friend rather than as a potential meal. Following a suggestion in 1969 by Lorenz, the authors concluded that "an entire behavior system [is] conditioned by the procedure of classical conditioning, not just an isolated response."[59] A similar conclusion was reached by Jenkins and his collaborators on the basis of sign-tracking experiments with dogs: "The concept of the CS-US episode as a model, or simulation of a natural sequence which induces a corresponding preorganized action pattern, accommodates a number of otherwise puzzling aspects of conditioned motor responses in a Pavlovian experiment. It makes understandable the observation that induced actions are often signal-centered and are often *appetitive*, or reinforcer-procuring, rather than *consummatory*, or reinforcer-consuming."[60]

The behavior of the dogs studied by Jenkins and his collaborators was far more variable than the sign-tracking normally seen in birds. Such individual variability was also seen in hungry rats when illumination or insertion of a lever signaled food. While many animals would approach, sometimes gnaw and press the lever, some rats would head towards the magazine aperture, the place where the food pellet was about to arrive, a form of behavior that was labeled *goal tracking.*[61]

Firm confirmation that goal tracking, or conditioned magazine approach, was also largely controlled by Pavlovian contingencies was not obtained until nearly four decades later.[62] Thus, Zener's claims in 1937 concerning the behavior of his dogs were at last vindicated. In the meantime, whereas associative learning theorists studying Pavlovian conditioning in the 1960s and 1970s had relied upon preparations involving the delivery of electric shock to an animal and some measure of fear, many subsequently switched to using auto-shaped pecking by pigeons or magazine entries by rats as the outcome measure. The change made it easier to defend behavioral research from the increasing number of critics concerned with animal rights.

Interactions between Pavlovian and Instrumental Conditioning

In his contribution to the Cambridge *Constraints* conference Jenkins made the important point that a discriminated operant procedure can produce an implicit Pavlovian relationship, since the discriminative stimulus becomes a predictor of the reinforcer, and, likewise, a Pavlovian procedure can produce an implicit operant relationship in that the conditioned response becomes reliably followed by the reinforcer. The idea that instrumental and Pavlovian learning can interact led to understanding of some hitherto puzzling behavioral phenomena. One such example was a report by the Brelands of what they called the *misbehavior of organisms*. This referred to cases in which they had used operant methods with some initial success to train an animal to perform a complex behavior but then the behavior had broken down.[63] One case was that of raccoons who were trained to collect a wooden disc and then deposit it in a box in order to obtain food; after this chain of behavior was first established, instead of steady improvement, the latency to complete the sequence increased as the raccoons spent more and more time rubbing and kneading the discs. The idea that this was because the discs had become closely associated with food and, as a result, were treated as food-like objects was confirmed when rats were trained to obtain a ball-bearing and deposit it into a hole; like the raccoons, they also started to display food-related behavior towards the ball bearing and a consequent decline in performance, but only when the behavior was reinforced with food and not when water was used as the reinforcer.[64]

Animals not only approach localized signals for positive events, they also withdraw from signals for aversive events. Interactions between the latter type of Pavlovian conditioned behavior and instrumental behavior are also seen in avoidance learning. As described in Chapter 4, a shuttle box consists of two compartments separated by a small barrier; following a two-way avoidance procedure a rat is given a short-duration warning signal (WS) and, if it fails to move to the other compartment before the WS terminates, it receives a shock, whereas moving to the other compartment before WS termination prevents delivery of a shock. Avoidance learning with this procedure is rapid so long as a diffuse stimulus is used as the WS; it is even faster if a localized stimulus is presented in the compartment that the rat currently occupies, but extremely slow if the WS is located in the compartment that the rat now has to enter.[65]

Robert Bolles's career is described in Chapter 4 (Figure 6.12). What is relevant in the present context is his increasing concern with avoidance learning that came to dominate his research when he moved to the University of Washington. As detailed in Chapter 4, over a series of studies he and his students demonstrated that the readiness with which a rat learns some avoidance behavior depends on the degree to which the required avoidance response is compatible with the rat's *species-specific defense*

Figure 6.12 Bob Bolles in 1966, the year he published his *Theory of motivation*. From Bouton and Fanselow (1997). Reproduced with permission from the American Psychological Association.

reaction (SSDR). These are innate responses displayed by a frightened animal that vary according to the situation. An important example is that in a small, enclosed space that offers no escape a rat will freeze.

Taste Aversion Learning

The studies of constraints on learning, auto-shaping and related topics described in this chapter were carried out within mainstream traditions for studying learned behavior by experimenters who held academic appointments in university psychology departments. The final body of research to be described consisted of experiments that used what were seen at the time as bizarre methods and that were carried out in a research laboratory run by the US Navy. This research was initially either ignored, dismissed, or treated with suspicion.[66] One probable factor was that the lead author was believed to be a Mexican. This was not a belief that John Garcia (1917–2012) tried hard to counteract, even though his parents were immigrants from northern Spain who met in California; his father was born in the province of Asturias and his mother in Galicia. Another factor may have been that he was still a graduate student when his first paper on taste aversion learning was published in 1955. Although he had entered graduate school to work in Tolman's laboratory in 1949 – five years before Bolles arrived – Garcia was not awarded his Ph.D. till 1995.[67]

Garcia grew up on his parents' small farm near Santa Rosa to the north of San Francisco. On leaving school he trained as a mechanic and worked on trucks and ships. During World War II he became first a pilot in the US Army Air Corps and then an intelligence specialist. As with Hoffman, the GI Bill enabled Garcia to attend college after the war. The bachelor's degree he obtained from Santa Rosa Junior College led to enrolment as a graduate student at Berkeley.[68]

The reason for the 46 years it took Garcia to obtain his Ph.D. was that two or so years after starting at Berkeley, he needed to take a break to earn money. The position he obtained was at the US Naval Radiological Defense Laboratory in San Francisco, where he met Bob Bolles.

In the context of the radiation sickness experienced by survivors of Hiroshima and Nagasaki, Garcia was employed to study the effects of gamma- and X-radiation on rats. When given a choice in a T-maze between an arm that was lead-shielded and one that was not shielded, rats showed no sign of learning which side was safe. However, exposure to such radiation was not entirely ineffective. Garcia noticed that the rats would no longer drink water out of the plastic bottles they were given while being exposed to radiation, although they would still drink water from glass bottles. It seemed that they may have developed an aversion to the taste of plastic.[69] This observation prompted the first conditioned taste aversion (CTA) experiment: After rats had been given a novel saccharin solution to drink and then exposed to gamma radiation, they consistently avoided the saccharin when given two-bottle preference tests between the saccharin solution and water.[70] There was no mention of this study or of the several published reports on radiation-induced taste aversion that followed in the standard textbook on learning[71] that was published in 1961, six years after the first report of a CTA by Garcia and his colleagues.

It took eleven more years before the publication of two key papers prompted wide appreciation of the importance of CTA research. In 1966 Garcia and Koelling reported that, although a taste was readily associated with sickness so as to produce an aversion to that taste, they could not detect any aversion produced by pairing the taste with a shock; on the other hand a visual/auditory stimulus – 'bright noisy water' – produced when the rats contacted the spout of a water bottle was readily associated with the shock, in that the rats subsequently avoided this bottle; on the other hand, they did not avoid this bottle after the same stimulus was followed by sickness.[72] As hostile reviewers pointed out, these experiments lacked critical control conditions; it was not easy for Garcia and Koelling to find a journal that would accept their paper.[73] One reviewer allegedly commented that the results were as unlikely as finding bird-shit in a cuckoo clock. However, the basic stimulus selectivity – or *cue-to-consequence* – effect was later thoroughly replicated[74] and has never been in serious doubt ever since; see Figure 6.13.

Also in 1966, Garcia and colleagues reported evidence that taste aversion learning could take place despite a delay between experience of the taste and subsequent sickness that was far longer – up to several hours – than any previous example of an animal learning to associate two events that were separated in time.[75] Despite some initial suggestions that the result might be an artifact due, for example, to a persistent

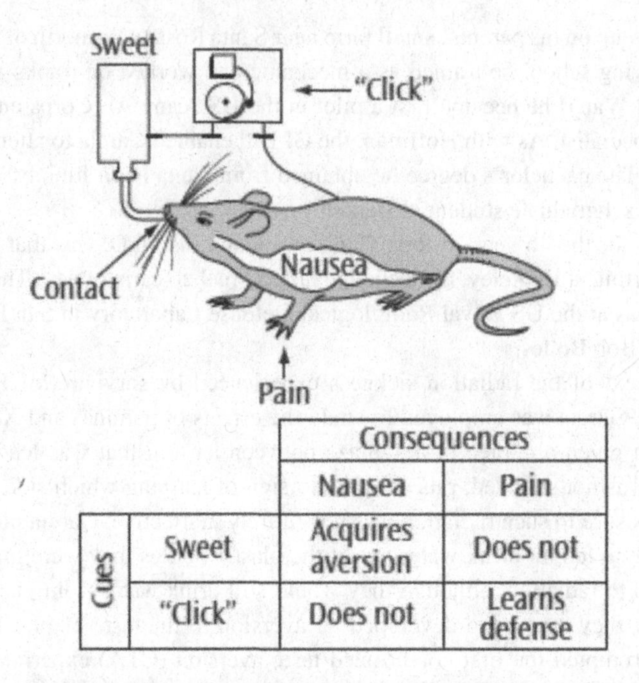

Cues		Consequences	
		Nausea	Pain
Cues	Sweet	Acquires aversion	Does not
	"Click"	Does not	Learns defense

Figure 6.13 Diagram illustrating the cue-to-consequence effect reported by Garcia and Koelling (1966). Rats readily learn to associate a sweet taste with nausea and to associate a click with pain. Taste-pain and Click-nausea associations are formed only under extreme conditions, if at all. The diagram was drawn by Garcia, who was proud of his training as a graphic artist at an early stage in his varied career.
From Garcia, Clarke and Hankins (1973). Reproduced with permission from Springer Nature.

aftertaste, again subsequent studies confirmed that acquisition despite a long delay is a very real characteristic of taste aversion learning;

Garcia's theory was that there existed a gut defense system containing learning processes with quite different properties from those studied by mainstream learning researchers. This system, he suggested, was a very primitive one that had evolved very early to defend animals from ingesting toxins. As further examples of special properties, Garcia pointed to one-trial learning – the acquisition of a taste aversion after a single pairing of a taste with sickness – and claimed that, once an aversion to a particular taste had been acquired, it was completely resistant to extinction.

A boost to the claim that poison-based aversion learning is unique came from the discovery – in the now conventional laboratory that Professor Garcia headed at the University of California, Los Angeles – of *taste-potentiated odor* conditioning. As his team first demonstrated, when a rat is exposed to a novel odor such as that of almond and later made sick, little or no aversion to the odor can be detected in a subsequent test. However, if during the conditioning trial the rat is given a novel taste, such as saccharin, at the same time as exposure to the odor, a strong aversion to the odor is established.[76] This was the opposite outcome from what was expected from experiments going back to Pavlov's early research. These had consistently found that the adding a strong stimulus to a weak stimulus *overshadowed* conditioning of the weak stimulus (see Chapter 9).

Garcia decided that taste potentiation resulted from the taste acting to 'gate' an odor into the gut defense system. Also in 1979, a similar effect in pigeons was reported that was related to color. Pigeons do not readily acquire an aversion to the color of their drinking water. However, if a pigeon is made sick after its drinking water is colored blue for the first time and is also given a novel salty taste, the bird will subsequently avoid tasteless blue water.[77]

By the time such examples of potentiation by a taste of various kinds of aversion learning were reported, taste aversion learning was looking far less unusual than it first appeared. For one thing, despite early claims to the contrary, such aversions could be readily extinguished. Also, one-trial learning was found in other types of learning; a rat given a single shock after being placed in a novel place or after exposure to a loud sound, will subsequently freeze when returned to that place or when the sound is repeated. Furthermore, some of the most striking features of taste aversion learning turned out to be prominent only if the sickness-inducing treatment – normally injection of lithium chloride – was strong and the taste entirely novel. Taste aversion learning is retarded by giving an animal previous exposure to the target taste; that is, it is subject to *latent inhibition* (or the *stimulus pre-exposure effect*) just like any other kind of conditioning.[78] And as for the tolerance of long delays between taste and sickness, Sam Revusky, who had collaborated with Garcia for a couple of years, argued that this is because other forms of associative learning are subject to interference from intervening events during the delay between a target stimulus and some outcome, but, in the case of taste aversion learning, the only effective source of interference during the delay period is exposure to another taste.[79]

Revusky noted that "general process approaches to learning presuppose that there are learning processes governed by principles general enough to apply to many species in wide varieties of learning situations." He went on to argue that such general principles also apply to taste aversion learning, but with a single exception: Taste aversion learning shows a degree of stimulus selectivity – the *cue-to-consequence* effect – that is far larger than that seen in any other preparation.[80] This still left taste potentiation of odor aversions to be explained. Then in 1980, Bob Rescorla reported a number of experiments demonstrating the immediacy and strength of associations formed between two stimuli that are presented together in a simultaneous compound; such simultaneous associations could be formed between two tastes or between a visual and an auditory stimulus. They could also be formed between an odor and a taste. This led him and Paula Durlach to successfully test the idea that such associations form during experiments on taste-potentiation of odor aversions and provide the basis for second-order conditioning of the odor aversion.[81] In other words, presentation of the odor in a subsequent test reminds the animal of the taste with which it had become associated and that taste has become highly aversive and so the animal avoids the odor.[82]

Concluding Comments

During the 1960s and 1970s, the issues described in this chapter were focal for researchers concerned to understand how animals learn. By the 1980s it had become widely accepted, for example, that the effectiveness of instrumental conditioning varies according to what kind of response is involved, that the consequences of Pavlovian conditioning are not limited to autonomic responses, and that – as Jenkins had argued – any conditioning procedure is likely to involve an interaction between Pavlovian and instrumental processes. Furthermore, it was clear that animals were capable of forming an association between events that were separated in time by many hours, as both Garcia and Hogan had claimed in the face of great skepticism.

As for imprinting, most learning theorists lost interest and it became a topic that at best was mentioned in a couple of pages towards the end of textbooks on 'learning and behavior.' Imprinting also became another of the phenomena studied by the classical ethologists that by the 1980s were increasingly ignored by researchers in animal behavior. Instead, foraging behavior became one of the rare areas of research in which psychologists and ethologists stayed talking to each other. See Figure 6.14.

Figure 6.14 In 1973 John Garcia paid a visit to Konrad Lorenz in Altenberg, Austria. Reproduced with permission from Riccardo Draghi-Lorenz. Also, thanks to the Garcia family for permission to reproduce it.

7 Discrimination Learning, Attention and Stimulus Generalization

Early research on discrimination learning: Karl Lashley and Isadore Krechevsky

In the early years of the twentieth century, there was widespread interest in how animals perceive their worlds. For example, can nonhuman mammals see color? John Watson and Robert Yerkes failed to find any evidence for color vision in their rats but did find that the one monkey they tested was sensitive to color. Unknown to them, at around the same time Pavlov asked the same question of his dogs and concluded that they lacked color vision. What about visual patterns? Again, unknown at the time to researchers in the USA, Pavlov, and his students had used salivary conditioning to train dogs to respond differentially between a circle and an ellipse. Meanwhile, a large number of experiments were carried out in American laboratories, mainly using albino rats, that "yielded little evidence that the rodents are sensitive to visual patterns."[1]

Karl Lashley became involved in such research as early as 1912 when he tried – and failed – to train rats to discriminate between patterns that differed in size and shape.[2] The problem – as he later recognized – lay in the equipment he and others used in their experiments. As shown in Figure 7.1, the discrimination box required rats to choose which of two alleys to enter on the basis of patterns displayed at the far end; as an example, two squares varying in size are shown in Figure 7.1. Choice of the correct alley could lead to food and choice of the wrong alley could result in the animal receiving a shock. Training rats on even a very simple discrimination such as between vertical and horizontal lines could take many hundreds of trials.

Many years later – in the summer of 1929 to be precise – Lashley hit on a solution. The training procedure for the jumping stand shown in Figure 7.2 required that on each trial a rat had to jump from the stand (*s*) through one of two doors, on which were displayed different visual patterns. The door with the positive stimulus was unlocked so that, when the rat made a correct choice, it could reach the platform (*f.p*) on the other side of the screen where some food was waiting. The door containing the negative stimulus was locked so that, if the rat made a bad choice, it would collide with this door and fall into the net (*n*) below. When trained on this equipment rats learned very rapidly to discriminate between a variety of visual patterns.[3]

Lashley attributed the success of this method to the way that it ruled out all but visual cues. Thus, when perched at least 20 cm from the doors containing the patterns, a rat was unlikely to be able to smell these stimuli and certainly could not touch the patterns. The equipment made the rats attend to the only stimuli that were relevant to

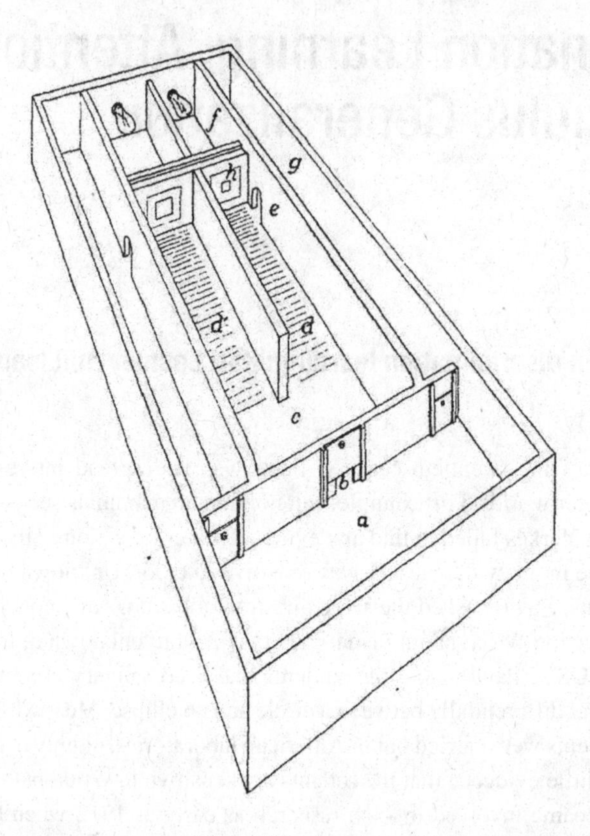

Figure 7.1 As reported in 1912, Lashley used this discrimination box to test whether rats could learn to enter the correct arm, d or d´, on the basis of shapes mounted on the doors at the end of the arms. Success was rare.
From Lashley (1912). Public domain.

their task. Subsequently, the Lashley jumping stand became a standard tool for the study of visual discrimination learning. The versions that succeeded the original of 1930 became more humane. Not surprisingly, many rats exposed to the equipment shown in Figure 7.2 showed considerable reluctance to jump. Lashley sometimes had to use a whip to get some rats to jump, a detail of his procedure that never reached a published page.[4] The later versions contained a ledge in front of each door; if a rat landed on the ledge in front of the correct door, it needed now only to knock the door down to reach the food and, if landing on the wrong ledge, then the only punishment was that of having to remain on the ledge and to go without the food reward until the experimenter picked up the rat and returned it to its home cage.[5]

Lashley and many of his fellow psychologists were mainly interested in discrimination learning as a tool for understanding an animal's perceptual world. Isadore Krechevsky became inspired to study the process. During his senior undergraduate year at New York University in 1929, a bad attack of flu meant Krechevsky was confined to bed and to read more thoroughly than usual. One book prompted a Eureka moment. The book was Lashley's *Brain mechanisms and intelligence* that had just been published.[6] The passage that inspired Krechevsky was one in which

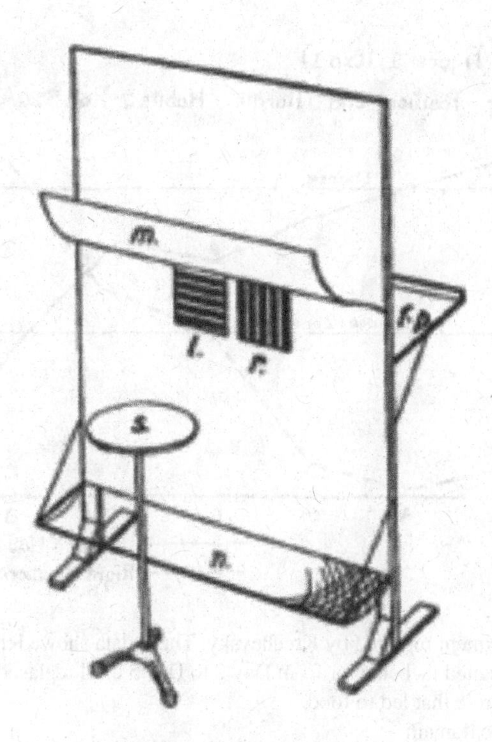

Figure 7.2 By 1930 Lashley had developed this jumping stand, in which a hungry rat on the pedestal ‘s’ was faced with two windows. If it jumped at the correct window, it would crash through to the food on the platform at the back. The wrong window was locked, so that, if the rat jumped at it, the rat would fall onto the ledge below. This equipment enabled Lashley to obtain good evidence at last for rats’ ability to discriminate between visual patterns. From Lashley (1930). Public domain.

Lashley speculated that his rats responded in turn to a variety of different cues in the process of acquiring a visual discrimination. Lashley noted that he did not know how to obtain objective evidence for such a process. Krechevsky decided that he knew how this could be done.[7]

After obtaining his bachelor’s degree Krechevsky remained for a further year at New York University to complete a Masters degree. His first experiment on discrimination learning used equipment he built that resembled the two-alley discrimination box shown in Figure 7.1. Some of his rats were given a brightness discrimination task and others a hurdle problem; the latter involved climbing over a rod to enter an alley and the height of the rod signaled whether this was correct or not. As he emphasized, it was important to analyze the records of individual animals. This was done in terms of whether on a given day a rat showed a preference for the left or the right alley or showed a preference for the hurdle or brightness that would lead to food. The data from each rat were presented as in the representative graph in Figure 7.3.

Most learning curves, whether the original ones Thorndike obtained from cats and dogs in his puzzle boxes or later ones based on rats learning to find their way through a maze, showed a typical pattern of performance that from the start steadily improved,

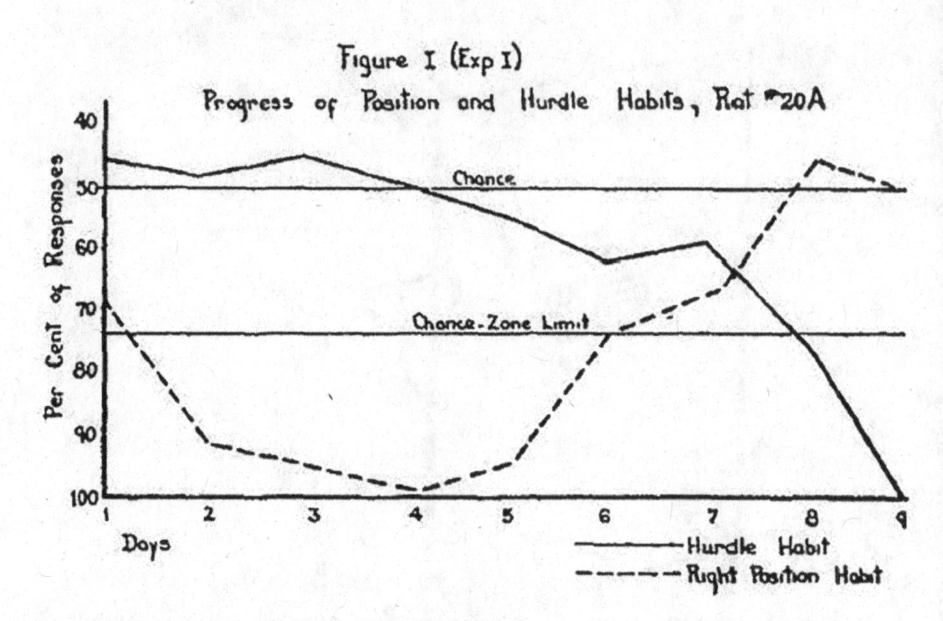

Figure 7.3 Record from an experiment reported by Krechevsky. These data showed that a rat's initial position preference dominated its behavior from Day 2 to Day 5 until replaced by an increasing preference for the hurdle that led to food.
From Krechevsky (1932). Public domain.

albeit at an increasingly slower rate. In contrast, the typical learning curve for an animal in a discrimination task showed a long period of chance performance before an animal made an increasing number of correct choices. Krechevsky's point was that, during the usual long period when performance remains at chance, the animal is not behaving randomly, rather it is behaving systematically with respect to some potential cue. In the example shown in Figure 7.3, the rat is showing a preference for the right alley and, since the correct hurdle is only half the time on the right, this produces chance performance. It is, as Krechevsky put it, the rat is trying out a series of 'hypotheses' as to what predicts the way to the food reward.[8]

'Hypotheses' was a term suggested by Edward Tolman on learning about these experiments when Krechevsky arrived at the University of California at Berkeley to enter the Ph.D. program there. It was a label that Krechevsky described as "dubious," a "confession of failure," and it was always placed within scare quotes in his published reports. The experiments that followed at Berkeley included one that reversed the conditions in a discrimination task: Once a rat had learned, for example, to enter the more brightly lit alley to obtain food, the less brightly lit alley was made the correct choice. The typical pattern produced by such a reversal was that the rat would first persist in choosing the previously correct alley, then perform at chance for many trials – usually displaying a strong position preference on these trials – and eventually begin to choose the newly correct alley. In a final summary of the report on these experiments, he stated: "…helter-skelter trial-and-error learning must go by the board as a valid description of the learning process. …. the learning process at every point consists of a series of integrated, purposive behavior patterns."[9]

Figure 7.4 David Krech (previously 'Krechevsky') on returning to Berkeley in 1947. From Carroll (2017). Reproduced with permission from Cambridge University Press.

On obtaining his Ph.D., Krechevsky obtained a number of short-term research positions, including a year spent with Lashley in Chicago, but during the Depression era, few long-term academic positions were available. This was an era of strong anti-semitism that – along with the reputation for being a 'trouble maker' and left wing – made it particularly difficult for Krechevsky to obtain or retain any position in a university. During World War II he served for a while in the US intelligence services, where he changed his name to David Krech. Partly as a result of this work he became increasingly involved in social psychological research and established a strong reputation in this area. This gained him a return to Berkeley as a professor and colleague of his former advisor, Tolman; see Figure 7.4.

In reflecting upon his early research on discrimination learning, Krech observed that: "My method of analysis and the resulting data were eventually to lead to the opening of the Continuity-Noncontinuity controversy in learning theory and to bring to me the enthusiastic support of Tolman, the saddened disapproval of Schneirla, the benign approval of Lashley, the vigorous wrath of Spence, and the unforgiving suspicion of all professing Behaviorists – Watsonians in their time, then Hullians, and finally Skinnerians."[10]

As for the *Continuity-noncontinuity controversy* that Krechevsky opened, the 'non-continuity' claim was that at any given time animals only learn about one aspect of their perceptual world, one hypothesis at a time, or in Lashley's later version, can selectively attend to only one aspect at a time.[11] The opposite claim was that animals learn simultaneously about all aspects of their perceptual world; this 'continuity' position was most vigorously promoted by Kenneth Spence, as related in the next section.

The Vigorous Wrath of Kenneth Spence

Kenneth Spence was born in born in Chicago in 1907, but his formative years were spent in Montreal, where his father was employed as an engineer. As a young man, Spence's main, very active, and injury-inducing interest was in competitive sport. At the age of 22, he completed an undergraduate degree at McGill University and then remained there to complete a one-year Masters in psychology in 1930. He then moved to Yale University to begin doctoral research on visual acuity in the chimpanzee, with Robert Yerkes as his advisor.

Following his appointment in 1924 as Professor of Comparative Psychology at Yale, Yerkes had set up a laboratory for studying chimpanzees. Appreciating that the New England climate was unsuitable for such animals, Yerkes obtained enough money from the Rockefeller Foundation to realize at the age of 54 his long-held dream of establishing in Florida a research center for the study of the great apes (see Chapter 5). Building started in Orange Park, Florida in 1930, the year Spence arrived in Yale. What with the development of Orange Park and a range of other commitments beyond Yale, it seems unlikely that face-to-face meetings between Yerkes and his new graduate student took place very often.

On the other hand, Spence rapidly formed a close relationship with Clark Hull, then 46 and, as Professor of Psychology at Yale and Head of the Institute for Human Relations, at his creative peak. Spence joined the impressive group that met weekly with Hull (see Chapter 2). Two years after arriving at Yale he published his first research paper. This reviewed previous studies that involved rats learning to find their way through complicated mazes that contained blind alleys. Spence aimed to explain why rats found it easier to learn to avoid some 'blinds' than others. He rejected explanations by other researchers in favor of a combination of two new concepts developed by Hull: The *goal gradient hypothesis* and *anticipatory goal reactions*. Hull had yet to publish the paper containing these ideas but had given Spence a pre-publication copy of the paper. For a graduate student in his mid-20s, this must have seemed great honor. In the spirit of Hull's aim to build a mathematical foundation for psychology, Spence even devised a formula for measuring the relative likelihood that a rat would enter a particular blind alley.[12] The respect became mutual. In his influential *Principles of Behavior* of 1943 Hull acknowledged his great intellectual debt to Spence. Their relationship continued to have a profound impact on both of their subsequent careers.

Just like Krechevsky and many other brilliant students with Ph.D.s from prestigious universities and with highly regarded professors providing references, Spence discovered that there were very few academic positions available in this Depression era. His Ph.D. of 1933 gained him only a research fellowship to remain at Yale for a further year. Fortunately, he seems to have impressed his advisor, Yerkes, who appointed him in 1934 to a three-year research assistantship in the newly opened primate facility in Florida. Here Spence continued his studies of the chimpanzee visual system. These experiments required that the animals had to discriminate between different stimuli and – predictably given his Yale experience – Spence became as interested in the processes by which such learning occurs as in how chimpanzees see their world.

Spence explained his ideas on discrimination learning in what became a classic paper. This starts by acknowledging the value of the data that Krechevsky had reported but is scornful of the claim that these revealed the inadequacy of 'trial-and-error' theories of learning and the involvement of more complex processes of 'hypothesis testing.' Spence decided that Krechevsky's problem is that he does not understand current theories of trial-and-error learning, meaning Hull's latest theory. "It is the purpose of the present paper to attempt a theoretical account of discrimination learning founded on 'trial-and-error' principles similar in nature to those recently exploited by Hull in the field problem box and maze learning."[13]

Spence's theory was expressed in terms of equations expressing changes in the strength of S-R connections that were either increases following reinforcement or decreases due to the absence of reinforcement generating "inhibition or frustration." It became known as *continuity theory*. The basic idea was that, when, for example, a rat in a Lashley jumping stand jumps towards the correct stimulus, A, that happens to be on the left, the subsequent reward strengthens both the 'A-approach' connection and the 'left-approach' connection. On the next trial, A might be on the right and B on the left; if the rat jumps to the left once again, this will weaken both the 'B-approach' and the 'left-approach' connections, as well as any tendency to alternate between the two doors. If a position preference or a tendency to alternate is particularly strong at the start of training – due to the rat's past history – the 50% reinforcement schedule for such tendencies in a typical discrimination learning task could result in them remaining the dominant behavior, thus masking the steady increase in strength of the A-approach connection until this becomes the dominant habit.[14]

Spence's next target was Gestalt psychology or, more specifically, the claim by Wolfgang Koehler that the perception of relationships between sensory events is fundamental and perception of the absolute properties of individual events is secondary. The idea goes back to the nineteenth century analysis of melodies, whereby the relationship between notes is more important than their individual frequencies. Koehler's *The Mentality of Apes* was published in English translation in 1925. Nazi persecution of scientists with Jewish origins led to the emigration from Germany to the USA of many Gestalt psychologists, eventually including Koehler – who was a Gentile – in 1935, while Spence was still working in Orange Park, Florida. The arrival of these immigrants further increased the impact of Gestalt ideas among English-speaking psychologists (see Chapter 2).

The *Mentality of Apes* has remained famous for the experiments it reported on Koehler's tests of insight in his chimpanzees. Among other less well-known experiments were ones that tested the chimpanzees' capacity for *transposition*, a name that evokes the fact that a melody can be transposed from one key to another. The present example of transposition comes from the tests Spence first carried out with his chimpanzees in Orange Park. Three individuals were trained to respond to a large (160 square inches) white square and not to respond to a small (100 square inch) square. The squares were mounted on the front of food boxes with the larger square indicating that this box was unlocked and so the chimp could retrieve the piece of banana it concealed, whereas the smaller square was mounted on a locked food box. Once these chimps were performing at an almost perfect level, they were given a set of three tests. One of these offered a choice between an even larger square of 256 square inches and the original positive square of 160 square inches. All three chimps showed a strong preference for the larger square, avoiding the previously rewarded square. Similar results were obtained when a choice between 320 and 200 was given. Thus, the chimps were showing *transposition* from their training pair to these test pairs of stimuli and were apparently responding on the basis of the *relative* size of the squares, as having learned to 'choose the bigger of the two.' However, when the size of the test pairs was increased yet further to 409 vs. 256, transposition became weaker. Spence found less evidence for transposition from two chimps that were trained to choose the smaller of the original squares, 160 over 256.

This initial experiment was followed by a second, involving six new chimps, that yielded the same pattern of results.[15] "The general conclusion that transposition can be found as long as a test pair of stimuli does not differ too much from the original training pair was consistent with what a range of other researchers – some strongly adhering to Gestalt psychology and others more skeptical – had found using a number of different species." What Spence set out to do was to explain both how transposition could be explained without appealing to an animal's perception of the relationship between two training stimuli and why the effect was not found when more extreme test stimuli were used.

His analysis was again expressed in mathematical terms. It included the assumptions in his theory of the previous year that reward reinforced the connections between all stimuli present and the response that produced the reward, whereas non-reward weakened all such connections. These were combined with assumptions based on Pavlov's concept of stimulus generalization: Excitation spreads from a positive conditioned stimulus to similar stimuli, as does inhibition to stimuli similar to an inhibitory stimulus, and the extent of either kind of generalization decreases as the distance increases from the conditioned stimuli. Spence's theory was illustrated by diagrams of the kind shown in Figure 7.5.

In this figure the numbers above each stimulus indicated on the x-axis represent the effective strength of the connection between the stimulus and the approach response. This is calculated by subtracting the amount of inhibition generalizing to the stimulus from the amount of excitation generalizing from the positive stimulus. Where two test stimuli differ considerably in effective strength, as between 409 and

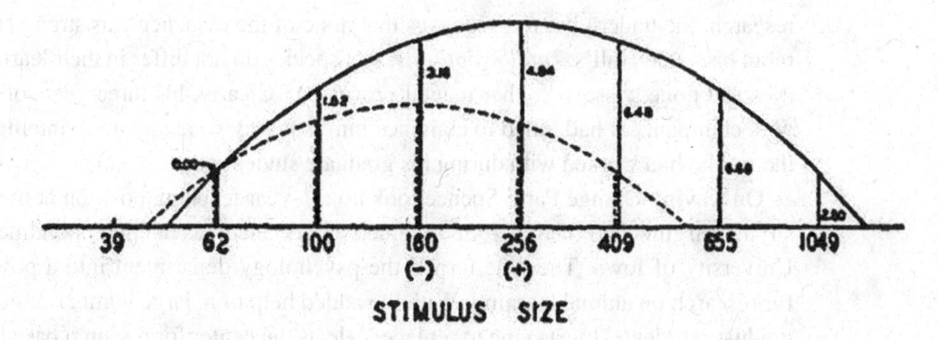

Figure 7.5 Hypothetical functions illustrating generalization of both excitatory and inhibitory connections in Spence's model of how transfer can occur from discrimination training – in this case between a square of 256 square inches, serving as the positive stimulus, and a smaller one of 160 square inches serving as the negative stimuli – to various pairs of test stimuli. From Spence (1937a). Public domain.

256, then a clear preference for the larger stimulus will be obtained. In contrast, when a more extreme pair of test stimuli is used, 655 vs. 409, the effective strengths are very similar and transposition will no longer be obtained. Furthermore, when an even more extreme pair, 1,049 and 655, are used, the prediction is now that the reverse of transposition will be observed, that is, preference for the smaller of the two stimuli.[16]

The generalization functions produced by Spence were entirely hypothetical. As described in the final section of this chapter, peak shift and other predictions from Spence's model were first confirmed – over 20 years later – in pigeons by Hanson.[17] However, a further effect that Spence predicted, namely, reversal of transposition has never been found.

The paper on transposition was written during the final year of Spence's 3-year contract of employment at Orange Park. Spence never wrote a biography, not even the short sketch that almost all of his eminent contemporaries contributed to the series of volumes of the *History of Psychology in Autobiography,* and none of the few brief biographies say much about this stage in Spence's career. We do not know whether Yerkes would have wanted Spence to remain at the primate colony. Besides the research described here, Spence had published several excellent reports on apes' visual abilities based on the experiments he was employed to carry out. On the other hand, his ideas on comparative psychology were entirely opposed to those of Yerkes.

Ever since his early collaboration with John Watson, Yerkes had devoted his research career to delineating the ways that species differ in their learning and cognitive abilities. He had increasingly concentrated on investigating the ways that the great apes appeared to be superior to any other nonhuman species. In his final year at Orange Park, possibly following Yerkes' suggestion, Spence had published a review of research over the previous decade on "learning and the higher mental processes in infrahuman primates."[18] Without being explicitly scathing about the quality of much of this research, the underlying message was that none of the evidence was strong enough to rebut his – and Hull's – *null hypothesis*: that species do not differ in their learning abilities and none possess 'higher mental processes.' Clearly, his three years of working with chimpanzees had failed to convince him that they were any more intelligent than the rats he had worked with during his graduate student years at Yale.

On leaving Orange Park, Spence took up a 1-year teaching position at the University of Virginia and then in 1938 obtained a permanent academic appointment at the University of Iowa. There he turned the psychology department into a powerhouse for research on animal learning. With the added help of a. large number of productive graduate students, Iowa came to replace Yale as the center for research based on what became known as the Hull-Spence theory of behavior.

Spence and His Students at Iowa (1938–1964)

In terms of academic geography, the University of Iowa was a long way from the top US universities in the 1940s and 1950s. Since Spence was not yet a well-known figure in psychology, it is not at first obvious why so many able students sought him out as a graduate advisor and subsequently became effective champions of his theories. However, when Spence arrived, the Department of Psychology at Iowa already had a strong tradition and reputation.[19] A memoir found on a computer after the author's death suggests one explanation as to how in his early days at Iowa Spence attracted and then inspired so many bright and highly productive students.

The title of the memoir was *A woman's struggle in academic psychology (1936–2001)*.[20] Tracy Seedman was raised in New York City as the daughter of recent immigrants escaping anti-semitism in Hungary and Russia and received a superb education from the city's excellent public school and public library systems; see Figure 7.6. In 1936 Tracy entered Brooklyn College as an 18-year-old undergraduate, supporting herself by working at Macy's and benefiting from the college's then policy of not charging qualified students any tuition. In her junior year, she met the man who later became, not just her husband, but a lifetime collaborator, Howard Kendler. In her sophomore year, she was persuaded by Abraham Maslow – well before he became famous – to major in psychology. A course with Solomon Asch confirmed her ambition to become an academic psychologist. As someone working within the Gestalt tradition, Asch recommended that she apply to study with the distinguished Gestalt theorist, Kurt Lewin, at the University of Iowa. An added attraction for Tracy, as well as for Howard Kendler and two other Brooklyn College students, was that one could live a lot more cheaply in Iowa City than in New York.

On arrival at the University of Iowa in 1940 the four would-be graduate students found that enrolling in psychology was not straightforward. Lewin was based in the Iowa Child Welfare Station rather than in the Department of Psychology, while the Head of the Psychology Department was John McGeoch. The latter was establishing himself as the leading exponent of the paired-associate approach to the study of human memory and of interference theories of forgetting, an approach very

Figure 7.6 Tracy Seedman Kendler's high school graduation photo of 1936. From Kendler (2003). Reproduced with permission from the American Psychological Association.

compatible with Hull-Spence theory and antagonistic towards Gestalt ideas. McGeoch made it clear that he would rather not have any more Jews from New York in his department, and especially not a woman, and told them that his class on the history of psychology was closed.

"Howard and I, confused and unhappy, sat on a bench in the student union trying to decide what to do, when a young man approached. He introduced himself as Kenneth Spence, as associate professor of psychology, and enquired about our problem. We told him only that the history class was closed. He offered to register us for his second-year graduate seminar in learning theory. We both registered for his class and so began our graduate training."[21]

Tracy Seedman completed an initial Masters thesis under the supervision of Lewin, as she had originally planned. However, Spence's influence eventually prevailed and both she and Howard Kendler undertook Ph.D. theses with Spence as their advisor. Her topic was discrimination learning in rats. In her memoir, she reported that Spence "admittedly enjoyed converting Gestaltists" to his behaviorist theories and commented that "what made his theory so attractive was that it resembled theories in the natural sciences in its capacity to generate specific testable predictions." As Spence's first female Ph.D. student, she noted that "while he enjoyed talking to his male students, he was very uncomfortable with me." In their second year at Iowa, Tracy and Howard married. "Spence told me soon after to confine my ambitions to being a good wife to Howard."[22]

As now Tracy Kendler, she completed her Ph.D. in 1943. During this period Spence published a further paper on transposition – on the *intermediate size problem*[23] as described later – and thus it is not surprising that this was the topic of Tracy Kendler's thesis experiments. She tested predictions from Spence's analysis of transposition. Her experiments used hungry rats trained to discriminate between two levels of brightness: One group of rats found a food reward after entering the brighter of two fairly dimly lit alleyways, while a second group found food after they entered the dimmer of two brightly lit alleyways. After hundreds of training trials, tests were given between pairs of brightnesses. The results extended previous findings that transposition effects decrease in size as the values of the test pairs increasingly differ from those of the training pairs, an important element of Spence's analysis. It failed to confirm his prediction that a reversal of transposition would occur once the test pair of brightnesses were sufficiently remote from the training pair. This failure meant either that very different generalization functions are needed from those sketched by Spence or that, as Hull conceded, the 'S' in S-R theory does not need only to refer to elemental stimuli but could stand for a pattern of stimuli, that is, a relationship.[24] In other words, Koehler's claim that animals can learn about relationships between elemental stimuli may be correct after all.

These thesis experiments were not published until seven years after Tracy Kendler had obtained her Ph.D.[25] Around the latter time, her husband began the project that eventually made their partnership famous. These experiments used a design that allowed a comparison between *Reversal shifts* and *Non-reversal shifts*. These can be explained using diagram shown in Figure 7.7.

During the first stage the participants – rats, children, or undergraduates – are trained on a set of four stimuli that in this example differ both in color and size. As shown in Figure 7.7, size is the critical cue in the first discrimination, with choice of the larger objects being rewarded, and color is irrelevant. Once this has been learned, a second discrimination is introduced. For participants given a *reversal shift*, size is still relevant but now reward follows choice of the smaller objects. For participants given a *non-reversal shift*, color has become relevant, in that choice of black objects is rewarded, and size has become irrelevant.

Unless they are over-trained on some discrimination – as discussed later in this chapter – rats generally take longer to learn a reversal shift than a non-reversal shift. This outcome is entirely compatible with Spence's continuity theory of discrimination learning. This predicts that a neutral cue is more rapidly converted into a positive cue than is a cue that is inhibitory at the start of the reversal stage. But, when undergraduates are given these tasks, the opposite result is obtained. The Kendlers explained this difference in terms of a 'mediation theory'; undergraduates realize that size, for example, is the important dimension for solving the problem and decide to stick with this belief when conditions change. This was not the way that the Kendlers expressed this idea. They used a very different vocabulary: 'Orienting responses,' 'symbolic

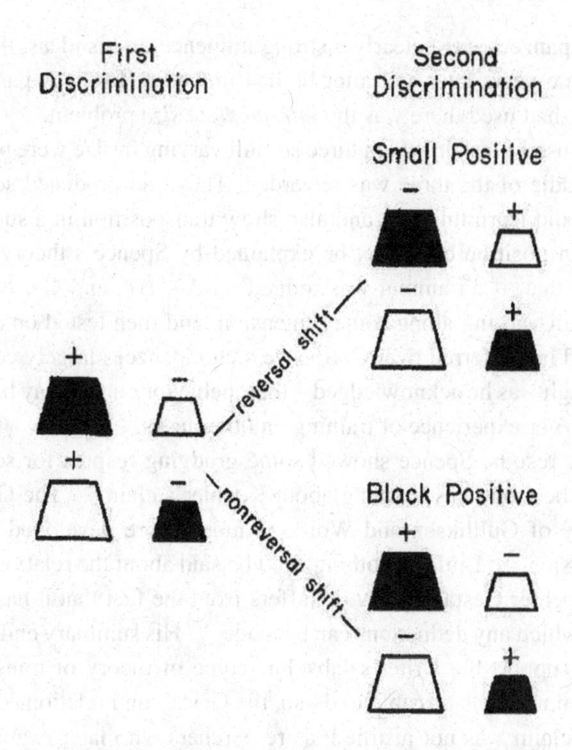

Figure 7.7 Diagram illustrating experiments designed to compare reversal and non-reversal shifts. In this example of a 'reversal' shift a rat or child first learns to choose the larger of two objects and then is shifted to learning to choose the smaller of the objects. In the 'non-reversal' shift shown here, after learning to choose the two larger objects, the task in the next stage is to choose the two black objects.
From Kendler and Kendler (1962). Public domain.

responses' and 'symbolic cue,' terms that must have reassured Spence that his former students were not slipping into becoming 'cognitivists.'[26]

When children ranging in age from three years old to 10 were tested, it turned out that the younger children, like rats, found the non-reversal shift easier, whereas beyond the age of seven, a child was more likely to learn the reversal shift more rapidly. For the Kendlers, such discrimination learning tasks provided a way of studying the evolving role of language, from being simply a means of communication to becoming – in addition – a tool for solving problems.[27] Again, this is not quite how the Kendlers expressed this idea. Tracy Kendler went on to become a leading researcher in the area of cognitive developmental psychology. After a series of temporary research positions and short-term academic appointments, in 1966 she finally achieved her ambition of obtaining a tenured appointment at a university where she had her own laboratory and own graduate students to advise, the ambition that many of the male students from her Iowa days had fulfilled very many years earlier.[28]

Now to return to Spence's early years in Iowa. His classic paper of 1937 on transposition was published while he was still working in the primate center in Florida and was entirely theoretical. Although the experiments on discrimination learning he carried out with chimpanzees were clearly a strong influence on his ideas, they were not reported until some years later and after he had moved to Iowa. A particularly interesting test that he had used there was the *intermediate size* problem.

Earlier studies had used a task in which three stimuli varying in size were presented and choice of the middle of the three was rewarded. They had produced some evidence that animals could learn this task and also show transposition in a subsequent test. This form of transposition could not be explained by Spence's theory. On the contrary, it predicted that, if an animal was trained on A–, B+, and C–, where the letters represent stimuli varying along some dimension, and then tested on an array, B, C, and D, B should be preferred over C. Spence's chimpanzees largely confirmed this prediction, although – as he acknowledged – their behavior on test may have been influenced by their earlier experience of training on other tasks.

In discussing these results, Spence showed some grudging respect for some earlier American researchers but was scathing about Koehler's claims. "The Gestalt or configurational theory of Gullliksen and Wolfle is much more developed than the primitive Koehler version..... Little or nothing need be said about the relational interpretation. Like the Koehler Gestalt theory, it suffers from the fact that it has no specific postulates from which any deductions can be made."[29] His summary ended: "The results are shown to support the writer's 'absolute' type of theory of transposition and to contradict, insofar as it is possible to do so, the Gestalt and relational interpretations."[30] This final claim was not justified; as researchers who later examined the intermediate size problem pointed out, "the fact that an animal responds to certain absolute features of a stimulus object presented in some arbitrary context does not justify the conclusion that it is incapable of response to the relations."[31]

When studying an animal's ability to learn about the relationships among three visual stimuli that, in the above example, differ in size, the way that the stimuli are presented is clearly important. Animals are more likely to learn about the relationships between the stimuli – 'larger than' or 'intermediate in size' – if the stimuli are presented in a way that the animal can see all three within a single glance. They are less likely to perceive such relationships if they need to shift their gaze from one position to another. It is not clear that the arrangement Spence used with his chimps was optimal for detecting relational learning and also not clear that he ever obtained a clear transposition effect. When stimuli are presented in a way designed to facilitate learning about the relationship between them, then transposition, including that of intermediate size learning, can be found in chimpanzees.[32]

Spence does not appear to have continued with an active interest in transposition. His main focus in Iowa became the motivational side of Hull-Spence theory. As for the general topic of discrimination learning, he published two further papers on this topic. One reported a further test of his continuity theory. In his 1936 paper on discrimination learning he proposed an experiment directed towards the following question: "Does the animal before it learns a discrimination problem, that is, before it responds systematically to the positive stimulus cue and while it may be responding systematically to other stimulus cues, learn anything about (form any associations with) the positive stimulus cue?"[33]

In 1936 he had been able to cite a study of *pre-solution reversal* in support of his continuity theory. Subsequently, Krechevsky had criticized this earlier study and had run his own experiment which – sure enough – supported Krechevsky's non-continuity theory, in that his rats learned a reversed discrimination as quickly as controls and thus appeared not to have learned anything about the previously positive cue. The details of Spence's new experiment on this issue are not important here, except to note that it was better designed and the analysis more appropriate than Krechevsky's. The results "refuted the assumption of this type of theory that the animal learns nothing with respect to the relevant stimulus cue before it begins to respond systematically to it and while it is exhibiting some other 'hypothesis'.... and are considered to be in agreement with the continuity type of theory."[34]

Subsequent research of this kind has supported Spence's conclusion that animals can learn about the significance of cues that are not currently controlling their behavior. However, it has also indicated that systematic responding to irrelevant cues, such as position, can interfere with learning about the relevant cues.[35]

Spence's opponent in the next debate over his theory of discrimination learning was an even more junior researcher than Krechevsky. Where the latter was provocative, Jeff Bitterman (see Chapter 5) was as aggressive and became as good at experimental design as Spence. Having earlier criticized Spence's claims on transposition, Bitterman and one of his students at the University of Texas published an experiment that appeared to show that successive discriminations are learned more rapidly than simultaneous discriminations, a result that was both generally surprising and contrary to Spence's theory. To take as an example an experiment that Spence designed to answer this challenge: In a simultaneous discrimination using a simple T-maze a rat might need to learn to turn left at the choice point on half the trials to enter the black arm and to avoid entering the white arm, and on the other trials to turn right to enter the black arm and avoid the left arm; a successive discrimination would be one in which a rat has to turn left when both arms are black and to turn right when both are white.

Consistent with previous experiments by other researchers, Spence found that the simultaneous discrimination was on average learned in less than half the number of trials needed for the successive discrimination. He noted that this kind of experiment raises the issue of what response is learned: 'enter the black arm' or 'turn left when the black arm is on the left'? The opposite result reported from Bitterman's laboratory probably reflected the use of an unusual and particularly complex task.[36] This prompted a further experiment in Bitterman's laboratory which is illustrated in Figure 7.8. Their rats learned the successive discrimination more rapidly than the simultaneous version. However, the simultaneous task was quite atypical. The conclusion to the paper reporting this result implied, although very indirectly, that Spence's analysis was generally appropriate but could not account for animals' responses to certain configurations of stimuli.[37]

By the time he left Iowa in 1964 to take up an appointment at the University of Texas, Spence had directed 75 Ph.D. theses, which must be close to a record for an experimental psychologist, and it was once said that "almost every major psychology department in the United States had a Spence PhD on its faculty."[38] He was

Figure 7.8 Bitterman and Wodinsky used a Lashley jumping stand to train rats in either of two kinds of discrimination. In the simultaneous version, the rat needed to jump to one of the gray cards when the white card was on the right and to jump to the other gray card when the white was on the left. In the successive version, the stimulus card was either all white, signaling that, say, the left window was correct, or all black, signaling that the other window was correct. From Bitterman and Wodinsky (1953). Public domain.

Figure 7.9 Kenneth Wartenbee Spence.
From Amsel (1995). Reproduced with permission from the Briscoe Center for American History, University of Texas.

remembered by Allan Wagner, who studied under Spence from 1956 to 1959, as "a person of extraordinary presence, who appeared naturally to be at the center of things …. (and who) would 'hold court' (to) students who were generally intimidated in his presence. He was always 'Dr. Spence,' never 'Kenneth'."[39] From 1962 to when he died in 1967 Spence was the most frequently cited author in publications by the American Psychological Association; see Figure 7.9. Nevertheless, during this period his influence was already declining. The cognitivists he had attacked for decades were increasingly influential within American psychology, as were – on the other extreme – the Skinnerians he had mocked for their obsession with cumulative records and lack of any theory.[40] A major contribution that has survived is his 1936 theory of discrimination learning or, more precisely, the idea of interactions between generalized excitatory and inhibitory processes, as illustrated in Figure 7.5. And probably his lasting influence is most clearly seen in the much later work of Allan Wagner (see Chapter 9).

Douglas Lawrence and the Acquired Distinctiveness of Cues

While Clark Hull was still alive, Yale remained the most important center for his neo-behaviorism. During the late 1940s, Hull's health deteriorated and his former student and now colleague, Neal Miller, became increasingly influential and ultimately Hull's successor at Yale when Hull died in 1952 (see Chapter 2). Well before then, Miller had acted as advisor to a graduate student whose ingenious experiments for his Ph.D. played a major role in the later downfall of Hull's theory of learning.

Douglas H. Lawrence was born in Canada in 1918 but was raised in Washington State. After obtaining a B.S. and then an M.S. in 1942 from the University of Washington, he served as an officer in the Amy Air Corps, where his duties were mainly those of selecting personnel for specialized training. At the end of World War II, he enrolled as a graduate student at Yale and obtained his Ph.D. in 1948.[41]

The hypothesis he set out to test was: "if the S (an animal) is trained to respond to some aspect of the situation other than the initially dominant one, then this new aspect will tend to become more distinctive and the other aspects relatively less distinctive." This implies that two kinds of learning take place when an animal is trained in some discrimination task: "(a) the learning involved in the acquisition of the correct instrumental responses in a discrimination situation and (b) the learning involved in modifying the initial order of distinctiveness among the cues."[42] The experiments he ran to test this idea used two pieces of equipment that he designed and built. As shown in Figure 7.10, one was used for simultaneous discrimination training, where a rat had to choose which of two goal boxes to enter, and the other for successive discrimination training, where a rat had to choose which of two arms of a T-maze to enter.

For his first experiment, Lawrence divided his sixty hungry rats into three groups. These differed as to the initial training they were given in the simultaneous apparatus. One group was trained on a black-white discrimination, with half the rats given food after entering the black goal box and the other half given food after entering the white box. A second group was trained to discriminate on the basis of the wire mesh covering the floors of the goal boxes, where one was a fine mesh and the other course. The discrimination task faced by the third and final group was to choose between a narrow or a broad goal box.

(a)

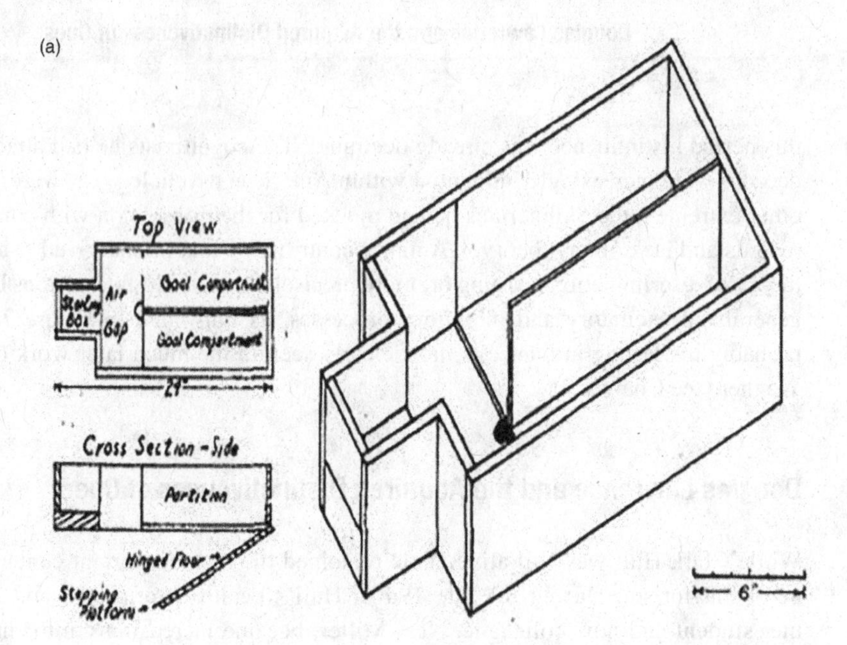

(b)

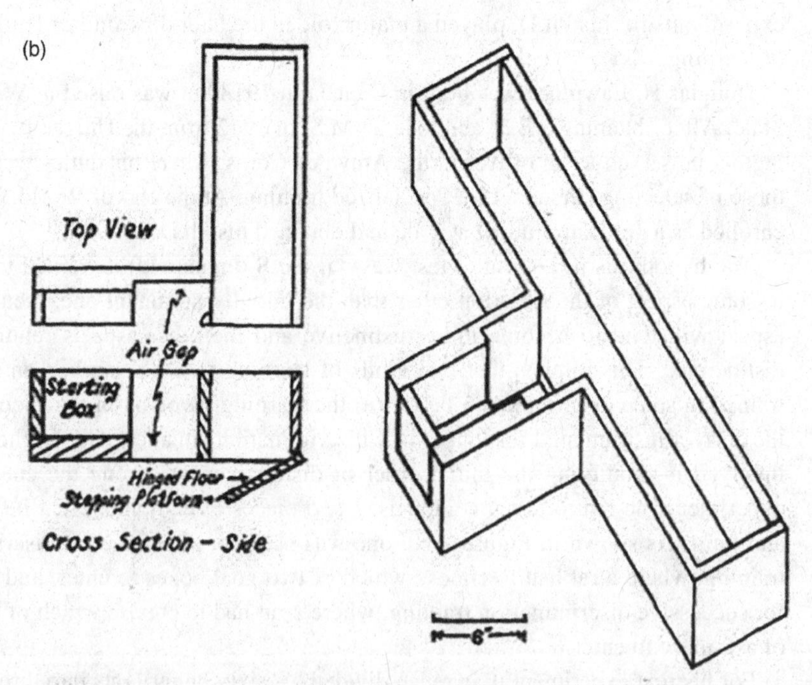

Figure 7.10 Lawrence used this set of T-mazes to study transfer from a simultaneous discrimination to a successive discrimination task. He used three pairs of cues: black vs. white, fine vs. coarse mesh floor, and wide vs. narrow goal boxes and T-maze arms. The maze above was used to train the initial simultaneous discrimination; that below to test how rapidly rats would learn a successive discrimination based either on the same cues or different cues. From Lawrence (1949). Public domain.

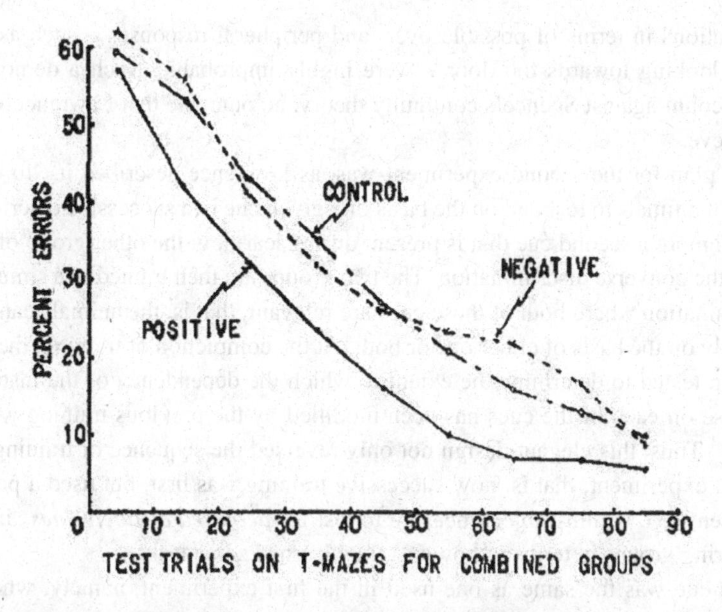

Figure 7.11 Learning curves showing the decrease in errors with repeated trials from three groups rats trained on a successive discrimination, using the maze shown in Figure 7.10. The groups differed in terms of the training they had previously been given in the simultaneous maze. The *Positive* group of 18 rats were given the same cues in this successive discrimination as those that had been used in the previous simultaneous discrimination task (positive transfer). They learned the successive discrimination more rapidly than both a *Control* group previously trained on cues that were no longer present and a *Negative* group previously trained on cues that were now irrelevant.
From Lawrence (1949). Public domain.

In the successive discrimination task that followed there was a relevant cue and irrelevant cue. For example, some subgroups of rats were required to learn to turn left when the cross alley was black and turn right when it was white, while whether the floor of this alley was covered in fine or course wire mesh was irrelevant.

The results were very clear. As shown in Figure 7.11, there was positive transfer when the relevant cue in the successive discrimination was the same as that in the previous simultaneous discrimination. These rats learned the successive discrimination more rapidly – in terms of a decrease in errors – than either the rats given a successive discrimination task in which the irrelevant cue had been the one they had previously been trained on (negative transfer) or the rats given a relevant cue they had not encountered previously (control group). No difference between these latter two conditions was detected, as can also be seen in Figure 7.11.

Lawrence explained these results in terms of mediating processes that "are conceived of as having the characteristics of instrumental behavior, i.e. they are assumed to be learned, unlearned and to show all the other functional properties that have been demonstrated for instrumental behavior.... They can equally well be overt or covert peripheral or central."[43] In the report of a second experiment, it became clearer that Lawrence was mainly interested in such processes when they were central and when explanations in terms of possible overt and peripheral responses – such as looking or not looking towards the floor – were highly improbable. Such a demonstration would count against Spence's continuity theory, an outcome that Lawrence was keen to achieve.

The plan for the second experiment was, as Lawrence described it, "to train one group of animals to respond on the basis of a given cue in a successive discrimination and to ignore a second cue that is present during learning; the other group of animals learns the converse discrimination. The two groups are then trained in a simultaneous discrimination where both of these cues are relevant, that is, the animals can respond correctly on the basis of either one or both. At the completion of training, the animals are then tested to determine the extent to which the dependency of the instrumental response on each of the cues has been modified by the previous training with those cues."[44] Thus, this elegant design not only reversed the sequence of training used in his first experiment, that is, now successive training was first, but used a potentially more sensitive within-subject measure to test for *acquired distinctiveness* instead of comparing groups in terms of how fast they learned some task.

One cue was the same as one used in the first experiment, namely, whether the walls of the goal boxes were black or white. For a second cue, Lawrence used curtains hanging within the goal boxes, with one near the entrance and the other further in; these curtains consisted of chains hanging an inch apart, so that they were a negligible obstacle for a rat making its way through a goal box and also did not make it difficult to see whether the box was white or black. The use of these curtains reflected the idea that there was no way that the rats could fail to notice them on their way to the food at the end of the correct goal box: "The cues, black-white and chains-no chains, were of such a nature that the animal was certain to be physically stimulated by them on each choice."[45] Thus, during the initial successive discrimination training one group of rats would have to learn to go left when both goal boxes had curtains near their entrance and to learn to ignore whether both boxes were black or both were white; another group would learn to respond on the basis of white vs. black and ignore the presence or absence of curtains.

It turned out that Lawrence's rats learned the successive discrimination when the curtains were the relevant cue as rapidly as when brightness was the cue. Similarly, when faced next with the simultaneous discrimination, in which brightness and presence/absence of curtains were both relevant, the rats' rapid learning was not affected by their previous training. The critical results came from the series of tests. In the first one, the rats needed to choose between, say, a white box without a curtain – where white had been a positive stimulus – and a black box with a curtain – where the presence of a curtain had been a positive stimulus. It turned out, as predicted, that rats' choices were made on the basis of the cue that had been relevant during their initial successive training. This showed that initially learning that, for example, brightness was the relevant cue in the successive discrimination and the presence or absence of curtains was irrelevant, ensured that brightness was the more distinctive cue in the simultaneous discrimination that followed. Further tests yielded similar outcomes and further confirmation that 'distinctiveness' could be 'acquired.'

In describing Lawrence's experiments, it is difficult to avoid using terms like 'learning to attend' or 'paying attention to.' He admitted that his analysis was very similar to that of Lashley, who in 1942 had discussed the role of attention in discrimination learning. However, while arguing that his choice of cues meant that the results of this second experiment could not be explained in terms of the 'orienting responses' favored by Spence, Lawrence also argued for the use of 'mediating responses' rather than appealing to attentional processes. Appeals to attention "tend to emphasize the all-or-none shifts in attention during the pre-solution period of learning. The present formulation assumes that the changes in the mediating process are of a gradual and continuous nature."[46] It is intriguing to notice that, as well as acknowledging a debt to his advisor, Neal Miller, in this second paper Lawrence also thanked "Dr. Clark L. Hull."

The third experiment that Lawrence published from his Ph.D. was based on an idea that went back to the early experimental psychologists of the nineteenth century and probably even further. The idea became known as *transfer along a continuum.* William James gave the example of being touched simultaneously at two points on the skin; when they are far apart, they are perceived as two separate events, but, when close together, normally they produce what appears to be a single impression. However, even two points close together can be perceived as two separate events if the observer is given training that starts with a large separation and progressively brings the points closer together.[47] Pavlov reported several experiments of this kind using salivary conditioning with dogs. The dimensions used by various students in his laboratory included shades of gray, ellipses with varying ratios between their semiaxes, frequency of beats from a metronome, and tones produced by organ pipes.[48] Although the reports are, as usual, very sketchy, it seems that in each case a dog discriminated between two stimuli very close together after being trained to discriminate between two distant points on the same dimension.

Lawrence's interest in producing a clear demonstration of this phenomenon with rats arose because he saw that it could not be explained in terms of a continuity theory such as Spence's but required an explanation in terms of acquired changes in distinctiveness. His experiment used the simultaneous discrimination apparatus shown in Figure 7.10, with the walls of the goal boxes lined with cardboard painted with six different levels of gray, ranging from (a) "Lightest Gray" to (g) "Darkest Gray."

The target discrimination for his four groups was between two close shades, (d) and (e). One group (Hard Discrimination Group) were from the outset given all 80 trials of simultaneous discrimination training on this pair. Their performance slowly improved but never reached the level achieved by other groups that were first trained on the very easy discrimination – (a) vs. (g) – and were either abruptly shifted to the hard discrimination or allowed a gradual transition – (a) vs. (g), then (b) vs. (f) and so on. This gradual group ended up performing with the smallest number of errors. The results were not surprising but nonetheless provided the clearest demonstration to that date of transfer along a continuum.[49]

After completing his Ph.D. in the Fall of 1949, Lawrence took up an appointment at Stanford University. It was an obvious choice. Miller, his advisor, had graduated there and the Head of the Psychology Department was Ernest Hilgard, by then recognized as a top expert on animal learning following his co-authorship of the standard text in this area.[50] It seems that Lawrence set up a new animal laboratory at Stanford where, supported by a grant from the Ford Foundation, he collaborated on a series of experiments with students. The most interesting one returned to the topic of transposition. This used a jumping stand and a successive discrimination procedure whereby a rat was rewarded for jumping to the left door when, for example, both doors displayed panels in which the top half was a lighter gray than the bottom half, and for jumping to the right when both displayed the opposed arrangement. Lawrence and his student employed a procedure based on transfer along a continuum to train his rats, first, to perform on the basis of a large contrast between the two extreme grays and then between the middle gray, no.4, and all the intermediates. Transfer tests revealed for almost every possible combination of grays that rats continued to respond on the basis of whether to top half was lighter or darker than the bottom half. The authors did not claim on the basis of these results that Koehler was right and Spence wrong. Instead, they suggested – very reasonably – that the finding that their rats were responding on the basis of the relational aspects of the situation "does not imply that this mode of responding is more basic or fundamental than responding to the absolute or particular characteristics of the situation. .... Rather, this study indicates that when these particular aspects are minimized as reliable cues .... and the relational properties are emphasized, S is perfectly capable of responding to the latter."[51]

Lawrence began a collaboration with a social psychologist in his department, Leon Festinger, on the topic of *cognitive dissonance* and on the remarkable persistence of some partially rewarded behaviors. Interest in this topic seems to have eventually swamped his interest in the acquired distinctiveness of cues. After this phase, he remained an active researcher, but now concerned with the role of attention in human perception, until he retired in 1977. He died in 1999, aged 81.

Discrimination Learning at Oxford: Analyzers, Octopuses, and the Overtraining Reversal Effect (ORE)

After 1932, when Grindley's last research report appeared, for two decades there was no further active interest in animal learning in Britain. Then, starting in Oxford, such research blossomed in a way that led to important advances in understanding discrimination learning. This occurred mainly because of a confluence of two developments. The first was increasing interest in human selective attention. Whereas among American behaviorists 'attention' was almost a taboo word, this was seen as a key topic among the British psychologists who had studied the performance of personnel such as radar operators during World War II and afterwards worked at the Applied Psychology Unit in Cambridge. An important book that summarized this research and described an early theory of human cognitive processing was Donald Broadbent's *Perception and communication* of 1958.[52] His theory conceived of attention in terms of filtering information.

The other development was due to J. Anthony Deutsch (1927–2016). Deutsch was born in Czechoslovakia and arrived in the UK in 1939 – as a 12-year old on his own. He obtained First Class Honors in the unique Oxford PPP degree (Psychology, Philosophy, and Physiology). In contrast to what was – and still is – an essentially impossible career path within the American university system, at Oxford and many other British universities it was quite acceptable for someone to start as an undergraduate, continue as a graduate student and, immediately after completing a Ph.D., obtain a faculty position, all at the same university. This was Deutsch's early career. His outstanding Ph.D. thesis gained him an appointment as 'Lecturer in General Psychology' in 1952, when he became a member of the Institute for Experimental Psychology. The thesis contained a critique of Hull's S-R habit theory as to how rats learn to find their way through a maze. Deutsch proposed an alternative theory that was based on the idea that rats learn to approach a succession of stimuli. His theory was expressed in terms of the development of connections between a series of stimulus *analyzers* and became possibly the first theory of animal learning to be implemented in a mechanical system.[53] This was something to which Hull had aspired but never achieved.

To demonstrate the superiority of his theory over those of Hull and his fellow behaviorists, Deutsch set up a rat laboratory in Oxford, constructed a number of mazes, and obtained a research grant from the UK Medical Research Council that allowed him to employ a research assistant. The mazes were designed to test, for example, a Hullian theory of extinction[54] or a theory developed by Spence and Howard Kendler as to how rats can choose the appropriate arm of a T-maze according to whether they are hungry or thirsty.[55] For Deutsch, the results from these experiments confirmed the superiority of his theory. A particularly complicated maze is shown in Figure 7.12. This was intended to overcome some of the objections that had been lodged against the original experiment from Tolman's laboratory on *insight* in rats.[56] The details of the experiment are not important here, except to note that Deutsch's theory uniquely predicted the test condition under which rats would choose to take the long arm after finding that choice of one of the two short arms no longer led to food. The comment following the results section was: "The prediction from Deutsch's theory is thus strikingly confirmed."[57]

In 1959 Deutsch spent a year as a Fellow of the Center for Advanced Studies in California. That led to a series of appointments at American universities. In 1966 he obtained a full professorship at the University of California, San Diego, where he remained for the rest of his life. Once in America he appears to have left behind his interest in animal learning and concentrated instead on what was then known as physiological psychology. However, his influence on animal learning research in Britain persisted. When a lecturer at Oxford he had taught two undergraduates who were to make major contributions to the study of discrimination learning. The first was N. Stuart Sutherland (1927–1998) and the second was Nicholas Mackintosh (1935–2015).

The description by a friend and colleague of Sutherland as having a "colorful personality"[58] was an understatement; of which, more later. Although born in the same

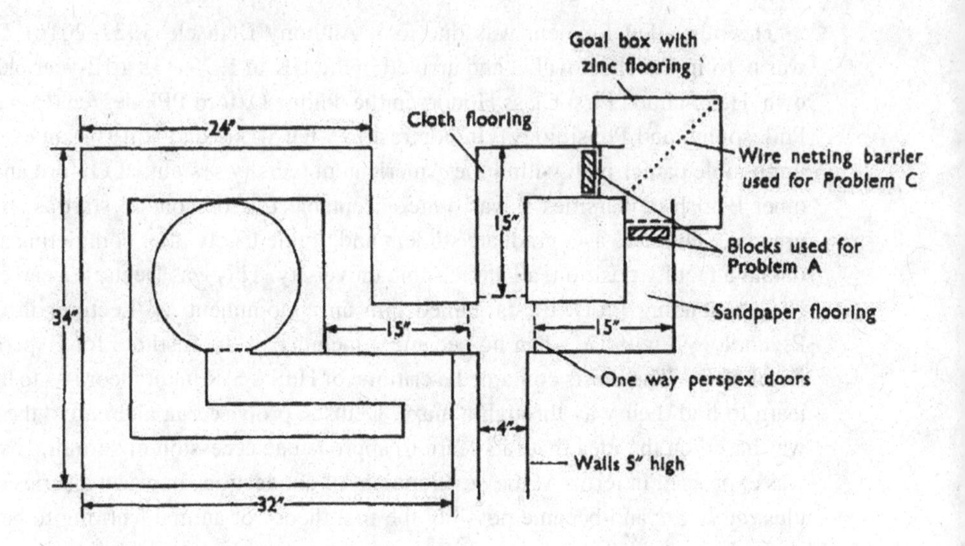

Figure 7.12 Maze constructed to test Deutsch's theory of reasoning in the rat. From Deutsch and Clarkson (1959b). Public domain.

year as Deutsch, his interest in psychology was slower to develop. On leaving school Sutherland first studied classics at Oxford and then was conscripted into National Service – as was still universal in the UK until the late1950s – with the result that he spent two years with a training unit in the British Army. His interest in psychology seems to have been sparked by this experience, since he returned to Oxford to read psychology and philosophy. This led in 1954 to a fellowship at Magdalen College, Oxford, and eventually an appointment in 1960 as a Lecturer in the Oxford Institute of Experimental Psychology.[59] This may have been as a successor to Deutsch.

Sutherland was later to establish at the University of Sussex the largest center ever in the UK for the study of animal learning, and indeed the largest in Europe. However, early in his career, this was not his core interest. This was in visual perception and, in particular, the perception of shapes by octopuses. His doctoral research was carried out at the Zoological Station in Naples, Italy. Sutherland's supervisor, the London-based biologist, J.Z. Young, had already set up a research program in Naples in order to study the octopus's visual abilities when Sutherland started to spend every summer in Naples.

At first, Sutherland used a discrimination procedure that Young's team had already developed. An octopus was placed in a tank that contained at one end a small cave, made of bricks, in which the animal would sit and look down the length of the tank. Shapes of the type shown in Figure 7.13 were cut out of plexiglass and mounted on a rod so that they could be lowered into the side of the tank opposite to the cave. The animals were trained on successive discriminations involving two shapes, one designated as positive and the other as negative. If the octopus shot out a tentacle to attack a positive shape, it was given a piece of sardine; if it attacked a negative shape, it was given an electric shock.

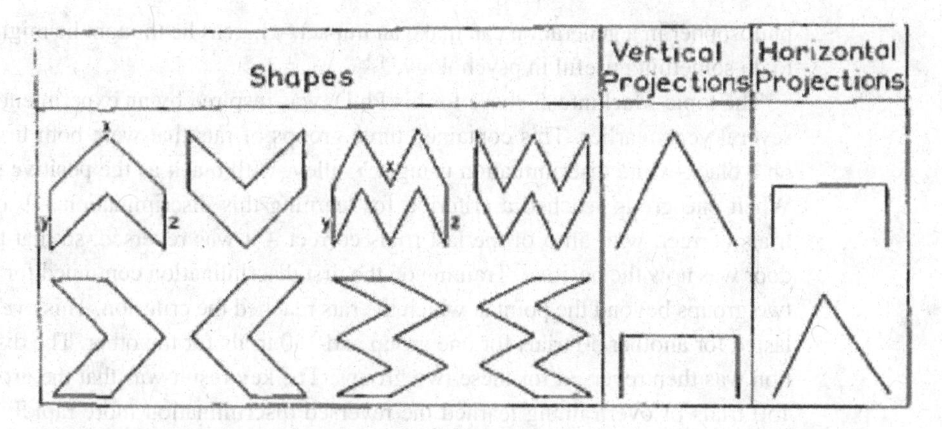

Figure 7.13 Sutherland trained octopuses to discriminate between these pairs of shapes that were presented either vertically or horizontally.
From Sutherland (1959). Public domain.

Sutherland used the results from a series of such experiments over several summers to develop a theory of octopus shape perception based on hypothetical neural units, *analyzers*, that were tuned to detect either the vertical or the horizontal extent of a shape. Not every prediction from this theory was confirmed. The ability of octopuses to discriminate between mirror images of the same shape was a controversial issue that distinguished Sutherland's theory from one that Deutsch had proposed.[60] The two fell out over this controversy but apparently became good friends again years later.[61]

While discrimination learning was at first simply a tool as far as Sutherland was concerned, it was from the start of his research career a topic of great interest in its own right for his Ph.D. student, Nick Mackintosh. At Mackintosh's funeral, three accounts of his life were provided by his brother-in-law. The short account was: "A very clever man who liked a drink." The slightly longer version was: "One man; two degrees; three wives; four names; five children; six visiting professorships."[62] Instead of summarizing Mackintosh's marital history, a psychologist might have included "three very important books."

A striking aspect of the early part of Mackintosh's life is how small a role his parents played. His family was based in Malaya, while he and his sister generally remained in the UK, looked after by a grandmother, until in 1940 both Mackintosh and sister were sent to join a group of children that were evacuated to Canada for three years. After the war, he attended a series of private schools that provided him with an education that was effective enough for him to obtain at the age of thirteen a scholarship to Winchester College, one of the few elite English private schools to prize intellectual achievements above all others. Afterwards, he spent what later would be called a 'gap year,' working and traveling, until he had to spend two years of National Service, spent mainly with a tank regiment based in Germany. In 1957 he entered Magdalen College, Oxford to study philosophy and psychology and, on completing his undergraduate degree, continued there as a Ph.D. student in psychology. "He chose to study psychology over philosophy because he believed that only one philosopher in a generation can make an impact, whereas he thought he might be able to do something useful in psychology."[63]

The topic Mackintosh chose for his Ph.D. was inspired by an experiment reported several years earlier. This contained three groups of rats that were both first trained on a black-white discrimination using a Y-alley, with black as the positive stimulus. When one group reached a criterion for learning this discrimination – 9 out of 10 trials correct, with all 5 of the last trials correct – it was reversed, so that the white door was now the positive. Training on the first discrimination continued for the other two groups beyond the point at which the rats reached the criterion. This *overtraining* lasted for another 50 trials for one group and 150 trials for the other. The discrimination was then reversed for these two groups. The key result was that the group given 150 trials of overtraining learned the reversed discrimination more rapidly than the other two groups. This result was the opposite of what would be expected in terms of a simple S-R theory but was what the author had predicted.[64]

For the next 15 years or so this *overtraining reversal effect* (ORE) remained one of the hot topics in the study of discrimination learning. By 1969 the first ORE experiment had inspired over 50 experimental reports, some of which also found an ORE, while many found no effect of overtraining or, in a few cases, even that overtraining hindered reversal learning.[65]

The first of many experiments on the effects of overtraining that Mackintosh published was an ingenious attempt to pit two types of explanation for the ORE against each other. One type suggested that overtraining promoted a *general* improvement in an animal's ability to learn discrimination tasks. One explanation of this type was based on Harlow's *error factor theory*. This proposed that the increasing speed with which monkeys solved discrimination after discrimination in one of his learning set experiments resulted from them learning what errors to avoid – for example, a bias towards one position;[66] see Chapter 5. The other type, emphasizing *specific* effects of overtraining, followed Lawrence's idea of *acquired distinctiveness*. Here Mackintosh included as an example the theory that his supervisor had proposed. Sutherland had extended the idea that the octopus's visual system contains analyzers that are tuned to detect different features of an object to the proposal that all animals possess analyzers that select different aspects of their whole visual field.[67] Thus, overtraining could strengthen an analyzer tuned to detect differences in the color of the two doors that a rat needed to choose between.[68]

Only the specific type of explanation for the ORE predicts that overtraining on one discrimination should benefit learning of the reversal of this discrimination but not learning of a new discrimination. This was the prediction that Mackintosh tested in his first published experiment. His rats were initially trained on a black-white discrimination using a modified Lashley jumping stand; half were trained with black positive and the other with white positive. On reaching criterion rats were divided – as in the original ORE experiment – into three groups: They were given either zero, 75, or 150 further trials on their black-white discrimination. Subsequently, each group was divided into two subgroups, one of which was given a reversal of their original discrimination and the other a new discrimination, between a door that contained three

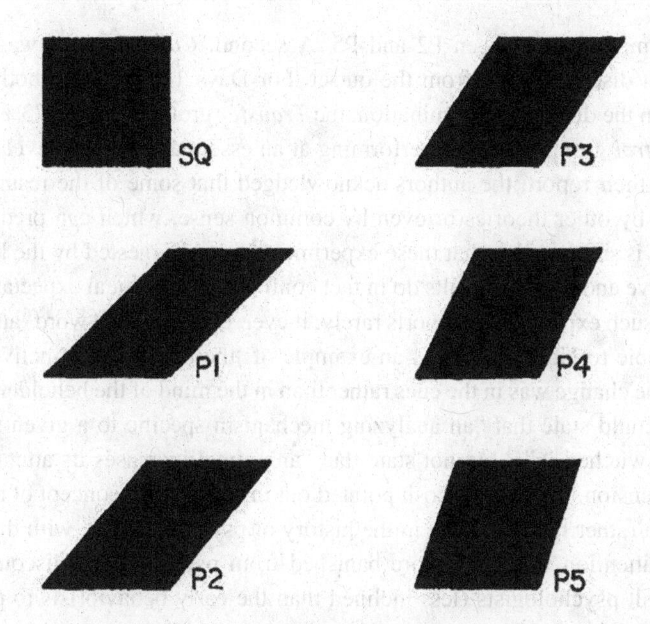

Figure 7.14 These shapes were used by Sutherland and his two coauthors to demonstrate *transfer along a continuum* in octopuses. The animals were initially trained to discriminate between the square and P3 and then shifted first to P1 *vs.* P4 and then to P2 *vs.* P5. From Sutherland, Mackintosh, and Mackintosh (1963). Reproduced with permission from the American Psychological Association.

vertical bars and a door that contained three horizontal bars. Mackintosh referred to the latter condition as a "non-reversal shift"; in doing so, he followed the terminology used by Kendler and Kendler[69] that was described earlier in this chapter.

The results were very clear and completely in line with the specific theory that Mackintosh favored. Rats given overtraining prior to a reversal of their original black-white discrimination displayed the ORE in that they learned more rapidly than rats reversed after reaching criterion. In complete contrast, rats given overtraining and then shifted to the new, and more difficult, vertical-horizontal discrimination learned this more slowly than rats given this second discrimination immediately after reaching criterion.[70]

When still an undergraduate Mackintosh had begun to spend his summers in Naples to work with Sutherland on discrimination learning by octopuses. He and another Oxford undergraduate found that octopuses could also display the ORE.[71] In 1960 the other undergraduate became his first wife, Janet Mackintosh. For their next published experiment, they joined Sutherland in testing a further prediction from the latter's analyzer theory. Their main aim, that of replicating in octopuses Lawrence's report of transfer along a continuum, was achieved despite a series of mishaps. One group of animals, the *Transfer* group, were first trained to make a simultaneous discrimination between the square and the parallelogram, P3, shown in Figure 7.14, then trained to discriminate between the two parallelograms, P1 and P4, and finally on the most difficult discrimination, between P2 and P5. A second, *Control* group was trained on the difficult discrimination from the outset. For Days 14–17, when both groups were trained on the difficult discrimination, the *Transfer* group averaged 73% correct, while the *Control* group were still performing at an essentially chance level of 54%. In concluding their report, the authors acknowledged that some of the results could "be explained by other theories or even by common sense, which can predict most things; all that is suggested is that these experiments were suggested by the hypotheses set out above and that the results do in fact conform to theoretical expectations."[72]

Until 1965 such experimental reports rarely, if ever, contained the word 'attention.' It was acceptable to label a result as an example of 'the acquired distinctiveness of cues,'[73] as if the change was in the cues rather than in the mind of the beholder. As late as 1963, one could state that "an analyzing mechanism specific to a given stimulus dimension is switched in"[74] but not state that "an animal increases its attention to a particular dimension." As Mackintosh pointed out in 1965: "The concept of attention has occupied a rather humble place in the history of psychology. .... with the rise of behaviorism, attention was once more banished from psychological discourse, and even the Gestalt psychologists (less inclined than the early behaviorists to prejudge the value of a concept from its introspective overtones) had little room in their system for a selective determinant of perception."[75]

The acceptance of selective attention as an acceptable concept by all but extreme behaviorists dated from the publication of a major review article by Mackintosh. Its title was *Selective attention in animal discrimination learning*. The core argument echoed one that Broadbent had made in 1958 regarding human information processing.[76] "Animals have nervous systems of limited size and therefore of limited capacity for processing and storing information. Thus, they are confronted with the problem of selection. At some stage, they must discard irrelevant or redundant information so as not to interfere with the storage of important information. .... If animals do not respond to all features of their stimulus input, then a sharp distinction must be drawn between the physical stimuli impinging on an animal in any given situation and the effective stimulus which controls the animal's behavior in that situation."[77]

The review contains over 150 references and suggests that Mackintosh had not only read every paper on discrimination learning in nonhuman animals ever published, as well as many involving humans, but had thought hard about every result they reported. On the basis primarily of research on three phenomena – acquired distinctiveness of cues, transfer along a continuum and the ORE – he concluded that only a two-stage theory could account for discrimination learning: "Animals must learn both to attend to the relevant cue and to establish appropriate choice responses."[78]

After completing his Ph.D. in 1963, Mackintosh remained at Oxford for another four years. One year as a postdoctoral researcher led to an appointment in 1964 as a university lecturer, presumably filling the position that Sutherland had left upon the latter's appointment as Professor of Experimental Psychology at the University of Sussex. In the mid-1960s Mackintosh held a series of visiting appointments at various universities in the United States. One of these (1965–1966) was the University of Pennsylvania where interactions with graduate students there, such as Rescorla,

LoLordo, Shettleworth, Maier, and Seligman, persuaded him of the importance of Pavlovian conditioning. This was followed by a spell at the University of Hawaii, where he formed a friendship with Louis Herman who studied cognition and communication in dolphins; see Chapter 5. After a brief return to Oxford, he and his family commenced a more settled life in Canada upon his appointment in 1967 as a Research Professor at Dalhousie University in Halifax, Nova Scotia.[79]

Despite his several relocations, Mackintosh continued to be as productive as ever. He continued with experiments on the ORE and also tested whether attentional theory could be expanded to account for other phenomena, such as the increased resistance to extinction produced by partial reinforcement. This research and the theoretical ideas behind it were described in a book he co-authored with his former supervisor. *Mechanisms of animal discrimination learning* by Sutherland and Mackintosh was published in 1971.[80] Three years earlier Mackintosh, together with a colleague, Vern Honig, had organized a conference at Dalhousie University that changed the direction of his research and that of many others; see Chapter 9.

The final experiment to be described in this section demonstrates that Mackintosh was not the only one to come up with ingenious experiments on the effects of overtraining. This was the topic of another Ph.D. thesis that involved testing rats on a modified Lashley jumping stand. Most of these doctoral experiments were carried out at the University of Cambridge in the UK, but the one described here was run when the graduate student, Geoffrey Hall, was on a working visit to Mackintosh's laboratory at Dalhousie.

As described earlier, Mackintosh had reported in 1962 that overtraining on one discrimination task, although benefiting reversal learning, produced slower learning of a second task – a *non-reversal shift* or *extradimensional shift* (EDS).[81] Hall's study started with the observation that Mackintosh's results had not been generally confirmed. Instead, several studies found the opposite effect, that overtraining helped new learning. Consequently, the evidence from EDS experiments "cannot provide a crucial test which distinguishes between attentional and other theories of transfer."[82] The basic idea behind his critical experiment was to demonstrate a general benefit of overtraining that could not be explained in terms of it strengthening rats' tendency to attend to a particular dimension.

There were three stages to the experiment. In Stage 1 rats were trained on a horizontal-vertical problem, half to criterion and the other half overtrained by 150 trials beyond criterion. Stage 2 was an EDS, in that the two groups were now trained to criterion on an easy black-white discrimination. In Stage 3 this second discrimination was reversed. The results from the second stage showed no effect of overtraining in Stage 1; this failure to replicate Mackintosh's finding of overtraining interfering with learning an EDS was not unexpected, given the use of an easy black-white discrimination in Stages 2. The important finding was that overtraining in Stage 1 had a major impact on learning the reversal of the black-white discrimination; the overtraining group required an average of 70 trials to reach the criterion, while the criterion group took more than twice as many trials. These results "demonstrated that the appearance of a helpful overtraining effect does not depend on the continued presence of the dimension upon which overtraining was given. This result is opposite to the prediction of Mackintosh's version of analyzer theory and calls for an account of overtraining in which transfer effects are not specific to the overtrained dimension."[83] The paper ends with an admission that there is no well-developed alternative to attentional theory: "No explanation is available that can deal easily with all the available experimental evidence, but a possibility that might be usefully investigated is that overtraining has its effects by producing useful observing responses or more complex 'response strategies' described by Mandler (1966) and Hall (1973)."[84] Spence would have been half pleased by such a conclusion.

This criticism of Mackintosh's ideas on the effects of overtraining did not prevent Hall from obtaining a postdoctoral position in Mackintosh's laboratory. How both Mackintosh's and Hall's ideas on attention were subsequently modified when introduced into their theories of associative learning is described in the final chapter.

Discrimination Learning and Generalization in the Pigeon: Stimulus Control

From 1950 Charles Ferster was employed at Harvard University as a postdoctoral researcher on a project that was directed by Skinner. Ferster ran dozens of experiments that exposed pigeons to a wide range of reinforcement schedules; see Chapter 8. In 1953 Ferster described what proved to be a highly influential example of how these techniques could be used to study discrimination, using what became known as a *multiple schedule* of reinforcement.[85] In an initial baseline stage pecking at a response key is reinforced intermittently whichever of two colors are projected onto it. Once a pigeon is responding steadily and at an equal rate to both colors, then in the subsequent discrimination stage pecking at one of the colors ($S^{delta)}$ is no longer reinforced, while reinforcement continues at the same rate for pecking at the now positive color (S^D). "Despite the alternation of the reinforced and non-reinforced stimuli, the decline in rate in S^{delta} is extremely orderly. Moreover, after two or three hours of training, the rate under the nonreinforced color will be less than 1/100 the rate under the reinforced color."[86]

One theoretical issue behind the sudden burst of experiments that are described in this section was the question: What is the true shape of a stimulus generalization function? As Spence had shown, it is possible to explain a phenomenon such as transposition in terms of interactions between generalization to a test stimulus from a positive stimulus, S+, and generalization from a negative, that is, inhibitory, stimulus, S-. However, predictions from such a theory depend on the assumed shape of these generalization functions. Various suggestions on this topic were made in the absence of good experimental data. A particularly influential suggestion was that of Lashley and Wade who had argued in 1946 that generalization results from a failure to discriminate between the S+ or S- and some test stimulus.[87] Their theory predicted that a generalization function can be predicted from data on an animal's ability to discriminate between close pairs of stimuli along some sensory continuum.

The obstacle blocking good data on generalization was that, after a traditional procedure was used to condition responding to some S+, during testing with a series of related stimuli in the absence of any further reinforcement the conditioned response would rapidly extinguish. A possible way around the extinction obstacle had been shown by Skinner during his wartime work that initiated the use of pigeons in behavioral research. In a series of experiments that were not reported for two decades, Skinner and his students reinforced pecking, for example, at a green triangle on an intermittent reinforcement schedule and then tested in extinction for responding to triangles of a different color or to different shapes.[88] Norman Guttman was one of the students involved in these experiments. Later, after establishing his own pigeon laboratory at Duke University, he teamed up with Harry Kalish to develop what became a standard method for studying stimulus generalization. The method relied on the fact that responding that is reinforced on a lean variable-interval schedule can persist in extinction for a very long time. "The obtaining of generalization gradients for individual Ss in this experiment (is) an outcome of the fact that aperiodic reinforcement greatly increases resistance to extinction, such that the introduction of a test stimulus for a brief interval during experimental extinction reduces the response strength by a small fraction of its total extent, for example in a 30-seconds test, 50 responses may be subtracted from a 'reserve' of several thousand."[89]

In their classic experiment, Guttman and Kalish used 24 pigeons that were divided into four groups. The mildly hungry birds were trained to peck on a variable-interval reinforcement schedule (grain became available at variable intervals), with a mean interval length of 1 minute, at a response key that was illuminated by a near-monochromatic light whose wavelength differed from group to group. Once the birds were responding at a steady rate, testing began under extinction conditions. Within a test session, eleven different wavelengths were projected onto the response key, each for 30 seconds at a time. The data were remarkably orderly. As shown in Panel (a) of Figure 7.15, in each group responding declined in a regular manner as the test stimuli became increasingly distant from the training stimulus. Panel (b) shows how discrimination between two adjacent pairs of wavelengths changes with wavelength for both pigeons and humans. The absence of any correspondence between these functions and the generalization data shown in Panel (a) is not what Lashley and Wade would have predicted.

The next major experiment on this topic was carried out in the same laboratory by a Ph.D. student, Harley Hanson; Guttman was his research adviser. The aim of Hanson's experiment was to test for the effect of discrimination training on generalization gradients. In doing so, this would test a number of implications of a Spence-type theory. These included predictions that "the post-discrimination gradient will be steeper than the generalization gradient in the region of S- … (and that) … the mode of the post-discrimination gradient will be displaced away from S- in relation to the mode of the generalization gradient."[90] Hanson pointed out that these predictions did not follow from Hull's assumption that the 'true' shape of generalization gradients was an exponential function.

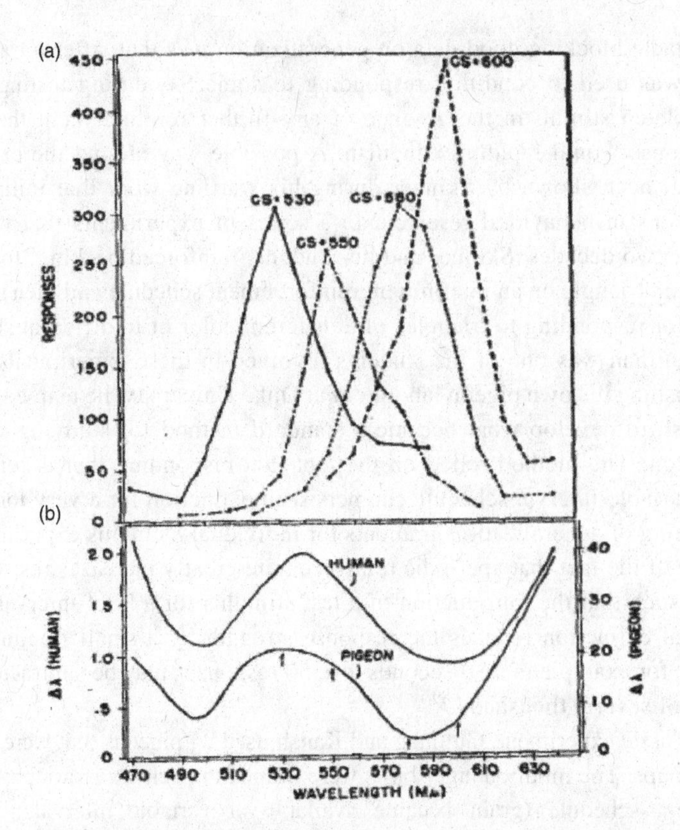

Figure 7.15 Panel (a) shows stimulus generalization gradients obtained on test after four groups of pigeons were trained to peck at different wavelengths of colors projected onto the response key. Panel (b) Hue discrimination as a function of wavelength in both pigeons and humans. From Guttman and Kalish (1956). Public domain.

All thirty-three pigeons in this experiment were first trained to peck at a response key that was illuminated by light of the same wavelength, 550 mu, using a 1-minute variable-interval schedule. They were then divided into five groups, of which one – the control group – was given no further training, while the other four were given discrimination training on a multiple schedule. This arranged that 1-minute periods in which pecking at the S+ was still reinforced on the variable-interval schedule alternated with 1-minute periods in which the response key was illuminated by some other wavelength, the S-, and pecking at it was not reinforced. The conditions for the discrimination groups differed in terms of how close their S- was to the S+; the S- values used ranged from very close, 555 mu, and thus a difficult discrimination for this group, to quite distant, 590 mu. Generalization testing followed the procedure used by Guttman and Kalish, in that under extinction conditions each pigeon as presented with a series of 30-second exposures to a wide range of different wavelengths.

The results were very clear in confirming the Spence-type predictions quoted above. As seen in Figure 7.16, compared to the function produced by the control group, at least in the discrimination group given the most difficult discrimination, 550

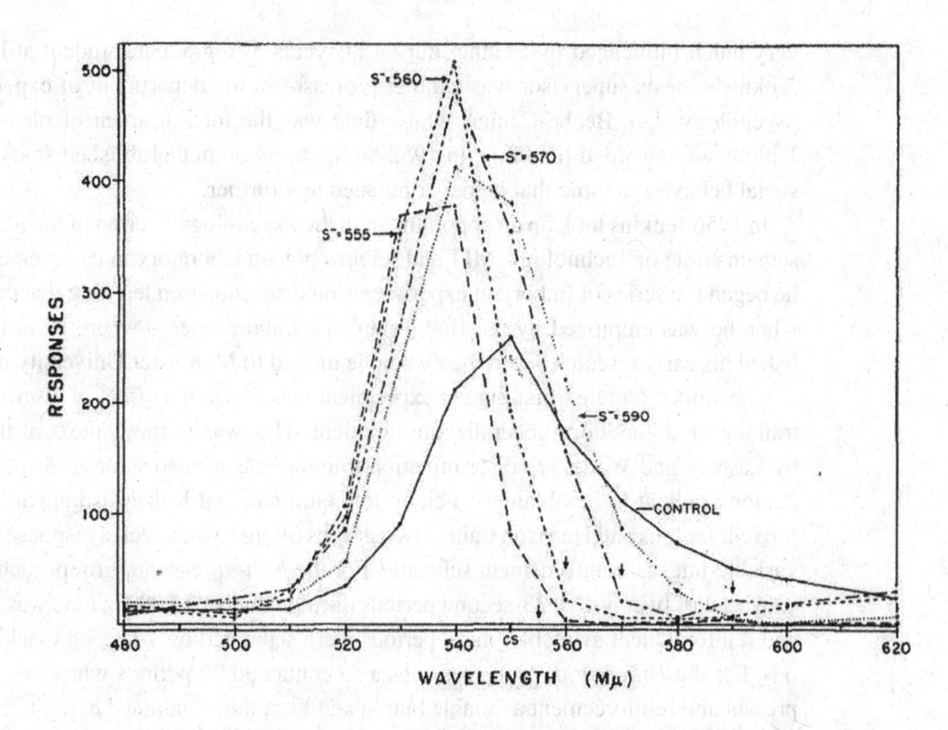

Figure 7.16 These data are from the first experiment to demonstrate a peak shift effect, that is, peak responding shifted to a color other than the positive stimulus and away from the negative stimulus. These generalization functions were obtained on test after four groups of pigeons had been trained using a multiple schedule that reinforced pecking when one color (positive stimulus) was present but did not reinforce pecking to a second color (negative stimulus). The control group was simply trained to peck at a single color without any negative stimulus being presented. From Hanson (1959). Public domain.

mu vs. 555 mu, the slope of the generalization functions was steeper in the region of the S- than on the other side. Most importantly, all four discrimination groups displayed what was subsequently named the *peak shift*: Maximum responding was displaced from the S+ to a stimulus on the side away from the S-.

The generalization data shown in Figure 7.16 differ in one glaring respect from what would be predicted from a Spence-type analysis: In the discrimination groups rates of responding to the S+ and to similar colors were around twice as high as response rates to S+ in the control group. Any theory of generalization of inhibition would predict that discrimination training should produce response rates to the S+ that are *lower* than in animals not given discrimination training. Pavlov had documented such an effect for salivary conditioning in dogs and named it *negative induction*.[91] As described in the next section, attempts to understand what became known as *behavioral contrast* continued to generate a large number of experiments after Hanson's original discovery of this effect until it was finally understood some 15 years later.

Another important experiment on stimulus generalization inspired by the Guttman and Kalish study was carried out by Herb Jenkins as the lead researcher. Although very much influenced by Skinner during his years as a graduate student at Harvard, Jenkins's thesis supervisor was another professor in the department of experimental psychology, J.G. Beebe-Center, whose field was the measurement of pleasantness. Jenkins was awarded his Ph.D. in 1952 on the basis of an unpublished study of rats' social behavior, a topic that he never pursued any further.

In 1956 Jenkins took up an appointment in the Psychology Section of the Massachusetts Institute of Technology, MIT and set up a pigeon laboratory in its basement. Here he began the series of important experiments on discrimination learning that continued when he was employed by the Bell Telephone Laboratories – where he at last published his early research – and then when he moved to McMaster University in 1963.

The aim of Jenkins' first pigeon experiment was to test the effect of discrimination training on an auditory generalization gradient. This was in the context of the claim by Lashley and Wade that discrimination training was needed in order for a generalization gradient to be obtained, a claim that Guttman and Kalish had apparently disproved. Jenkins and Harrison trained two groups of pigeons to peck a response key on a variable-interval reinforcement schedule. For the *Non-differential* group, each session consisted of twenty-five 33-second periods during which a 1,000 Hz tone was sounded and reinforcement available; these periods were separated by 7-second blackout periods. For the *Differential* group sessions also contained 25 periods when the tone was present and reinforcement available but, in addition, they contained at least 25 periods when the tone was absent and pecking was not reinforced. In the subsequent generalization test eight tones, ranging from 300 Hz to 3,500 Hz, were presented.

As shown in Figure 7.17, generalization gradients were essentially flat in the Non-differential group but sharp in the Differential group, an outcome consistent with Lashley and Wade's suggestion that the slope of a generalization gradient indicates the extent to which the subject paid attention to the feature of the training situation. The authors argued that Guttman and Kalish's success in obtaining steep generalization gradients after non-differential training was a special case, in that, when visual stimuli are projected upon the response key that a pigeon has to peck, this inherently involves differential training.[92]

The next development was direct measurement of generalization of inhibition. To achieve this, Jenkins and Harrison used very similar training to that given to the Differential group in their previous experiment. Thus, in one experiment three pigeons were trained on a multiple schedule in which pecking was reinforced in the presence of white noise (S+) but not reinforced in the presence of a 1,000 Hz tone (S−). The results obtained when they were tested for generalization to tones ranging from 300 Hz to 3,500 Hz are shown in Figure 7.18. It can be seen that responding was least to the 1,000 Hz tone and greater to the higher tones that had never been reinforced.[93]

Given that there were now abundant data on stimulus generalization, the question of the true shape of such functions could be addressed. Some "concluded that the question of the shape of the gradient of generalization is meaningless because there is really no independent measure of dissimilarity";[94] in other words, almost any function could be produced by selecting a particular transformation of the x-axis. The brilliant solution was pointed out by Roger Shepard, another employee of Bell

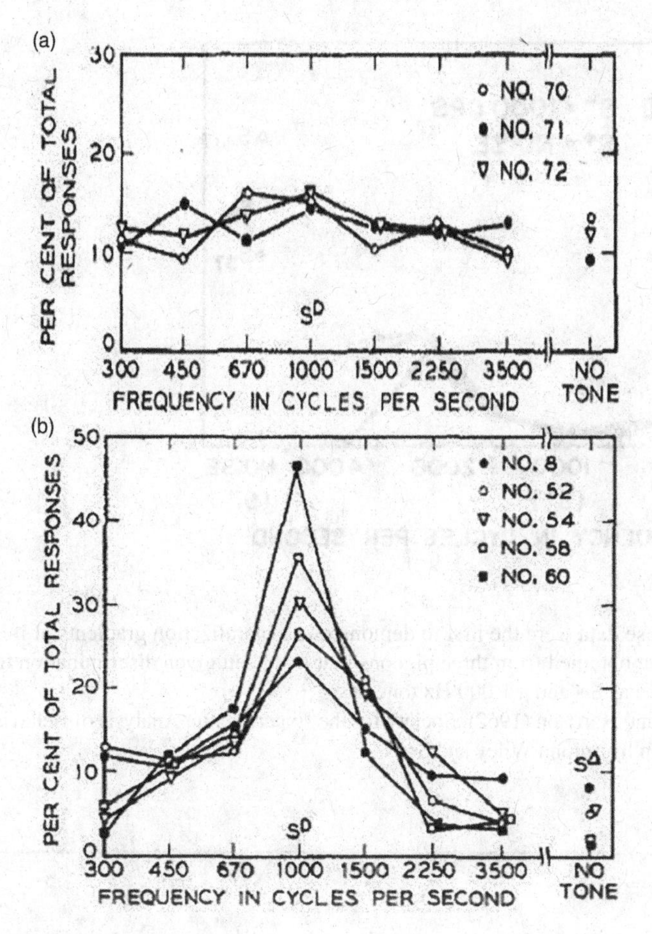

Figure 7.17 Panel (a) Generalization gradients from a group of pigeons following non-differential training to peck a response key in the presence of a 1,000 Hz tone. Panel (b) Generalization gradients from a group of pigeons following differential training in which pecking was not reinforced unless a 1,000 Hz tone was present.
From Jenkins and Harrison (1960). Public domain.

Telephone Laboratories. (One cannot but wonder how he and Jenkins persuaded the top executives that research on stimulus generalization in pigeons was relevant to human communication!). Shepard showed that, when there are a number of generalization functions along some continuum, then there is a unique transformation of the x-axis that will produce the same shape for each function. "The discovery of that transformation yields two quite different kinds of information in a single stroke. First, it tells us what the true psychological spacing of the test stimuli is in the underlying continuum or 'psychological space.' And, second, it tells us what the uniform shape is for all generalization gradients in this underlying space."[95]

The transformed gradients based on data from Guttman and Kalish and from a subsequent experiment are shown in Figure 7.19. In order to adjust the interstimulus spacing on a trial-and-error basis until a uniform set of gradients was obtained,

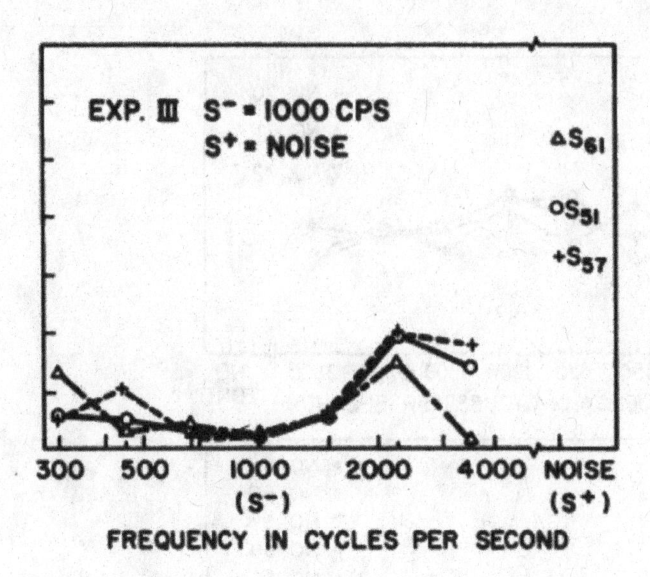

Figure 7.18 These data were the first to demonstrate generalization gradients of inhibition. They were obtained from three pigeons that had been given discrimination training between a noise as S+ and a 1,000 Hz tone as S-.
From Jenkins and Harrison (1962). Society for the Experimental Analysis of Behavior, used with permission from John Wiley and Sons.

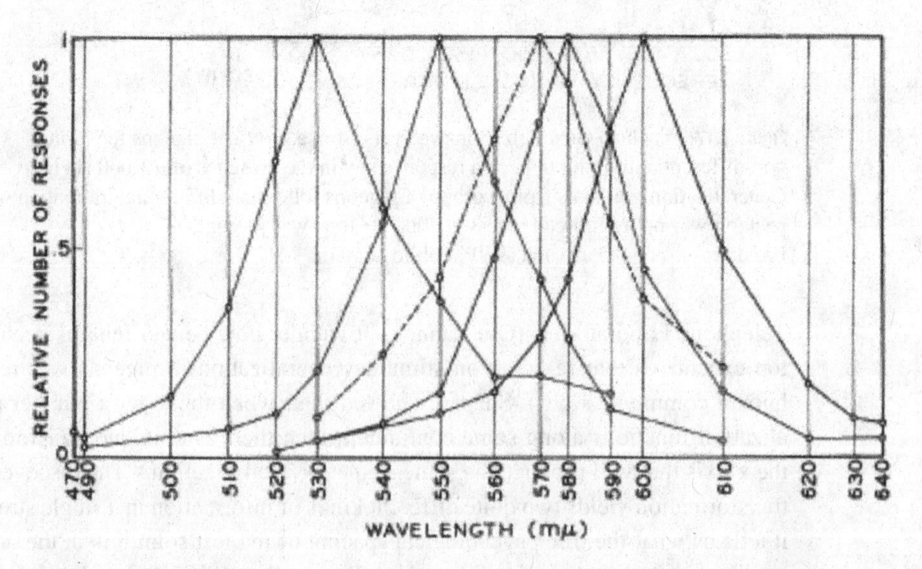

Figure 7.19 These graphs show the 'true' shape of stimulus generalization gradients. The data are from Guttman and Kalish (1956) – see Figure 7.15 – and from a later experiment. The shapes of these functions have been transformed by manipulating spacing along the wavelength dimension until all five are of the same shape.
From Shepard (1965). Reproduced with permission from Stanford University Press.

Shepard constructed a mechanical device. Later he was able to use the new digital computer that Bell Telephone acquired. It may be seen that the 'true' shape of generalization gradients revealed by this method were quite different from those assumed 20 years earlier by Spence (cf. Figure 7.5).

Behavioral Contrast and Errorless Discrimination Learning

The pigeon lab that Skinner and Ferster had set up at Harvard attracted an increasing number of graduate students. They arrived to discover that Skinner had long since given up active participation in research and spent most of his time at home writing books. In place of Skinner, his former Ph.D. student, Richard Herrnstein became adviser to many of these students; see Chapter 8. One such student was George Reynolds. His thesis experiments concentrated on the question of whether the increase in pecking at, say, a red key that occurs when interrupted by periods in which pecking at a green key is not reinforced, that is, *behavioral contrast*, is a function of the *decrease* in responding that occurs to the green key. On the basis of several experiments, he rejected this possibility. Instead, Reynolds proposed that the rate at which a pigeon pecks at a given stimulus depends on the relative rate with which this behavior is reinforced: "The frequency of reinforcement in the presence of a given stimulus, *relative to the frequency during all the stimuli that successively control an organisms' behavior*, in part determines the rate of responding that the given stimulus controls."[96] This was a conclusion consistent with his adviser's emerging theory; Herrnstein was thanked in print.

Intriguingly, Herrnstein was not thanked in print by a later student who also obtained his Ph.D. on the basis of research carried out in the Harvard pigeon lab. In his account of behavioral contrast, Herb Terrace championed the factor that Reynolds had rejected, namely, the decrease in responding that occurs to the S- stimulus when a discrimination is introduced. Terrace did so on the basis of experiments that became highly influential, particularly for applications of operant conditioning. Once again following a brief report by Skinner,[97] Terrace's *errorless discrimination* training worked in the following way. First, pigeons were trained to peck at a red key and then, when steady responding on a variable-interval schedule of reinforcement had been obtained, a second stimulus was faded in. The fading procedure started with insertions of 5-second interruptions in which the response key was dark and then proceeded in several steps over three sessions; these interruptions became progressively longer and the key became progressively greener, until the stage was met when 3-minute periods in which pecking at the red key was reinforced alternated with 3-minute periods when the response key was "a fully bright green" and no reinforcement was available. In addition, during the first session of this fading procedure, the first 25 interruptions were introduced when Terrace judged that a pigeon was not in a good position to peck at the key.

In the first of many such experiments, the three pigeons given such training soon after steady pecking to the red key had been obtained made a total of 5–9 'errors'; that is, pecks to the green key. These compared to the total of 1,922–4,153 errors made by a comparison group of three pigeons that had been trained in the standard way of abruptly introducing three-minute periods of when the key was a bright green and no reinforcement was available. The latter group tended to show increases in responding to the S+, whereas this was not seen in the pigeons given errorless training. Consequently, Terrace noted that "the data from the present study, especially the close correlation between the occurrence of responding to S- and the increase in the S+ rate of responding, suggest that behavioral contrast results from the occurrence of unreinforced responding to S-, rather than the acquisition of a discrimination *per se*."[98] The small number of pigeons and the variability of the data make it appropriate that this was just a 'suggestion' and not a firm conclusion.

His experiments at Harvard gained Terrace an academic appointment at Columbia University in New York, where for many years there had been a productive pigeon lab in the Psychology Department; see Chapter 8. Terrace ran further experiments on errorless learning that confirmed his earlier suggestion that behavioral contrast does not occur when a discrimination is acquired in this manner. In addition, he found that his errorless pigeons did not display the emotional behavior that conventionally-trained pigeons displayed when the S- appeared and that they did not produce a peak shift in a generalization test. "These differences in performance led to the conclusion that S- functions as an aversive or inhibitory stimulus following discrimination learning with errors and as a neutral stimulus following discrimination learning without errors."[99]

The flurry of experiments in other laboratories that were inspired by Terrace's reports of errorless learning did not consistently support his theory. In particular, studies carried out in Mark Rilling's lab at Michigan State University suggested that the way Terrace interpreted his results was misleading. "The research by Rilling and his students demonstrates that the manner of presentation of conditions during discrimination learning, not the production of errors, determines the so-called by-products of discrimination learning."[100] Furthermore, the apparent advantages of errorless learning, as championed, for example, by Skinner in the context of programmed instruction, may not be real. "The organism that has acquired the discrimination without errors is retarded in detecting changes in the response-reinforcer relationship in the presence of S- as compared to an organism that mastered the discrimination with errors. When inhibition is measured with a resistance-to-reinforcement procedure, more conditioned inhibition is obtained for the birds that acquired the discrimination without errors than for the birds that acquired the discrimination with errors. Therefore, errorless learning is clearly not the best learning for an organism exposed to a changing environment."[101]

The increasing number of psychology departments in which behavioral contrast or related phenomena were studied included the new Laboratory of Experimental Psychology at the University of Sussex. Here Terry Bloomfield ran experiments to test his 'response suppression' theory, one closely related to Terrace's account.[102] Around the same time, Sebastian Halliday and I ran experiments in which pigeons needed to discriminate between a stimulus signaling that they needed to peck at a key to obtain the reinforcer on a variable-interval schedule and a second stimulus signaling that the reinforcer would be delivered at the same rate independently of their behavior, a schedule that became known as a *variable-time* (VT) schedule. No behavioral contrast was obtained under these conditions. We suggested that this was because the decline in responding produced by a VT schedule does not involve response inhibition.[103]

Whereas in almost all pigeon experiments related to behavioral contrast used pecking at an illuminated response key as the target response, a Sussex graduate student, Fred Westbrook, trained many of his pigeons to press a lever to obtain the standard 3-second access to grain. There were two groups of pigeons in his most important experiment: A lever-pressing group and a key-pecking group. Both groups were trained on multiple schedules in which a noise was present during one 30-second component and a tone during the second such component. In Stage 1 (baseline) responding, whether key-pecking or lever-pressing, was reinforced on a variable-interval 1-minute schedule in both components. Stage 2 consisted of discrimination training in which responding in one component was no longer reinforced; that is, a *mult* VI-1-min Extn schedule. In Stage 3 baseline conditions were restored by resuming reinforcement during the stimulus that had signaled the absence of reinforcement in Stage 2.

The key result was that the lever-pressing birds displayed negative induction in Stage 2, in that response rates in the S+ component declined. This effect was not seen in the key-pecking birds. If anything, this group showed a behavioral contrast effect, albeit a very small one compared to that normally obtained when the discrimination is between visual stimuli displayed on the response key.[104] The performance of bird No.1718 from the lever-pressing group is shown in Panel (a) of Figure 7.20 and can be compared to data from the five birds in the key-pecking group shown in Panel (b).

By the mid-1960s there was even a pigeon lab at the University of Pennsylvania. However, the research performed there did not focus on operant behavior but rather on the newly-discovered phenomenon of *autoshaping* that later acquired the more appropriate and general name, *sign-tracking*. As described in Chapter 6, a key finding by Williams and Williams was that a hungry pigeon would develop pecking to a visual signal displayed on a response key if that signal was reliably correlated with the delivery of grain, even when an omission schedule imposed from the start of training meant that pecking was never immediately followed by the reinforcer.[105] This and subsequent experiments from the same laboratory firmly established that sign-tracking in the form of pecking at the signal on the response key was based entirely on Pavlovian conditioning. This provided the basis for two of Williams' students, Elkan Gamzu and Barry Schwartz, to explore the possibility of interactions between operant and Pavlovian conditioning in the kind of multiple schedule that produces behavioral contrast. They came up with the *additive theory* of behavioral contrast.

Additivity theory is best illustrated by an experiment reported by Schwartz that expanded on a neat idea from Keller.[106] This was to compare the standard procedure for discrimination learning in pigeons, one using visual stimuli superimposed on the response key, with a procedure in which these stimuli are superimposed

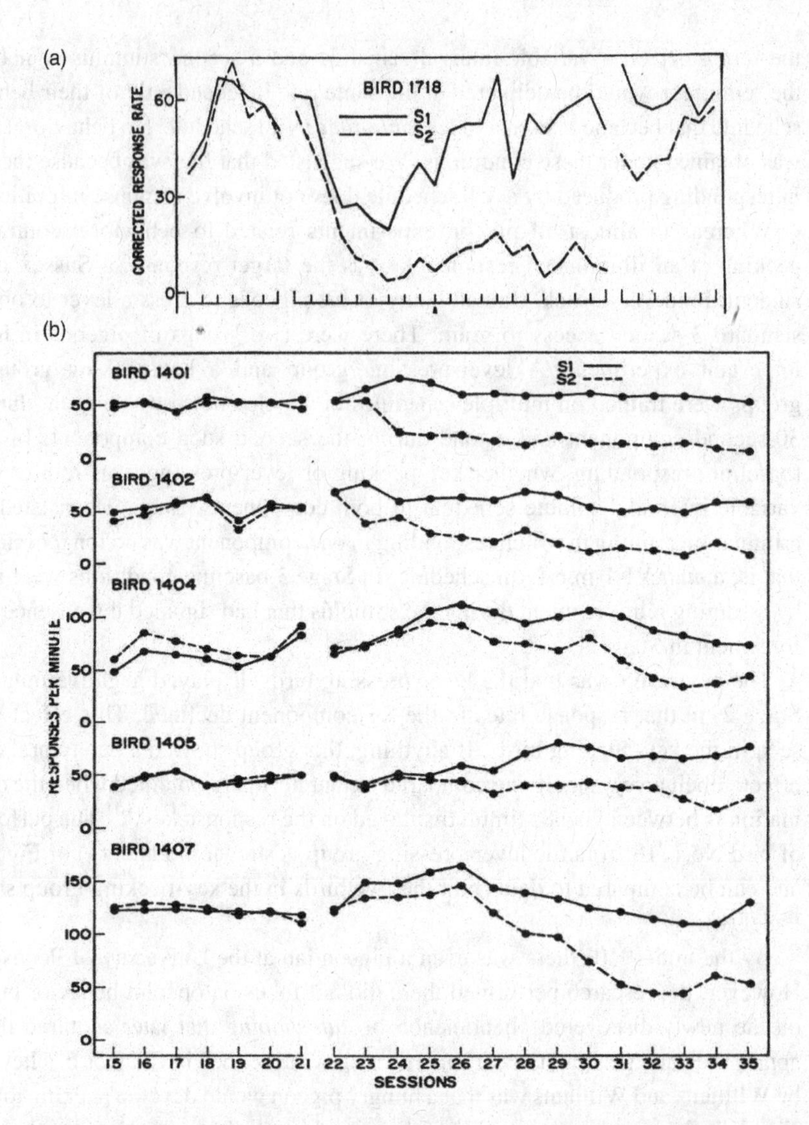

Figure 7.20 These data from individual pigeons revealed that a behavioral contrast effect fails to result from discrimination training when auditory stimuli are used instead of the standard visual stimuli projected on a response key. Panel (a) Data from a pigeon first trained *to press a lever* to obtain food on variable-interval schedule and then given discrimination training, whereby responses were reinforced in the presence of S1 (noise) but not in the presence of S2 (tone). Panel (b) Data from five pigeons trained to peck a response key for food and then given the same discrimination training between S1 (noise) and S2 (tone). From Westbrook (1973). Reproduced with permission from John Wiley & Sons.

on a second key. Pecking at this second, *signal* key has no consequences. Schwartz used a multiple schedule in which, for example, the center response key – the one that the pigeon needed to peck at to obtain the reinforcer – was always white and the signal key to the side was either green, indicating that pecks to the center key would be reinforced, or blue, indicating that such responses would *not* be reinforced. Such conditions produced considerable pecking at the signal key when it was green which was accompanied by a reduction in responding to the center key. For Schwartz signal-key pecking was produced by Pavlovian conditioning, that is, was an example of sign-tracking, and this behavior interfered with instrumentally conditioned responding on the center key. When under otherwise identical conditions the green and blue stimuli were projected onto the center key, a standard behavioral contrast effect was obtained. For Schwartz, this confirmed the idea that behavioral contrast resulted from Pavlovian-based pecking being superimposed on instrumentally based pecking.[107]

Additivity theory explained a wide range of evidence from experiments on this type of discrimination learning, as reviewed by Schwarz and Gamzu.[108] The theory has not been seriously challenged since then and their review marked a steep decline in experiments on behavioral contrast. The broader contribution of this body of research was to raise the important possibility that many of the phenomena that had suggested the need for species-specific laws of learning might – like behavioral contrast – be explained in terms of interactions between instrumental and Pavlovian processes. "In many areas, the study of key pecking in pigeons, bar pressing in pigeons, and bar pressing in rats yield different results. Nevertheless, these differences can be interpreted and understood in terms of learning principles that are already well established. They require no new formulation – only new combinations of old formulations. Whether similar analyses will provide sufficient explanations of other biologically constrained phenomena is an open question."[109]

8 B.F. Skinner and the Experimental Analysis of Behavior

Skinner (1904–1990) is still famous. Turn to the chapter on learning in any introductory textbook on psychology and you are likely to find the names of only two scientists who have carried out experiments on the learned behavior of animals: Ivan Pavlov and B.F ('Fred') Skinner. Skinner's fame is partly based on his pioneering experiments on operant conditioning and the invention of what he named an *operant chamber* and is widely known as a *Skinner box*. One enduring legacy is a behavioral technology that is in use in hundreds of behavioral neuroscience, behavioral pharmacology, and other kinds of laboratories around the world. This technology consists of both the Skinner boxes, now controlled by computers instead of the makeshift electro-mechanical devices that Skinner himself used, and the range of reinforcement schedules that he and his students first introduced.

A possible legacy with enormous consequences is the suggested use within the algorithms that respond to people's responses on social media of the reinforcement schedules and shaping techniques that Skinner first developed. The extent to which these algorithms have deliberately incorporated the principles of operant conditioning is not clear at present.

A further reason for Skinner's fame is that he established a distinctive research tradition, the imperialistically-titled *experimental analysis of behavior*. This label continues to be applied to a certain approach to experimental work and the label, *applied behavior analysis*, to applications of operant conditioning, most of which Skinner first suggested. Worldwide, there are many thousands of practitioners whose therapeutic approach is based on Skinner's ideas. In Australia, the most common form of 'applied behavior analysis,' or 'ABA,' is that applied to the behavior of children with autism. At the start of training for such work in Australia, would-be therapists are required to learn about Skinner's principles of operant conditioning, with little mention of other research that could be relevant to their future practice.

The peak of Skinner's fame among the general public was reached in the 1970s as a result of controversy over claims in his best-selling book of 1971, *Beyond freedom and dignity*,[1] which led to him being featured on the front cover of *Time* magazine. The key claim was that the idea that humans possess free will is an illusion; instead, their behavior is shaped since birth by a long series of reinforcing events. The book spelled out ideas that no other behaviorist since John Watson in the 1920s had tried to communicate to the general public, despite holding similar beliefs to Skinner's.

It is often difficult to find accounts of the early or personal lives of scientists working in the middle of the twentieth century. An internet search may uncover nothing more than a brief obituary notice. However, we know far more about Skinner's life than any other contemporary psychologist. The nineteen books he wrote included three volumes of autobiography. In addition, he published several articles about aspects of his life and work, topics that many other authors have written about, including a full-length biography of Skinner.[2]

The present chapter concentrates on Skinner's scientific work and that of researchers who worked within the tradition that he established. Nevertheless, it also includes an account of his life, his proposals as to how the principles of operant conditioning can be applied outside of the laboratory, the reasons for his huge influence in the middle of the twentieth century, and for its subsequent decline. What this chapter does not attempt is any analysis of his philosophical views or of his suggestions as to how society needs to change as first described in his novel, *Walden Two*, as well as thirty years later in *Beyond Freedom and Dignity*.[2]

Skinner's Youth

Fred Skinner grew up in the small railroad town of Susquehanna in northeast Pennsylvania. While his father worked as a lawyer, his mother made sure that Fred behaved in a way that would not prompt disapproval by the neighbors. Outside the house, Fred had great freedom, both to roam the surrounding countryside and to make an endless series of gadgets from the discarded bits and pieces he found around the house and in the neighbors' yards. Susquehanna was a god-fearing and hell-fearing town. On one very memorable occasion Fred's grandmother, wanting to make sure that the 5-year-old never told a lie, showed him "the coal fire in the heating stove and told that little children who told lies were thrown into a place like that after they died."[3] Nonetheless, as a teenager Fred let go his earlier Christian beliefs. He was left, however, with a strong Protestant work ethic.

Fred was lucky that two of his teachers in the local school appreciated his intelligence and potential, inspiring him both with a love of reading and to put effort into studying mathematics. On leaving school he was enrolled in Hamilton College, a small, very traditional males-only college near the town of Scranton in New York State, to which the Skinner family had moved after World War I. Intellectual accomplishment was a low priority at Hamilton College but Fred was again lucky in that in his second year he was invited to join a circle of men and women that met regularly to discuss books, art, and philosophy.

His interest and knowledge of literature grew throughout his final two years at Hamilton College. From a young age, Fred had continued to write stories and poems. Attending a summer school in Vermont on literature confirmed his interest in writing. There he met the poet, Robert Frost, who later read two stories that Fred had sent him. In reply, Frost encouraged Fred to pursue his ambition to become a writer.

Graduation in 1926 was followed by what Skinner later described as his Dark Year. Unemployed and frequently bored, Skinner lived with his worried parents and tried

Figure 8.1 B.F. Skinner on entering Harvard as a graduate student. From Skinner (1976). Reproduced with permission from the BF Skinner Foundation.

to write a novel. He subscribed to a number of literary magazines and, at a time when Skinner was reaching the conclusion that he had nothing to say, one magazine, *The Dial*, included some articles by the philosopher, Bertrand Russell. These led Skinner to Russell's 1927 book, *Philosophy*,[4] which discusses Watson's behaviorism. "Inspired by Russell, I bought Watson's Behaviorism. I lost interest in epistemology and turned to scientific issues."[5] That led to him buying the English translation of Pavlov's *Conditioned reflexes* which was also published in 1927. Skinner was hooked. He decided to carry out graduate study in psychology and applied successfully for enrolment in Harvard's Department of Psychology; see Figure 8.1.

Skinner's conversion to behaviorism was as immediate and complete as Saul's conversion on the road to Damascus to becoming Saint Paul. From his arrival at Harvard in the fall of 1928 until his death Skinner did not waver from his core belief that psychology should be based on a science of behavior that excluded all forms of 'mentalism' – or 'cognitivism' as the enemy was later named – and also made no reference to events within the brain. "The argument against physiology is simply that we shall get more done in the field of behavior if we confine ourselves to behavior. When we rid ourselves of the delusion that we are getting down to fundamentals, when we get into physiology, then the young man who discovered some fact of behavior will not immediately go after 'physiological correlates' but will go on discovering other facts of behavior."[6] Skinner also did not lack confidence in his ability to change psychology in this direction. Towards the end of his first year as a Ph.D. student, he wrote: "I have almost gone over to physiology which I find fascinating. But my fundamental interests lie in the field of Psychology, and I shall probably continue therein, even, if necessary, by making over the entire field to suit myself."[7]

Skinner as a Graduate Student and Postdoctoral Researcher

There was little sympathy for any form of behaviorism among Skinner's Harvard professors in the Department of Philosophy and Psychology. Its best days under William James were long past and now its most influential member was Edwin Boring, who directed its Psychological Laboratory. Boring had been a student of Edward Titchener, whose emphasis on private mental events had been the main focus for Watson's attack in 1913. Titchener in turn had studied in Leipzig with Wilhelm Wundt, widely acknowledged as the founder of experimental psychology as a scientific discipline. During Skinner's time as a graduate student, Boring both defended the value of introspection in psychology and constructed a graduate program that followed Titchener's training in Leipzig. Many of the elements, including Boring, were still in place over 30 years later, when I entered Harvard in 1963.

In several of the weekly colloquium meetings – research seminars – the assertive young graduate student refined his arguments for behaviorism in debate with Boring. Within Psychology Skinner's only support came from a fellow graduate student, Fred Keller, who was equally committed to behaviorism but more diffident than Skinner in promoting his beliefs.[8] Keller remained Skinner's lifelong friend and in the 1950s the most influential promoter of Skinner's ideas, as described later in this chapter.

Skinner's other support came from the Department of Biology. William Crozier had been appointed as head of a new Department of General Physiology in 1925 and had been promoted to a full Professorship two years later. Ambitious to expand his empire, Crozier encouraged Ph.D. students from psychology, as well as students in biology and medicine, to carry out experiments in his department. As well as research on the sensory systems of various species, Crozier encouraged students to carry out purely behavioral experiments. Thus, one of Skinner's early experiments, one that led in 1929 to his first research publication, was on the movement of large black ants when placed on a plane that was placed at an increasingly steep angle.

At least two features of the Psychology Laboratory meant that this was where Skinner spent most of his time. One was its workshop. The temporary absence of a technician meant that graduate students had complete freedom there to hone their practical skills. Skinner was already good at using hand tools to construct various gadgets and had even built detailed model sailing ships during his Dark Year. Now he could learn how to use power tools. The second feature was a small lab where another graduate student had started to test rats in mazes just before Skinner arrived at Harvard. Soon Skinner too was working with rats, as well as on a project that tested problem-solving in squirrels.

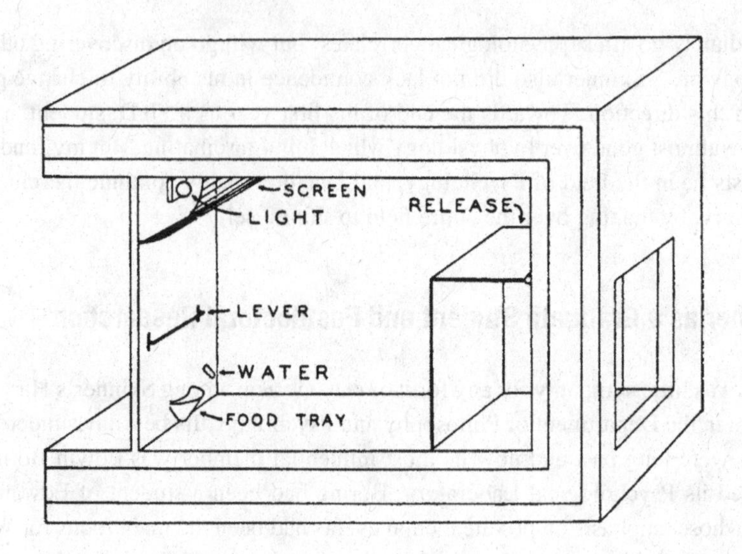

Figure 8.2 The first Skinner box.
From Skinner's *The behavior of organisms* (1938). Reproduced with permission from the BF Skinner Foundation.

Testing rats in a maze can require placing an animal in the start box, measuring the time it spends using a stopwatch, and retrieving the rat at some point after, say, it has had time to consume food in the goal box. Partly to minimize and eventually remove any need for handling, Skinner devised a series of gadgets and modifications, including a recording device built with the help of a newly arrived technician, Ralph Gerbrands. Skinner's aim was to develop a piece of apparatus that would produce highly reproducible data. Following Pavlov, he wanted to ensure that no external events influenced his rats' behavior and so he placed his equipment inside a sound-attenuating chamber. Unlike the variable behavior obtained from rats in a maze, yielding data that required averaging in order to find meaningful results, Skinner wanted to be able to obtain results from a single rat that would be essentially identical to those from every rat tested thereafter – or from any other kind of animal. His tinkering eventually resulted in a small box with a lever mounted on one wall and a device that would deliver a food pellet when the lever was operated. This was the first Skinner box; see Figure 8.2.

After working with just a few rats in the new equipment and still early in his third year as a Ph.D. student, Skinner came to believe that the kind of learning he was studying could not be fitted within the framework of Pavlov's conditioned reflex. What especially impressed him was the speed with which a rat that had already been fully acclimatized to the chamber and to receiving a freely-delivered food pellet would then learn to press the lever to obtain a pellet; in some rats, a single experience of a lever press being followed immediately by the click of the mechanism that delivered a pellet was all that was needed to produce a long series of presses when no more pellets were made available; see Figure 8.3. Early in 1932 the first of a series of papers by Skinner were published that reported his experiments on this new kind of 'reflex.'[9]

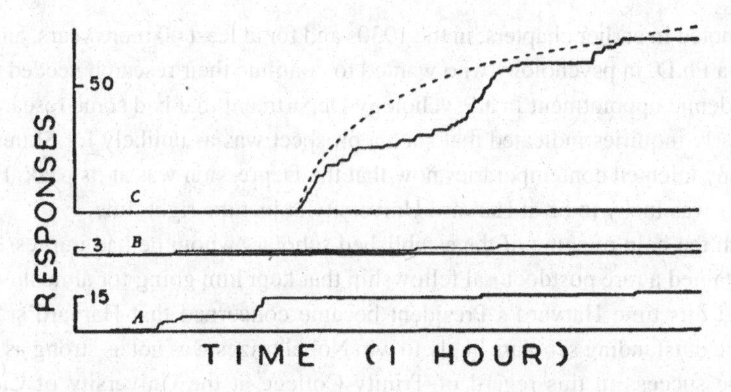

Figure 8.3 This example of a cumulative record from *The behavior of organisms* shows responding in extinction after a single lever-press had been reinforced.
From Skinner (1938). Reproduced with permission from the BF Skinner Foundation.

Many of the early studies of conditioning carried out by Pavlov and his students concentrated on extinction. In contrast, most researchers in the USA in the 1930s that were inspired by Pavlov continued to study how rewarded behavior was learned but not what happened when reward was discontinued. Skinner was an exception following a happy accident: While a rat was being tested, a failure in the mechanism for delivering its food pellets produced a steady decline in the rate at which it pressed the lever. Skinner immediately realized the importance of this phenomenon and further – now intentional – studies of extinction led to a major discovery. According to his account, a shortage of pellets meant that he needed to eke out his supply over a weekend when he continued to train his rats: "If I reinforced only an occasional response, my supply would last for many days."[10] The result, *periodic re-conditioning*, was the first of many reinforcement schedules to follow. As he reported to a meeting in September, 1932, with exposure to this schedule "the reflex assumes a constant strength, which is maintained without significant modification for as many as thirty experimental hours."[11] When extinction was at last introduced, he found that his rats would continue to press the lever for an unusually long time.

Skinner once met Pavlov. In 1929 the International Congress of Psychology was held in the Harvard Medical School. Pavlov was invited to give the principal address, even though he was now only a few weeks short of his 80th birthday. His lecture was in German, a language that Skinner and most of the audience could not follow. His main achievement from this meeting was to obtain Pavlov's signature.[12] Skinner tried to keep abreast of the experiments being carried out in St. Petersburg. He learned that Pavlov and his team were also studying 'periodic re-conditioning' by giving a dog access to food only on a proportion of the trials in which the conditioned stimulus was presented. They did not find unusual resistance to extinction as a result. This turned out to be typical of most forms of Pavlovian conditioning.[13] In contrast, partial reinforcement has been found to promote the persistence in extinction of all types of instrumental conditioning, as first accidentally discovered by Skinner.

As noted in earlier chapters, in the 1930s and for at least 40 more years, anyone finishing a Ph.D. in psychology who wanted to continue their research needed to obtain an academic appointment in a Psychology Department that had some research facilities. Early inquiries indicated that such a prospect was as unlikely for Skinner as for his many talented contemporaries now that the Depression was at its peak. However, Skinner was lucky to be at Harvard University at just the right time.

With the help of some of the established scholars whom he had impressed, Skinner obtained a rare postdoctoral fellowship that kept him going for almost two years. Around this time Harvard's President became concerned that Harvard's ability to produce outstanding scholars likely to win Nobel prizes was not as strong as it should be. The success in this regard of Trinity College at the University of Cambridge was presumed to be based on its tradition of giving generous fellowships to young scholars who showed outstanding promise. Harvard adopted this model by founding its Society of Fellows. There were Senior Fellows, who were already eminent in their fields, and Junior Fellows, who were supposed to be no more than 24 years old. Even though now 29, Skinner belonged to the first group to be appointed. Another member of this group was the to-become-eminent philosopher, Willard Van Quine.[13] The generous fellowship allowed Skinner complete freedom to do what he liked for a further three years.

Regular meetings of the Society of Fellows provided intellectual stimulation from a variety of sources. An argument about behaviorism with the elderly philosopher, Alfred North Whitehead, led to a challenge: How can a behaviorist explain the following sentence that Whitehead – and possibly no one else – had ever previously uttered, "No black scorpion is falling upon this table?" Skinner took up the challenge and started to draft a book he decided to call *Verbal behavior*. It took more than 20 years for him to complete this book. Most of his time as a Junior Fellow was spent in his rat laboratory.

John Watson had announced that the new science of behavior was to be based on the conditioned reflex. Skinner agreed: "I was convinced that the concept of the reflex embraced the whole of psychology."[14] But his concept of a reflex was very different from the traditional one. As a graduate student, many of the hours not spent by Skinner in his lab, were spent in the library instead. As well as major works on the reflex, his reading included books by two physicists, the nineteenth century German, Ernst Mach, and the twentieth century American, Paul Bridgeman. Both argued for a different view of science from the traditional one. Skinner adopted their *operationism*, the belief that science should be based on the description of natural phenomena, without any appeal to theoretical concepts, and applied it to the concept of a reflex. As he argued in his Ph.D. thesis, a reflex is not to be understood in terms of its presumed neurological basis, it is simply an observed "correlation between a stimulus and a response."[15]

Where Watson had proposed that the new science should be based on Pavlov's conditioned reflex, Skinner now proposed that it should be based on two types of conditioning: Type S was the kind Pavlov had studied and it applied to *respondent* behavior, which was of less importance than *operant* behavior that is produced by Type R conditioned reflexes. In this second type a discriminative stimulus, an S^D, does not *elicit* a response, but rather "sets the occasion upon which a response will be reinforced."[16]

The Behavior of Organisms of 1938

Skinner began his first book by spelling out his theories and particular version of behaviorism. Then, the bulk consisted of reports of the large amount of experimental work he had completed during his years as a full-time postdoctoral fellow. The contents owed a lot to Pavlov. The debt included terms like *reinforcement* and *induction*, as well as the idea that reliable and important behavioral data can be obtained from just a few representative animals, as long as they are tested in a closed environment that excludes most stimulation from the outside world. The debt also extended to the topics Skinner explored. His intention for *The behavior of organisms* can be seen as producing for operant conditioning the equivalent of Pavlov's *Conditioned reflexes* of 1927.

The skills Skinner acquired when an aspiring writer served him well in his later career. Most of the many books he wrote are very readable. His first is not one of them. Many sentences are decidedly clunky, as in this ending to the first chapter: "The rule that the generic term may be used only when its experimental reality has been verified will not admit the possibility of an ancillary principle, available in and peculiar to the study of behavior, leading to the definition of concepts through some other means than the sort of experimental procedure here outlined."[17] In addition, readability is not helped by constant requests that the reader examines figure after figure illustrating what an individual rat did on a particular day. A persistent reader will, however, discover in *The behavior of organisms* both seminal findings and most of the arguments that Skinner was to maintain throughout the rest of his life.

Of the many discoveries or developments Skinner described in 1938 that have been almost universally respected ever since, several stand out. A conceptual advance is to define responses and reinforcers in terms of their function. Thus, a lever-press response is any set of a rat's movements that result in depression of the lever; a reinforcer is any event that, when contingent of a response, increases the probability of that response recurring.

The most important theoretical contribution is the distinction between Pavlovian and operant (his term) or instrumental (the more general term) conditioning that, independently, Konorski and Miller were drawing at around the same time (see Chapter 1). An important empirical and methodological contribution is the first development of reinforcement schedules whereby a response is intermittently reinforced on the basis of either a fixed interval – termed *periodic reconditioning* in *The behavior of organisms* – or on the basis of a fixed number of responses between reinforcements, the *Fixed Ratio* schedule. Furthermore, the discovery that rats trained on an intermittent schedule continue to respond in extinction for far longer than when every response had been reinforced is of enduring theoretical and practical importance. Other important discoveries or developments include the chaining of responses and the capacity of a discriminative stimulus to serve as a secondary reinforcer.

Also very influential but less widely accepted in the long run are Skinner's dismissal both of any concept of inhibition and what he termed then as 'negative reinforcement' – more generally known as 'punishment'- as being the opposite of positive reinforcement. One of the justifications Pavlov had given for introducing the concept of inhibition was the phenomenon of disinhibition: After his dogs had undergone some extinction treatment, an unexpected event – for example, a door banging – could lead to a brief revival of salivation to whatever conditioned stimulus was then presented. Skinner put a lot of unsuccessful effort into obtaining a similar effect during extinction of previously reinforced lever-pressing. The unexpected events he introduced ranged from the mild – turning on lights above the lever – to the dramatic intervention of tossing a rat into the air; but nothing worked to reinstate lever-pressing. A related failure to document a phenomenon that Pavlov had described – that of transfer of the effect of a conditioned inhibitor to another conditioned stimulus –provided another reason for dismissing inhibition. As for punishing a rat for pressing its lever by delivering either a light mechanical slap or a brief electric shock, the limited data Skinner obtained convinced him that the resulting decrease in lever-pressing did not reflect the opposite process to that of positive reinforcement; instead, he suggested that such events produce an emotional state that interferes with pressing the lever.

The penultimate chapter of *The behavior of organisms* contains a sustained argument that was widely accepted for at least four decades by researchers studying the learned behavior of animals, even by the many who did not accept most of Skinner's other arguments. In place of what he saw as "the all but universal belief that a science of behavior must be neurological in nature," he argued that a science of behavior would make more progress if it avoided attempts to explain behavioral phenomena in terms of assumed events within what Skinner referred to as the 'Conceptual Nervous System.' He pointed out that, although the subtitle to Pavlov's 1927 book was *An investigation of the activity of the cerebral cortex*, no direct observations of the cortex are reported. Skinner was not just criticizing Pavlov but also contemporaries such as Hull. "I am asserting, then, not only that a science of behavior is independent of neurology but that it must be established as a separate discipline whether or not a rapprochement with neurology is ever attempted." He accepted that in some probably distant future, there could be a productive relationship between the two different sciences of behavior and of neuroscience. "I am not overlooking the advance that is made in the unification of knowledge when terms at one level of analysis are defined ('explained') at a lower level. Eventually, a synthesis of the laws of behavior and of the nervous system may be achieved, although the reduction to lower terms will not, of course, stop at the level of neurology."[18]

The behavior of organisms also contains proposals that were accepted only by dedicated followers. Until well into the 1960s to be a Skinnerian researcher meant employing just a few animals in an operant chamber, using their rate of responding as the only acceptable measure of behavior, reporting such data by reproducing samples of cumulative records, excluding any kind of statistical analysis and avoiding any temptation to try to 'explain' the results.

Toward the end of the book, Skinner argues that a science of behavior needs only to describe phenomena and should not attempt to explain them. He approvingly quotes the physicist, Ernst Mach, who had advocated such *radical empiricism.*[19]

From Rats to Words and Pigeons

As previously noted, in the mid 1930s obtaining an academic position remained a challenge, even for someone with a Harvard Ph.D., exciting new ideas, and an impressive research output. It did not help if one had little teaching experience, a reputation for arrogance, and a hypercritical attitude to others with different views on psychology. Nevertheless, with the unexpected help of a generous reference from Boring, in 1936 Skinner was appointed as an Instructor in the Department of Psychology at the University of Minnesota. The other major change to his life that year was getting married. Following a series of sexual and often intense relationships, he had met Yvonne Blue when she was on a visit to Cambridge from Chicago a few months before Skinner was due to leave Harvard. A few months later their wedding in Chicago began a marriage that lasted until Skinner's death in 1990, despite bringing only occasional joy to his wife. Yvonne later changed her name to Eve.[20] Seven years younger than Skinner, she died in 1997, also at the age of 86, after working for many years as a lecturer and gallery instructor for the Museum of Fine Arts in Boston.

Skinner brought with him to Minnesota four conditioning chambers. Their numbers were already increasing. Also built by Gerbrands following Skinner's design were two chambers that worked entirely mechanically; these were for Keller to take when he left Harvard. Clark Hull obtained Skinner's approval to have such chambers built at Yale and promptly named them 'Skinner boxes.'

In 1937 Skinner obtained a grant of $150 to study "the avoidance of undesirable consequences." This allowed him to develop a way of delivering shock to his rats via the grid floor of the conditioning chamber. Some rats learned that by pressing the lever more than once every 15 seconds they could prevent the delivery of a shock.[21]

Soon after arriving in Minnesota, Skinner began to collaborate with William Heron, a colleague within the department. Their experiments showed that both caffeine and Benzedrine increased the rate at which rats pressed a lever.[22] This was the first study to use responding on some operant conditioning schedule to assess the behavioral effects of a drug. They were completed in time to be reported towards the end of the *Behavior of Organisms*.

Skinner and Heron also collaborated in applying for a research grant. Their success enabled them to set up a set of twenty-four Skinner boxes. Heron was interested in testing potential differences between two strains of rats that he had selectively bred on how quickly they learned to make their way through a maze. The two strains were compared on their performance on a fixed interval schedule of reinforcement – still referred to as 'periodic reconditioning' – and subsequent extinction. No very interesting differences between the two strains were found. The study involved nearly one hundred rats and statistical analyses of the data. This was not to Skinner's taste.

An important concept described in *The behavior of organisms* was that of the *reflex reserve*. The basic idea is that, each time a response is reinforced, this increases the number of responses that an animal will make in a subsequent extinction stage. Skinner described this as a 'hypothetical entity'[23] and, as such, it can be seen as inconsistent with his argument that a science of behavior should be entirely descriptive. The large set of conditioning chambers now available in his Minnesota laboratory allowed him to carry out further tests of this hypothetical entity of a reflex reserve. The results convinced him to abandon this idea. Around this time a Ph.D. student at Stanford University, using what may well have been the first Skinner box on the West Coast, tested the idea of the reflex reserve and also decided it should be rejected: "The size of the reflex reserve was found to be unmeasurable, and statements expressing a relationship between the size of the reserve and other variables cannot be verified."[24]

In Minnesota what was now the biggest operant laboratory in the world was also used for a particularly important study. Although Skinner did not yet have the status that allowed him to act as the official advisor to graduate students, a number of them were attracted to work with him. One was Bill Estes, who, as Skinner described it, "began at once to do brilliant work." The "brilliant work" consisted of a series of experiments that used the electric shock device that Skinner had invented; Skinner himself disliked giving rats shocks.[25] The work followed some experiments reported in *The behavior of organisms* on what was initially labeled the *conditioned emotional response* or *CER*, and subsequently *conditioned suppression*.[26] These were followed by Estes' experiments on punishment which are described in Chapter 4.

Skinner himself spent less and less time in the laboratory. One distraction followed the birth of the couple's first daughter, Julie, in 1938. Yvonne did not cope very well with motherhood and, by the standards of that time, her husband became an unusually participant parent.[27] The other involvement that took him out of the lab was increasing excitement over his work on language which ultimately led to the publication of *Verbal Behavior* almost two decades later.

In 1940 the Second World War was underway in Europe and, although the USA was yet to become involved, this prospect seemed increasingly likely to many Americans, including Skinner. While on a train from Minneapolis to Chicago, Skinner was thinking about the horrors of bombing when "I saw a flock of birds lifting and wheeling in formation as they flew alongside the train. Suddenly I saw them as devices with excellent vision and superb maneuverability. Could they not guide a missile?" As soon as he was back in Minneapolis, Skinner began to work on the problem.[28]

Project Pigeon lasted three years and involved an enormous amount of work on the part of Skinner and also of the several students who joined him on the project. In addition to Estes, the other main participants were Howard Hoffman and Keller Breland. The project gave unlimited opportunities for Skinner to invent new gadgets and test out new methods of behavioral control. Funding was initially limited but it increased after Pearl Harbor and the entry of the USA into the Second World War. The General Mills company gave him a grant of $5,000 and the use of the top floor of a flour mill. Skinner began a sabbatical in January 1942 and spent almost all his time there, along with his select group of students. Elliot, the chair of Psychology

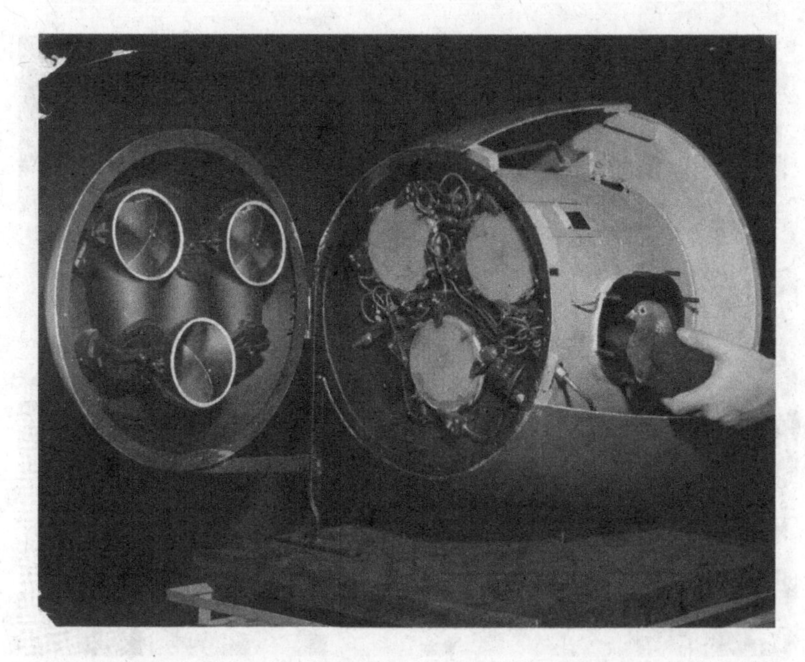

Figure 8.4 This is a mock-up of a missile that was intended to be guided by three pigeons pecking at the windows in front of them, as trained by Skinner and his team.
From Skinner (1979). Reproduced with permission from the BF Skinner Foundation.

noted: "The group lived, ate, and slept the project, and about that time Skinner seemed to some of us to become what might be called 'exalted' … The students … together with their leader … felt that ordinary pedestrian work of the kind we were doing in out war-ridden university just didn't count."[29]

Early in 1943 progress with the project was impressive enough for the National Defense Research Council (NDRC) to award $25,000 for the project. Development then reached a point where it was clearly feasible to have three pigeons, harnessed together and each able to peck at its screen when on a lean reinforcement schedule, guide a bomb to its target; see Figure 8.4. And then all financial support ended. The NDRC decided that developments in electronics had more of a future than the use of guidance by pigeons.

In the long run, what was more important than Project Pigeon was the experimental work Skinner and his students did during the pauses in the project while waiting for decisions on further finance. In one of his frequent letters to Fred Keller, Skinner wrote: "We are fast filling a sizeable book. Almost complete repetition of the B of O (with a different species) plus oodles of new stuff. Some work on discrimination that would put your eye out. A new trick with *variable* periodic reinforcement that eliminates the steplike effect of a temporal discrimination. Etc. Etc. and *beautiful* records. Everyone like an average of four to twelve rats."[30] The team developed their skills in shaping complex behavior sequences. One example was teaching a pigeon to play a simple tune on a four-key 'piano.' Another pigeon was trained to 'bat' a small wooden ball around a small wooden arena.

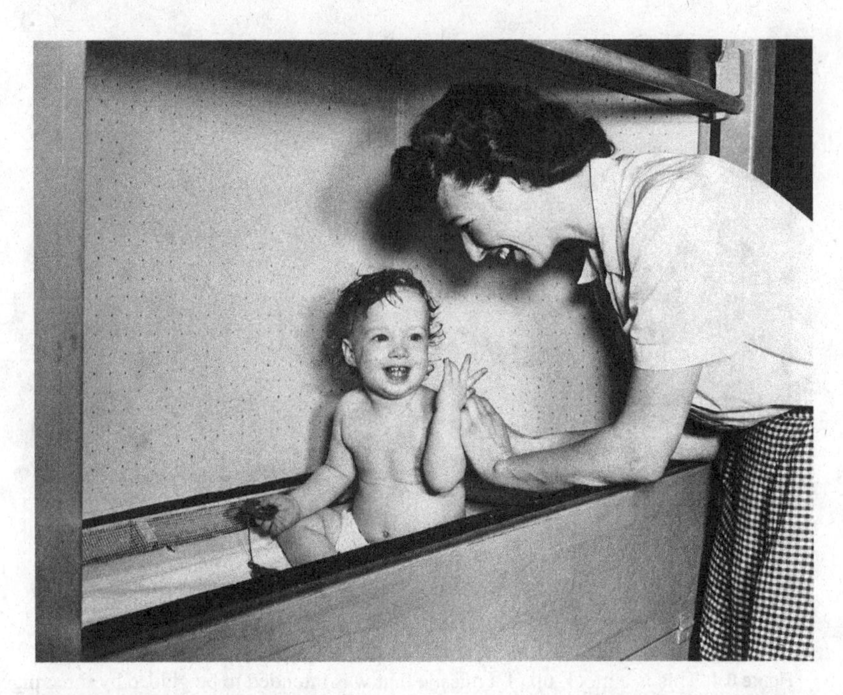

Figure 8.5 Skinner's wife, Yvonne, with their first daughter, Debbie, in the first 'baby tender' (later versions were named 'air cribs') that Skinner designed and built, as reported in Skinner (1945).
Image credit: Bettmann / Getty Images

Skinner decided that "the pigeon should be more widely used as an experimental subject. It has excellent color vision, it lives for a long time, its level of hunger or thirst can be easily controlled, and because it has been domesticated for several thousand years, it is resistant to human diseases."[31] To use pigeons was a further departure from the tradition of experimental medicine in which Pavlov had been trained. His use of dogs as subjects in his experiments on conditioning reflected the strong belief that his results would lead to insights into the workings of the human brain, just as his earlier research on the heart and then the digestive system of dogs would increase understanding of these organs in humans. The brain of the rat used by American learning theorists has at least many similarities to that of its fellow mammalian, *homo sapiens*. In contrast, the brain of a bird is structured in totally different ways from any mammal. If, however, like Skinner, one believes in a science of behavior that is entirely independent of neuroscience, it is of no consequence how an organism's nervous system is structured.

Skinner would have achieved a great deal of notoriety from *Project Pigeon* if it had not remained secret for several years. In the meantime, he achieved fame for constructing what he referred to as a 'baby-tender' that was used from the time his second daughter, Debbie, was born in 1944. This was essentially an air-conditioned space, maintained at a temperature that meant that Debbie had no need of any clothes, other than a diaper, with a large window that allowed her to watch what was happening in the room outside; see Figure 8.5. The article he wrote about this invention was re-titled by an editor as *Baby in a Box* and published in the Ladies Home Journal.[32] To his surprise, it prompted many critics to accuse him of subjecting his daughter to an experimental life in a Skinner Box and of increasing the risk that she would become a neurotic adult. He wrote later that "although the baby-tender would have been an ideal place for experiments, I did very few."[33] Debbie, like her older sister, grew up into a well-adjusted and healthy adult. Both sisters retained a warm relationship with their father throughout their lives.[34] There were more positive than negative responses to the article and Skinner was encouraged to make money from his invention. After many mishaps it was eventually marketed as an 'air-crib' and perhaps a thousand were sold.[35]

Skinner at Indiana University

Skinner's salary in Minnesota had remained low, when he learned of the possibility of a much higher income on becoming a full professor and chair of the Psychology Department at Indiana University. He moved there after a year on a Guggenheim Fellowship that was for developing his ideas on verbal behavior. "I also tackled theoretical issues, including the redefinition of some traditional terms. *Remembering* could mean either responding to a fragment of a discriminative stimulus or to the original stimulus after a period of time. *Believing* referred to the strength of one's behavior with respect to a less-than-adequate controlling stimulus."[36] He also managed to write an utopian novel, *Walden Two*, in seven weeks, run experiments with pigeons that tested their long-term retention of a discrimination, and spend a lot of time with his daughters; see Figure 8.6.

A highly influential report of an experiment he carried out at Indiana was titled *Superstition in the pigeon*. It was based on the idea that a reinforcing event that occurred immediately after a particular response would strengthen that response, whether or not the response had actually caused the reinforcer to arrive. Skinner placed a hungry pigeon in a conditioning chamber and simply arranged that food would be delivered at regular intervals. With an interval of 15 seconds between food deliveries, within a few such sessions six out of eight birds developed highly stereotyped repetitive patterns of behavior. "One bird was conditioned to turn counterclockwise about the cage, making two or three turns between reinforcements. Another repeatedly thrust its head into one of the upper corners of the cage. A third developed a 'tossing' response, as if placing its head beneath an invisible bar and lifting it repeatedly.... The conditioning process is usually obvious. The bird happens to be executing some response as the hopper appears; as a result, it tends to repeat this response. If the interval before the next presentation is not so great that extinction takes place, a second 'contingency' is probable. This strengthens the response still further and subsequent reinforcement becomes more probable. It is true that some responses go unreinforced and some reinforcements appear when the response has not just been made, but the net result is the development of a considerable state of strength." Skinner suggested that the same process was at work in the development of human superstitious behaviour; for example, that demonstrated by a bowler.[37]

Figure 8.6 Skinner with a demonstration pigeon at Indiana University, 1948.
From Skinner (1979). Reproduced with permission from the BF Skinner Foundation.

This idea was not seriously challenged until the early 1970s. Two discoveries, a theoretical development and a systematic replication of Skinner's original experiment were particularly important for the general abandonment of his concept of superstition.[38] One discovery was that of *schedule-induced drinking* or SID. John Falk was carrying out an otherwise standard experiment on lever-pressing by hungry rats that was reinforced by the delivery of a food pellet on a variable-interval schedule. What was unusual that the rats could drink at any time from the spout of a water bottle that was inserted into the chamber. Falk found that his rats would come to drink a large amount of water, even though they were not all deprived of water in their home cages. Under some conditions, the amount consumed was so excessive that Falk referred to the phenomenon as a form of *polydipsia*.[39] It seemed unlikely that his was a result of superstitious conditioning in that drinking was most likely to occur when a pellet was least likely to be delivered. Subsequent experiments by Falk and other researchers established that SID was a highly robust phenomenon. Other forms of schedule-induced behavior were also identified, though none as robust as SID. Falk named these, as well as SID, as examples of *adjunctive* behavior and drew parallels with the displacement activity of the kind identified by ethologists.[40] These studies appeared to demonstrate that not all behavior that an animal displays in a Skinner box is a result of operant conditioning.

The other discovery was that of *autoshaping* of the pigeon's key-pecking[41] and subsequent confirmation that this and other kinds of behavior by pigeons and other species were an example of Pavlovian conditioning; see Chapter 6. The theoretical development was based on demonstrations that, despite Skinner's assumption in 1948, temporal contiguity between two stimuli was not sufficient for conditioning to take place, what was required was a contingent relationship between them; as first demonstrated for fear conditioning;[42] see Chapter 4.

A careful replication of Skinner's 1948 experiment on superstitious behavior was carried out in 1968 by Virginia Simmelhag, a Masters student at the University of Toronto advised by John Staddon.[43] Instead of Skinner's impressionistic description of the behavior of his pigeons, Simmelhag recorded what each of her six pigeons was doing, second by second, in terms of sixteen predetermined categories of behavior. Her data provided the starting point for her advisor, John Staddon, to develop a theory that began with a distinction between *interim* and *terminal* behaviors, with the former roughly corresponding to what Falk had termed *adjunctive* behaviors, and *terminal* behaviors referring to those that occurred only when a reinforcer was shortly to occur. The latter were claimed to be based on Pavlovian conditioning. As for adjunctive behaviors, their origin was less clear.[44]

In a later review chapter on schedule-induced behavior Staddon suggested that interim behaviors, including schedule-induced drinking, are best understood as resulting from an animal learning when reinforcement is highly unlikely and that this leads to a motivational shift.[45] His account has been generally accepted ever since, although a few researchers have remained true to Skinner's theory of superstitious conditioning in attempting to explain schedule-induced behaviors.[46]

To return to Skinner at Indiana University, in 1946 he received a letter from Boring inviting him to give the prestigious William James Lectures the following year. The topic was to be language rather than Skinner's experiments and ideas on conditioning. *The Behavior of Organisms* had received mixed reviews and only eight copies were sold during the war years.[47] In contrast, the few articles that Skinner had published on verbal behavior had attracted wide and positive interest. At Harvard the following year, his lectures were well received and he made a good impression on the handful of experimental psychologists that now comprised the Department of Psychology; other psychologists had recently left the department to form a new department called 'Social Relations.' Skinner was offered, and immediately accepted, a full professorship at a salary of $10,000 per year, plus $4,000 to set up a laboratory and $1,000 a year to maintain it.[48] Twenty years after he had arrived as a graduate student, Skinner returned to Harvard and remained a member of the university for the rest of his life.

Operant Conditioning Research in the 1950s and the Influence of Fred Keller

A large proportion of the experiments on operant conditioning carried out in the 1950s were inspired by what can be seen as the pilot experiments that Skinner had described in the *Behavior of Organisms* or that had been run during Project Pigeon in between the main task of training pigeons to guide a missile. Even though Skinner still could not reveal what this main task had been, his 1950 paper with the title *Are theories of behavior necessary?*[49] included descriptions of several side studies that he and his dedicated team – Bill Estes, Norman Guttman, and Keller Breland – had carried out. The theme of his paper was that "it is possible that the most rapid progress toward an understanding of learning may be made by research that is not designed to test theories. An adequate impetus is supplied by the inclination to obtain data showing orderly changes characteristic of the learning process. An acceptable scientific program is to collect data of this sort and to relate them to manipulable variables, selected for study through a commonsense exploration of the field."[50] 'Common sense' was left undefined.

The skills and techniques that had been acquired during Project Pigeon provided the basis for an article Skinner wrote for *Life Magazine* in 1951 on *How to teach animals.* It heralded the increasing application of operant methods to training animals over the next few decades. The article began: "Teaching, it is often said, is an art, but we have increasing reason to hope that it may eventually become a science. We have already discovered enough about the nature of learning to devise training techniques that are much more rapid and give more reliable results than the rule-of-thumb methods of the past. Tested on animals, the new techniques have proved superior to traditional methods of professional animal trainers; they yield more remarkable results with much less effort."[51]

The article ended with a particularly persuasive example of how the principles of shaping can increase understanding of human interactions. It was based on the idea that an event such as one human being paying attention to another can act as a powerful reinforcer. "A scientific analysis can, however, bring about a better understanding of personal relations. We are almost always reinforcing the behavior of others, whether we mean to be or not. A familiar problem is that of the child who seems to take an almost pathological delight in annoying its parents. In many cases, this is the result of conditioning which is very similar to the animal training we have discussed. The attention, approval, and affection which a mother gives a child are all extremely powerful reinforcements. Any behavior of the child which produces these consequences is likely to be strengthened. The mother may unwittingly promote the very behavior she does not want. For example, when she is busy, she is likely not to respond to a call or request made in a quiet tone of voice. She may answer the child only when it raises its voice. The average intensity of the child's vocal behavior, therefore, moves up to another level – precisely as the head of the dog in our experiment was raised to a new height. Eventually, the mother gets used to this level and again reinforces only louder instances. This vicious circle brings about louder and louder behavior. The child's voice may also vary in intonation, and any change in the direction of unpleasantness is more likely to get the attention of the mother and is therefore strengthened. One might even say that 'annoying' behavior is just that behavior that is especially effective in arousing another person to action. The mother behaves, in fact, as if she had been given the assignment of teaching the child to be annoying!"[51]

Of the three students that worked with Skinner on Project Pigeon, Bill Estes was at the start of a very successful academic career, that included devising mathematical models of behavior; he published his *Statistical theory of learning* in the same year as Skinner's rejection of theories.[52] Keller Breland, together with his wife, Marion, had capitalized on the skills in shaping complex behaviors they had acquired during Project Pigeon by setting up an enterprise that trained various species for commercial purposes.[53] Norman Guttman had followed Skinner from Minnesota to Indiana, where he obtained his Ph.D., and eventually obtained an academic appointment at Duke University. There he set up a laboratory for both rat and pigeon experiments, while progressively moving away from Skinner's rejection of theory. Guttman trained a number of students who went on to become productive researchers within the operant tradition, while – like their advisor – not adhering to strict Skinnerian orthodoxy.[54] The most important work from Guttman's laboratory was on discrimination learning and stimulus generalization,[55] as described in Chapter 7.

The person most responsible for the expansion of operant research during the 1950s was Skinner's lifelong friend and correspondent, Fred Keller. Keller (1899–1996) was born on a farm in New York State, left school at a young age, and after several jobs that included work as a telegraph operator, enlisted in the US Army in 1918 in time to serve in France during World War I. After the Armistice, he remained in Germany until August 1919. On return to the States, he won a scholarship that allowed him in 1920 to enter Tufts College, located in a suburb of Boston, to major in English literature.

After reading Watson's 1913 manifesto, *Psychology from the standpoint of a behaviorist*, Keller decided to study psychology instead. He must have impressed his teachers at Tufts since, on graduating in 1926, he was awarded a teaching position there. Keller held this during the two years he studied for the Masters degree at Harvard that he obtained in 1928. He was also employed part-time by Harvard as a laboratory assistant and tutor until he, at last, found a full-time academic appointment at Colgate University in 1931. Seven years later he obtained an appointment in the Psychology Department at Columbia University, eventually becoming a professor in 1950 and remaining there until he retired in 1964.[56]

Keller's close friendship and intellectual interactions with Skinner during his Harvard days completely shaped Keller's subsequent career; see Figure 8.7. While at Colgate he invited Skinner to give a talk there in 1935 and asked him to bring with him one of the conditioning chambers that Gerbrands was now producing as a standard model in the Harvard workshop. Keller paid $45 for it and managed to get a cumulative recorder for an additional $20 or so. With the help of undergraduates, he began a series of experiments, starting with the effect on response rate of varying the degree to which his rats' access to food was restricted. He took the Skinner box to Columbia and started once again to run experiments with the help of undergraduates. This was in 1938, when *The Behavior of Organisms* was published.

Keller was ecstatic. He wrote to Skinner: "The book is great! … It comes up to, and goes beyond, everything I had anticipated, and I had Great Expectations. In my humble opinion, it is the most important single contribution that this century has seen in the field of psychology. As a beautiful example of inductive method and operationalism, it puts to shame the Hullites, the Titchenerians, et alia, with their high-powered

Figure 8.7 Keller and Skinner as graduate students at Harvard University in 1929. From Skinner (1979). Reproduced with permission from the BF Skinner Foundation.

deductions, their narrow applications, their physiologizing, and their vague dreamery of psychology-as-science."[56]

An early series of experiments run by Keller were influential in defining operant terminology. He found that his rats would consistently press a lever if the consequence was to turn off for a few seconds a bright light in their Skinner box.[57] He termed this kind of reinforcement based on the termination of some aversive state, *negative reinforcement.* Skinner adopted this usage. In *The Behavior of Organisms* he had used this label to apply to what became known as a *punishment* contingency.

Keller spent the Second World War developing a new training course in Morse code for the US Army Signal Corps. It was so successful that it became the standard course for many years. Thus, Keller was responsible for the first application of Skinner's principles to a real-world setting. After the war, Keller returned to Columbia where he was highly effective in conveying to students his enthusiasm for Skinner's ideas and research. It helped that in 1946 he was able to able to set up the first teaching laboratory containing Skinner boxes.

Undergraduates taking the introductory course in psychology given by Keller and his colleague, Nat Schoenfeld, were allocated a rat to carry out some simple operant experiment; see Figure 8.8. The text for the course was published in 1950 with a main

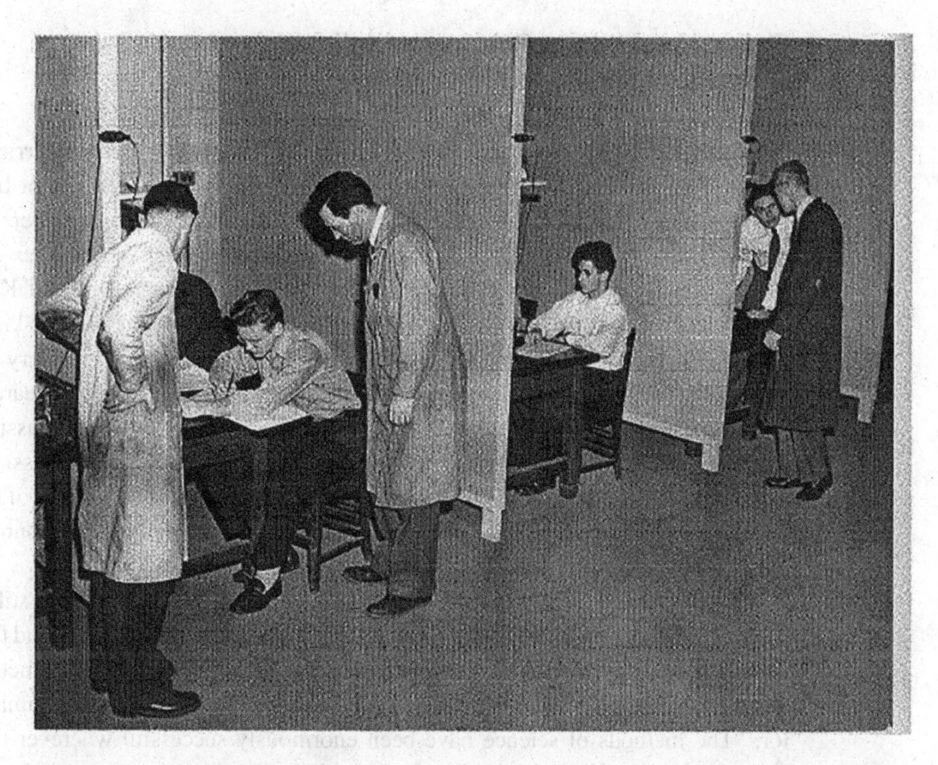

Figure 8.8 Keller and Schoenfeld supervising a rat lab class at Columbia University in the mid 1950s. Catania is the undergraduate at the far right, to whom Keller is talking. Schoenfeld is in profile looking at the work of a student on the left.
From Keller (2009). Reproduced with permission from Sloan Publishing.

title echoing that of eminent predecessors such as William James and, before him, Herbert Spencer: *Principles of Psychology: A systematic text in the science of behavior.*[58]

The book differed totally from traditional introductory texts. Instead of covering the topic of learning somewhere around the middle, sandwiched between early chapters on perception and later ones on cognition and social behavior, the sequence of chapters loosely followed that of the *Behavior of Organisms*, "our bible," wrote Keller. Thus, the first chapter was on the reflex, the second on Pavlovian conditioning, the third on operant conditioning, the fourth on extinction and reconditioning, and so on. Although the book described a large number of studies by researchers with no sympathy for Skinner's ideas, they were presented within the framework of the science of behavior defined by Skinner. There was no longer a need for the hard work required to read the *Behavior of Organisms*.

Keller was a skilled speaker as well as a good writer, with "low key, dead pan humor," and well-liked by his students.[59] As a result of his teaching, on graduating many students remained at Columbia or went elsewhere to undertake Ph.D.s in operant conditioning. By the end of the 1950s 37 graduate students had completed their Ph.D. at Columbia based on operant conditioning experiments.[60]

'Science and Human Behavior'; Applications of Operant Conditioning

Until the late 1950s most experiments on operant conditioning were carried out by researchers who had had direct contact with Skinner or had been taught by one of his or Keller's students. Few researchers without such direct contact were inspired by the *Behavior of Organisms*. In contrast, Skinner's *Science and Human Behavior* of 1953 made an immediate impact. Its origins were similar to those of Keller and Schoenfeld's book, in that it was written to accompany a course that Skinner was teaching at Harvard. However, *Science and Human Behavior* was a very different kind of book. Keller and Schoenfeld had incorporated a wide range of research within the Skinnerian framework of their text. In contrast, apart from some passing references to William James and many to Sigmund Freud, there was no discussion of any psychological research other than Skinner's own experiments. The aim of the book was to convince readers that applying the principles of operant conditioning would solve many of society's problems.

The book was written when the horrors of the Second World War were still fresh in peoples' memories and the risk was increasing that the Cold War would lead to nuclear destruction of the world. Skinner argued that the advances of natural science that had led to this situation needed to be matched by advances in understanding human behavior. "The methods of science have been enormously successful wherever they have been tried. Let us then apply them to human affairs. We need not retreat in those sectors where science has already advanced. It is necessary only to bring our understanding of human nature up to the same point. Indeed, this may be our only hope."[61]

What had held back understanding of human behavior was, Skinner argued, belief in 'free will' and appeal to inner causes. "When we say that a man eats *because* he is hungry, smokes a great deal *because* he has the tobacco habit, fights *because* of the instinct of pugnacity, behaves brilliantly *because* of his intelligence, or plays the piano well *because* of his musical ability, we seem to be referring to causes. But on analysis, these phrases prove to be merely redundant descriptions. A single set of facts is described by the two statements: 'He eats' and 'He is hungry'.... Thus we are unprepared for the properties eventually to be discovered in the behavior itself and continue to look for something which may not exist."[62]

Repeating the claim briefly introduced at the end of his 1951 *Life* article, Skinner argued that the results of operant experiments using pigeons or rats provide the basis for understanding human behavior. This claim was illustrated by further persuasive examples. In discussing the role of variable ratio schedules in real-life settings and their capacity for maintaining high rates of responding despite a low frequency of reinforcement, Skinner suggested that "the efficacy of such schedules in generating high rates has long been known to proprietors of gambling establishments. Slot machines, roulette wheels, dice cages, horse races, and so on pay off on a schedule of variable-ratio reinforcement.... The pathological gambler exemplifies the result. Like the pigeon with its five responses per second for many hours, he is the victim of an unpredictable contingency of reinforcement."[63]

In November, 1953 a Fathers' Day was held at the school that his daughter, Debbie, attended. Skinner and a few other fathers of 4th graders were allowed to sit in on an arithmetic class. "The students were at their desks solving a problem written on the blackboard. The teacher walked up and down the aisles, looking at their work, pointing to a mistake here and there. A few students soon finished and were impatiently idle. Others, with growing frustration, strained. Eventually, the papers were collected to be taken home, graded, and returned the next day. I suddenly realized that something had to be done. Possibly through no fault of her own, the teacher was violating two fundamental principles: the students were not being told at once whether their work was right or wrong (a corrected paper seen 24 hours later could not act as a reinforcer), and they were all moving at the same pace regardless of preparation or ability."

The solution to these problems was to be a technical one. "If students were to learn at once whether their response were right or wrong, and if the responses were to be those for which they were best prepared at the moment, instrumentation was needed." Within a few days, Skinner had built his first teaching machine.[64] He reported on subsequent developments in 1958. "Several machines with the required characteristics have been built and tested. Sets of separate presentations or 'frames' of visual material are stored on disks, cards, or tapes. One frame is presented at a time, adjacent frames being out of sight. In one type of machine, the student composes a response by moving printed figures or letters. His setting is compared by the machine with a coded response. If the two correspond, the machine automatically presents the next frame. If they do not, the response is cleared, and another must be composed. The student cannot proceed to a second step until the first has been taken. A machine of this kind is being tested in teaching spelling, arithmetic, and other subjects in the lower grade."[65]

For the next decade or more Skinner's major preoccupation was with the development of this teaching method and the attempt to have it incorporated within some educational system. He was able to obtain funding to set up a small research center at Harvard, the Center for Programmed Instruction, that was dedicated to developing these ideas.[66]

The arguments in *Science and Human Behavior* and those on the enhancement of educational practice were followed by suggestions as to how operant principles could be applied to the treatment of people with various kinds of mental disorder. Skinner obtained some grant money that allowed him to hire a former student, Ogden Lindsley, to run experiments on psychotic patients at the Metropolitan State Hospital in Waltham, Massachusetts. The experiments were designed to show that such patients would produce similar cumulative records to those of pigeons when the patients were pulling a plunger to obtain candy or cigarettes on various reinforcement schedules. The term *behavior therapy* was used for the first time in a report on this work, even though there was no evidence for the treatment to have been therapeutic for these patients.[67]

By the end of the Fifties the increasing range of applications of operant conditioning had convinced many who might have misgivings over Skinner's radical behaviorism that, nonetheless, operant conditioning was a powerful tool for improving human life. More generally, the belief that experiments on the learned behavior of animals was basic to understanding human psychology was held more widely than ever before.

Schedules of Reinforcement

In 1950 Skinner obtained a grant from the Office of Naval Research that was large enough for him to hire a postdoctoral research assistant. He sought Keller's advice on someone suitable who was about to obtain a Ph.D. from Columbia. Charlie Ferster was a perfect fit. He was highly committed to Skinner's views on psychology, experienced in running complicated operant experiments, and – most important of all – outdid Skinner in hoarding anything from mechanical gadgets, electrical relays from abandoned soda-dispensing machines, and an array of nuts and screws that might come in handy for setting up a new experiment, something he was routinely successful in doing. "For almost ten years, I carried around a 244-pole stepping switch (purchased from surplus for a dollar or two) before I finally threw it out."[68] The pigeon lab had been set up a few years earlier in the basement of Harvard's Memorial Hall. Its potential was greatly expanded as Skinner, who often made the first models, and Ferster, who perfected the equipment, had the means to study ever more complicated schedules of reinforcement.[69]

Ferster and Skinner worked feverishly on their pigeon experiments for two years; see Figure 8.9. Skinner later described their work: "With improved automation, Charlie and I were able to run nearly a dozen experiments twenty-four hours a day. Each morning we made our rounds, moving from apparatus to apparatus, unrolling the records which had been collected in baskets below the recorders, discovering occasional failures or in our planning, and making changes in the contingencies. There were many exciting moments, and scarcely a week passed without a surprise."[70] Then, Ferster was left with the challenge of translating the huge amount of data into the first draft of a book.

The result was a very unusual book. Ferster and Skinner's *Schedules of Reinforcement* of 1957 is over 700 pages long. Most pages contain at least one cumulative record. There is no summary of any of the results. The text is of the form: 'We did this with one, or sometimes two, pigeons or the occasional rat, and what happened can be seen in the accompanying record.' The book is best seen as a toolbox. If a researcher needs, say, a schedule that produces a very steady rate of operant responding and high resistance to extinction, then this is the schedule to use and this is what it should be called– a long duration *variable-interval* schedule, in this example. On the other hand, perhaps very slow responding is required? Then a *DRL* (= *differential reinforcement of low rate of responding*) could be the schedule to choose. And so on, for increasingly complicated arrangements, defined variously as *multiple*, *tandem*, *chain*, and several more.

For Skinner, now approaching 50 years old, this project was almost the last in which he became deeply involved in experiments. Whereas Pavlov had continued to enjoy the process of designing and running experiments well into his 80s, Skinner returned to his first love, writing. In 1953 Skinner handed over responsibility for the pigeon lab, plus supervision of the increasing number of graduate students running experiments there, to his two recent Ph.D. students, William (Bill) Morse and Richard (Dick) Herrnstein.

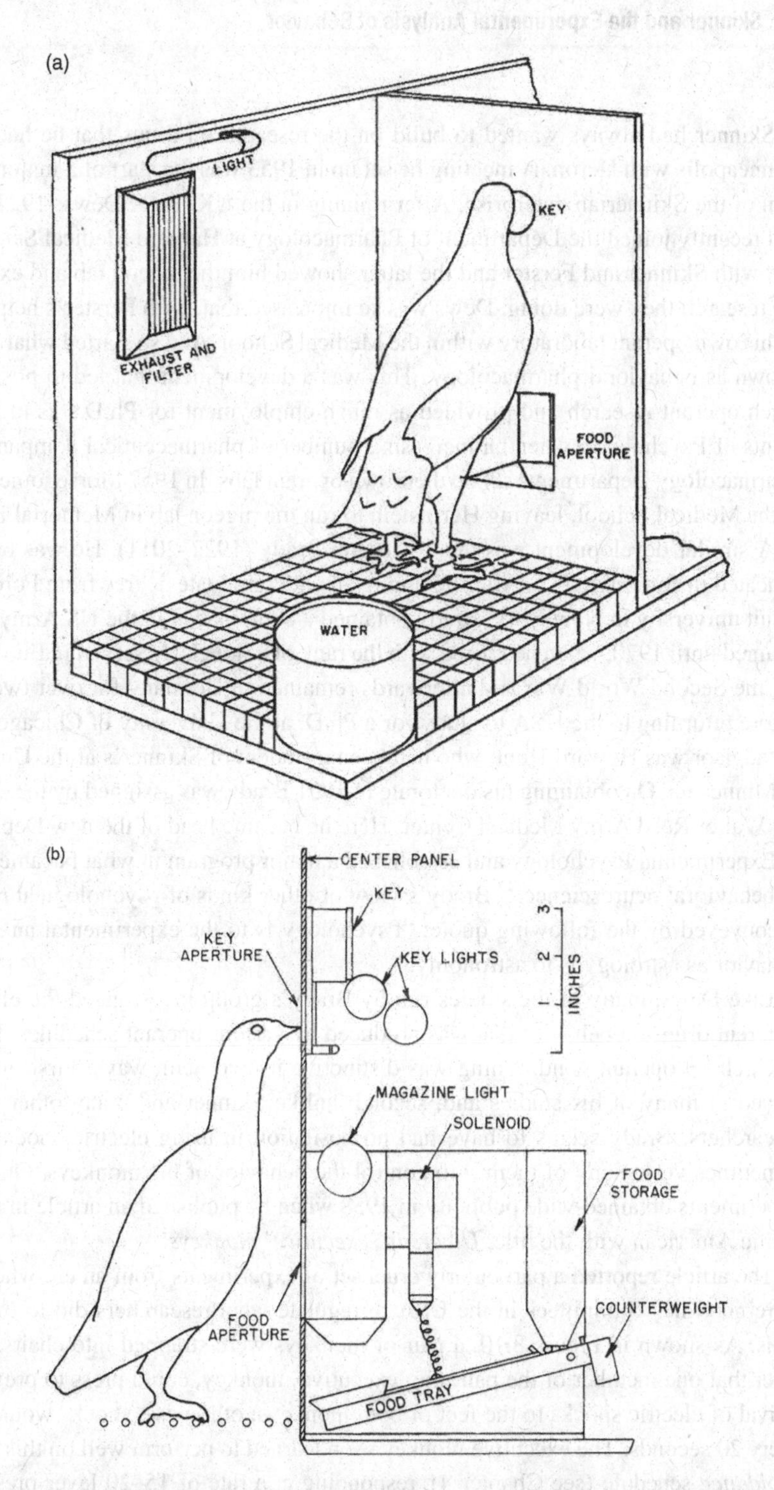

Figure 8.9 Sketches of the experimental chambers used by Ferster and Skinner in the early 1950s to study how rate of pecking at a response key by a pigeon was affected by various schedules of reinforcement. (a) Arrangement inside the operant chamber. (b) Details of magazine for delivering grain and other equipment.
From Ferster and Skinner (1957). Public domain.

Skinner had always wanted to build on the research on drugs that he had run in Minneapolis with Heron. A meeting he set up in 1953 was the start of a major expansion of the Skinnerian enterprise. After training in the UK, Peter Dews (1922–2012) had recently joined the Department of Pharmacology at Harvard Medical School. He met with Skinner and Ferster and the latter showed him the pigeon lab and explained the research they were doing. Dews was so impressed that, with Ferster's help, he set up his own operant laboratory within the Medical School. And so started what became known as behavioral pharmacology. This was a development that led to possibly as much operant research and provided as much employment for Ph.D.s as in Departments of Psychology, when an increasing number of pharmaceutical companies and Pharmacology Departments set up their own operant labs. In 1957 Morse joined Dews in the Medical School, leaving Herrnstein to run the pigeon lab in Memorial Hall.[71]

A similar development was due to Joseph Brady (1922–2011). He was born and educated in Brooklyn, NY. After obtaining an undergraduate degree from Fordham, a Jesuit university in New York, Brady obtained a commission in the US Army that he retained until 1970, when he retired with the rank of colonel. He served in Europe during the Second World War and afterwards remained in Germany for over two years, before returning to the USA to study for a Ph.D. at the University of Chicago. There his advisor was Howard Hunt, who had been a student of Skinner's at the University of Minnesota. On obtaining his doctorate in 1951 Brady was assigned by the Army to the Walter Reed Army Medical Center. Here he became head of the new Department of Experimental Psychology and established a major program in what became known as behavioral neuroscience.[72] Brady's view of other kinds of psychological research is conveyed by the following quote: "Psychology is to the experimental analysis of behavior as astrology is to astronomy."

Like Dews, many of the studies run by Brady's group investigated the effects of different drugs on behavior that was produced by various operant schedules. Brady's research on operant conditioning was distinctive in two main ways. First, monkeys served in many of his studies and, second, unlike Skinner and many other operant researchers, Brady seems to have had no hesitation in using electric shocks – and sometimes very many of them – to control the behavior of his monkeys. One of his experiments obtained wide publicity in 1958 when he published an article in the Scientific American with the title, *Ulcers in 'executive' monkeys*.

The article reported a particularly cruel set of experiments from an era when there were no ethics committees in the USA to regulate what researchers did to their animals. As shown in Figure 8.10, a pair of monkeys were strapped into chairs, with a lever that one member of the pair, the 'executive' monkey, could press to prevent the arrival of electric shocks to the feet of both monkeys; otherwise, shocks would arrive every 20 seconds. The executive monkey soon learned to perform well on this *Sidman avoidance* schedule (see Chapter 4), responding at a rate of 15–20 lever presses per minute, but nevertheless did not prevent every shock from arriving. As Brady wrote, the 'yoking' arrangement was one in which "both animals were subjected to the same physical stress (i.e., both received the same number of shocks at the same time), but only the 'executive' monkey was under the psychological stress of having to press the lever … After 23 days of a continuous six-hours-on, six-hours-off schedule the

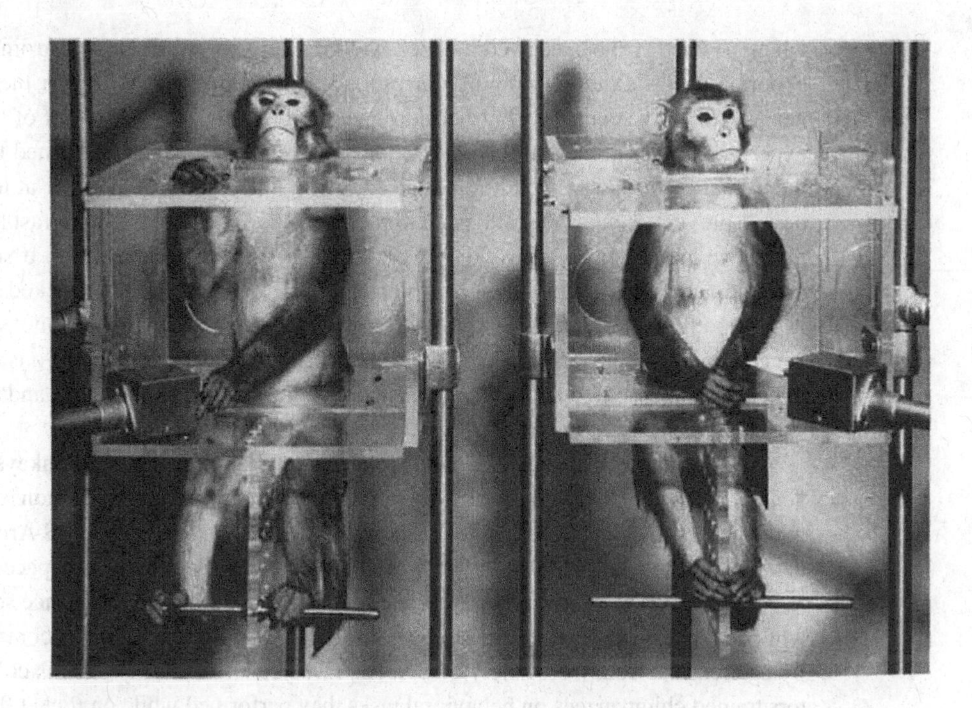

Figure 8.10 Monkeys serving in Brady's experiment related to ulcers in executives. The monkey on the left can press a lever to prevent arrival of an electric shock to both monkeys; for the one on the right the arrival of a shock is both uncontrollable and unpredictable.
From Brady (1958). Reproduced with permission from the Walter Reed Army Institute of Research, Library Special Collections/Archives, Silver Spring, MD 20910-7,500. Credit US Army Photograph.

executive monkey died during one of the avoidance sessions… An autopsy revealed a large perforation in the wall of the duodenum – the upper part of the small intestine near its junction with the stomach, and a common site of ulcers in man."[73] The control monkey remained in good health and, when sacrificed, showed no gastrointestinal abnormities. The experiment was repeated with a second pair of monkeys and this produced a similar result. Subsequent experiments used somewhat different conditions, such as an 18-hours-on, six-hours-off schedule, but failed to get the same effects.

Brady did not spell out the possible importance of his experiment for understanding the causes of human stomach ulcers. Nonetheless, the implication was clear from the title: Stress causes ulcers, and business executives have to make frequent decisions and as a result experience greater stress than their employees, thus running a greater risk of developing ulcers.

Although widely accepted for a time, the message was subsequently found to be nonsense. For one thing, the cause of human ulcers was later discovered to be bacterial and unrelated to stress.[74] In the context of animal experiments, a series of experiments by Jay Weiss, ones that antedated the better-known studies of learned helplessness in dogs (see Chapter 4), found that in a yoked design the rats that had no control over when or whether they received a shock were more likely to develop gastrointestinal abnormalities – although nothing corresponding to human ulcers – than their partner who did have control.[75] Furthermore, inspection of the original report of Brady's experiment revealed a fatal flaw. The monkeys were not randomly assigned to be in the executive or passive condition. Instead, the monkeys who were faster at learning how to prevent shocks were allocated to the 'executive' condition. Those fast learners may have learned fast because they were especially upset by the shocks. If so, what doomed them was possibly their heightened stress response to being shocked.

In the present context a lesson from this line of research is that, despite Skinner, generalizing from the results obtained from just a few animals can be very misleading. In contrast, Weiss used up to sixteen rats per condition in his experiments and applied statistical analyses to his data.

An important consequence of publishing his article on executive monkeys in the *Scientific American* was that Brady's work with primates caught the attention of a very wide readership. One reader was Wernher von Braun, then head of the US Army Ballistics Agency, who obtained the help of Brady in training primates for space flights. In 1959 two rhesus monkeys were trained to respond on a Sidman avoidance schedule while in the nose cone of a Saturn rocket during the first flight of a rocket containing a living creature. Both were safely recovered. Two years later Brady and his collaborators trained chimpanzees on behavioral tasks they performed while on rocket flights.[76]

The amount and range of operant research carried out in the 50s and early 60s can be appreciated by leafing through an important handbook published in 1966. *Operant behavior* was edited by one of Guttman's students, Werner Honig. His introductory remarks included the following: "The integration of operant methods into modern psychology is a recent as well as an important one. Skinner's earliest papers date back only three decades, and even much more recently, the use of operant methods was considered by many to be odd and rather fruitless. One professor of mine was of the opinion that putting an organism into a Skinner box taught us a lot about the box, but very little about the workings of the organism.... Some psychologists still think that ... operant methods are more suitable for the study of simple processes rather than more complex, higher-level behaviors. Much of the material in this book will counteract such an impression, if it is still current."[77] The list of twenty-three authors of the chapters that followed – still no women – is dominated by researchers who were either trained by Skinner, by Keller, or by one of their students. However, eight of the authors were men with a different background who had become convinced of the usefulness of operant methods to their independent research interests.

The process by which the 'experimental analysis of behavior' became a formal movement started when Skinner was still at Indiana University. He and Keller searched for a name for the conference they were organizing there and decided to derive one from the title and subtitle of Skinner's 1938 book: *The behavior of organisms: An experimental analysis*. There were only twenty or so contributors to this first conference in 1947 and only a few more to the second meeting in 1948 that was also held in Indiana. In 1949 the meeting was held at Columbia University in New York and at least 43 people were there. During the 1950s attendance at the successor meetings continued to grow.[78]

Many researchers faced a problem when imitating Skinner's use of just one or two experimental animals and the reporting of results simply in terms of representative samples of cumulative records: Mainstream psychology journals usually rejected their papers. Mainly due to the efforts of Ferster, who was then at Indiana University, the problem was solved in 1958 by founding a new journal, the *Journal of the Experimental Analysis of Behavior.*

The American Psychological Association (APA) has over 120,000 members and contains 56 Divisions at the time of writing. Many of these divisions, especially the earlier ones, reflect traditional categories within psychology: Personality and Social Psychology, Developmental Psychology, Clinical Psychology, and so on. In 1964 the APA was a lot smaller and had far fewer divisions. This was the year in which a new one was added: *Division 25: Behavior Analysis* was created with the aim of "promoting behavior analysis within the larger society." Thus, just over 30 years since Skinner first promised that the whole of psychology would be taken over by his radical behaviorism, a major beachhead had been established.

The Harvard Pigeon Laboratory in Its Heyday

From 1958 to 1962 the pigeon lab at Harvard was generously supported by a grant to Skinner from the National Science Foundation; see Figure 8.11. It made possible a particularly productive era.[79] The people who worked there at this time included, George Reynolds, whose experiments first documented the behavioral contrast effect described in Chapter 7, and Charlie Catania who concentrated on previously little studied *concurrent schedules* of reinforcement. As first described in the final chapter of Ferster and Skinner's book, a concurrent schedule is one in which an animal is offered a choice between two or more schedules of reinforcement. For example, a pigeon might be faced with two response keys, and pecking at one is followed by food on average once a minute – a variable interval (VI) 1-minute schedule, while pecking at the other is followed by food on average only every five minutes, a VI 5-minute schedule.

Generous technical support and production of electromechanical equipment was still provided by Ralph Gerbrands, who set up his own company producing operant conditioning equipment. Designing and constructing electronic devices was carried out by two further technicians, Rufus Grason and Steve Stadler, who also went on to form their own company to market products needed in operant conditioning labs. Additional staff, Mrs. Antoinette Papp and Walter Brown, were responsible for maintaining the pigeon colony; see Figure 8.12. A student would learn how to wire up a relay rack, using the recently introduced snap leads. When an experiment was ready to go and timers set, Mrs. Papp was informed; she would then weigh the pigeon selected for the experiment, place it in the operant chamber and press the start button. When the session ended, Mrs. Papp would remove the pigeon, weigh it once again and give it the required supplementary ration of grain when returning it to its home cage. This was operant conditioning research on an industrial scale![80]

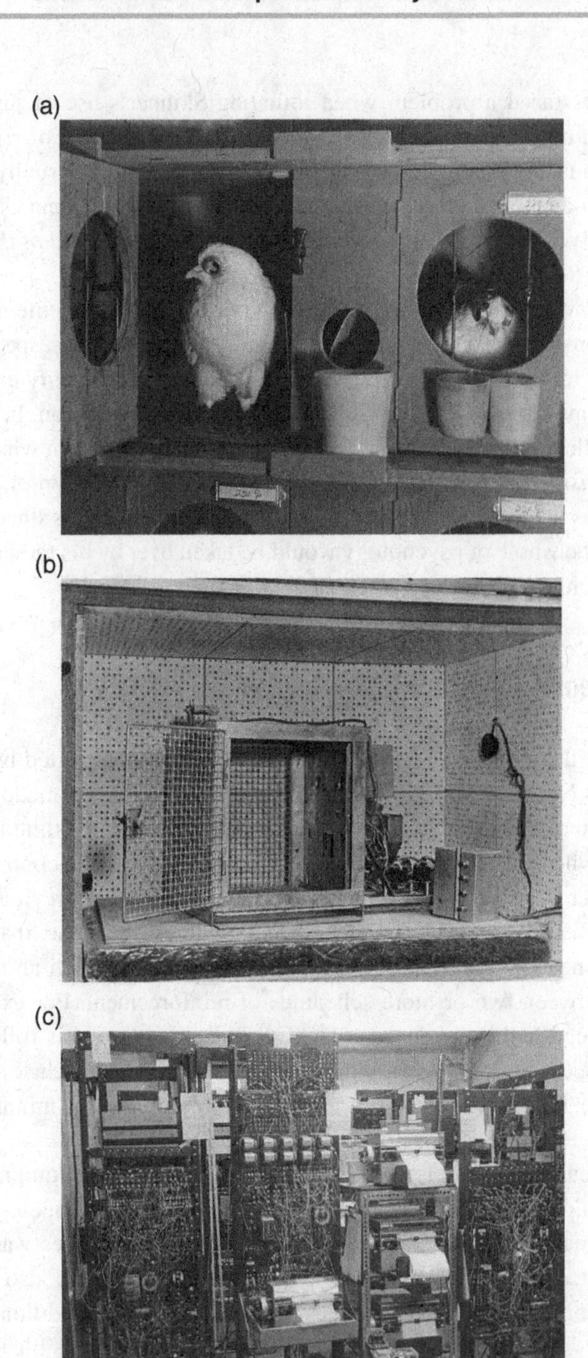

Figure 8.11 The Harvard pigeon lab around 1960. (a) Housing for pigeon. (b) An operant chamber, 'Skinner box,' fitted with two response keys for studying choice behavior (*concurrent reinforcement schedules*). (c) Rack-mounted electro-mechanical devices used to control such experiments, with cumulative recorders to record the data and monitor that all is well within an experimental session.
From Catania (2002). Reproduced with permission from John Wiley and Sons.

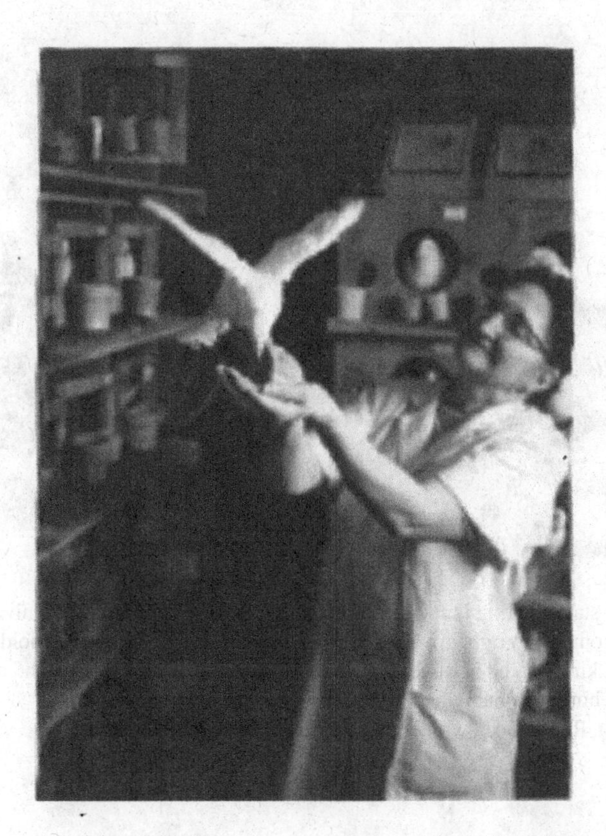

Figure 8.12 Mrs. Antoinette Papp, queen of the Harvard pigeon lab.
From Staddon (2016). Reproduced with permission from John Staddon.

Skinner was little involved in this feverish activity, other than attending the weekly pigeon meetings where he would comment on results that attracted his attention; see Figure 8.13. Presumably reflecting his own experience as a graduate student, he left his Ph.D. students alone to develop research projects. Consequently, most learned their skills and refined their ideas by interacting with other students and a postdoctoral researcher working in the lab. On one occasion, Skinner left for six months without warning any of his students.

In 1961 Skinner decided that he would apply for just a one-year extension of his grant. His application cited the past achievements of the lab, including the 1957 publication of *Schedules of Reinforcement* and almost 50 papers reporting research from the lab. He noted that "the laboratory has also served an important teaching function. More than 40% of Harvard Ph.D.s in Psychology during the past ten years have written theses connected with it." As for practical applications that were stimulated by animal-based research, he mentioned only teaching machines. In this context he provided an example of a current project within the lab, suggesting that the work of Herb Terrace on errorless discrimination learning (see Chapter 7) could be important for further developments of the teaching machine project.[81] The $50K he requested – and apparently was fully granted – ensured that the lab could continue to function at full speed.

Figure 8.13 A pigeon staff meeting in the Department of Psychology at Harvard University. Seated at the table from left to right are Mike Harrison, Bea Barrett, George Reynolds, and Charlie Catania. Walking behind is Roger Kelleher. Skinner is standing at the right and the seated figure behind him is probably Dick Herrnstein.
From Catania (2002). Reproduced with permission from John Wiley & Sons.

Skinner had contemplated retiring from Harvard in 1962, but instead was successful in obtaining a grant that released him from all teaching duties, except for a graduate seminar that he led on alternate years. Herrnstein took over the pigeon lab and became supervisor of the six to eight graduate students each year for whom the lab became the base for their Ph.D. research.

Richard Herrnstein (1930–1994) was born in New York City, the son of two recent immigrants from Hungary. His father was a house painter and, one assumes, the family were far from rich. Herrnstein's promising musical skills led him to the School of Music and Art in Manhattan. His subsequent studies at the City College of New York added an enthusiasm for psychology. In 1952 he enrolled as a graduate student at Harvard. It seems that music played a role in his choice of Harvard, in that initially he carried out experiments on sound perception in the Psychoacoustics lab headed by S.S. Stevens.[82] Why he decided to switch to the study of operant conditioning, with Skinner as his Ph.D. adviser, is hard to discover, since he never wrote any kind of autobiographical note and the few obituaries tell us little beyond the major milestones in his life. Herrnstein's Ph.D., awarded in 1955, was based on pigeon experiments that examined some properties of variable interval schedules. Afterwards, he served three years in the US Army, years that were spent running experiments in Brady's labs in the Walter Reed Army Medical Center. Despite the contrast between their backgrounds – a 50-year old WASP from a country town vs. a young Jew from New York – for many years Skinner and Herrnstein appear to have both liked each other

Figure 8.14 Skinner with Richard Herrnstein around the time that the latter was appointed as an Associate Professor in Harvard's Department of Psychology.
From Skinner (1983). Public domain.

and enjoyed a great deal of mutual respect; see Figure 8.14. They played chamber music together, with Herrnstein on the violin and Skinner at the piano.

Herrnstein was the only student of Skinner's to obtain an academic position at Harvard. This was not necessarily because Skinner regarded Herrnstein as his successor. In many ways, the latter departed even as a graduate student from Skinnerian orthodoxy. In the pigeon lab of the 1950s, there was a division between those who followed Skinner in using cumulative records as the primary way of reporting their results – mainly students who, like Catania, had been trained at Columbia University – and those, like Herrnstein, who relied on counters and timers and used cumulative recorders mainly as a way of checking that all was well within their pigeon chambers. What probably helped Herrnstein's appointment in 1958 was the good impression he had made when working in the psychoacoustics lab and the consequent likely support of Stevens, the most powerful member of the Psychology faculty. What may also have contributed to his appointment was that Herrnstein shared Boring's active interest in the history of psychology; later they co-edited a collection of key historical articles.[83]

At the time of his appointment, Herrnstein already had an impressive publication record and had acquired experience in a variety of operant methods. In 1955, the year he obtained his Ph.D., he published a paper reporting his graduate experiments with pigeons and was also co-author on a paper reporting the study of loudness judgments he had undertaken when working in the psychoacoustics lab. Herrnstein's three years spent in Joe Brady's lab were very productive. He collaborated with several other researchers, including Bill Morse with whom he published six papers, mostly on the effects of various drugs on the performance of rats on a range of reinforcement schedules. Another collaborator was Murray Sidman. Their work together, using both rats and monkeys, mainly involved the free operant avoidance procedure that Sidman had developed (see Chapter 4); this collaboration resulted in four more publications. Herrnstein also co-authored a paper with Brady on an experiment using a multiple schedule; this found that the response rate of one rat – deprived to an extraordinary 60% of its prior body weight – to obtain a condensed milk reinforcer in one component was affected by changes in the shock avoidance contingencies in another component. The second rat did not show this effect.[84]

On his return to Harvard Herrnstein continued with several novel and diverse projects. One study was almost the first to link placebo effects and Pavlovian conditioning; but see Chapter 1. After pointing out that placebo effects are commonly thought to result from suggestions by medical practitioners, he went on to suggest: "The placebo effect can, however, be viewed in a different way. The elicitation of a specific reaction by arbitrary agents, such as the abatement of a symptom after the mere sight of a physician and his medicines, may be nothing more than simple conditioning of the sort originally demonstrated by Pavlov with animals."[84] In an experiment that supported this idea a rat was trained to press a lever to obtain a sweetened condensed milk reinforcer during an occasional 5-minute period when its chamber was illuminated. When first injected with saline, no change in its performance occurred. However, after an injection of scopolamine that severely reduced the rate of lever-pressing, an injection of saline also slowed the rat's response rate, but to a far lesser extent. Thus, the injection procedure appeared to have become associated with the effects of scopolamine. In the 1960s the results from one rat could be enough to get a paper published in the prestigious journal, *Science*. In this case Herrnstein did at least note that "entirely analogous results were obtained from a second rat with an earlier version of the present procedure."[85]

Another of Herrnstein's highly influential studies from the early 1960s had echoes of Skinner's *Pigeon Project* two decades earlier. Herrnstein's project was not carried out in the Harvard pigeon lab but at the 'Limited War Laboratory' outside of Boston and partly supported by the US Army. Five homing pigeons were trained to peck at a translucent plate that could serve as a screen onto which colored slides could be projected. Herrnstein and his co-worker collected over twelve thousand slides of natural scenes, such as open countryside, cityscapes, or expanses of water, and these were divided into those that included at least one human being and those that did not. Pecking at the projection of a slide containing at least one human being was reinforced intermittently, whereas no reinforcement followed pecks at photographs lacking any human. All five pigeons learned within a few sessions to discriminate between the two sets, even though individual slides were very rarely repeated. Furthermore, the pigeons performed at a high level during tests in which entirely new slides were used. Where previously such discrimination training with pigeons and other animals had involved only simple sensory differences such as in the wavelength of lights or of frequencies of tones, the scientific point of this study was that pigeons could learn on the basis of a *complex visual concept*, as the title of the paper proclaimed.[86]

There was also a prominent applied aspect to this study. As Skinner later wrote: "As the war in Vietnam escalated, ambushes became a serious problem, and Dick Herrnstein taught pigeons to fly along jungle trails and stop when they found a person, their behavior reported by tiny transmitters."[87] A longtime friend and research collaborator of Herrnstein reported that they rarely disagreed on any topic but an exception was the Vietnam War: Unlike Skinner and most members of the Harvard community, Herrnstein was a supporter.[88]

At least in terms of subsequent citations, by far the most influential of the experiments that Herrnstein reported in the early 1960s was a study of how pigeons distributed their responses – pecking at one or other of two response keys – when the rate of reinforcement was systematically varied. For example, at one stage in the long experiment pecking at Key A might be reinforced every 2.25 minutes on average, while pecking a Key B was reinforced at the lower rate of every 4.5 minutes. As the relative rates at which reinforcements were varied between the two keys, the way that all three pigeons distributed their responses displayed a remarkably consistent and simple pattern. The percentage of responses made, say, to Key A, exactly matched the relative frequency of reinforcement available on Key A[89]; see Figure 8.15.

In the nineteenth century research on the relationship between subjective experiences, such as the loudness of a tone, and their physical basis had led to the adoption of a mathematical function known as the *Weber-Fechner Law*. Many experiments in the psychophysical laboratory at Harvard in the 1950s suggested that this 'law' needed to be replaced by a power function, leading to Stevens' *Power Law* of 1957.[90] Research focused on extending or testing this 'law' continued at Harvard for at least 10 years. In this context, it was not surprising that, the relationship between responding and reinforcement rates discovered by Herrnstein in 1961 became known as the *Matching Law*, rather than less grandiosely as say, the 'matching function.'

Some felt doubts about a law based on data obtained only under very specific conditions. As Herrnstein reported in his 1961 paper, matching between the relative rate of responding on Key A and the relative rate of reinforcement for these responses was found only when an extra contingency was inserted that was called a *Changeover Delay* or *COD*. In the 1961 experiment this prevented any reinforcement being delivered within 1.5 seconds of a pigeon changing over from pecking at one key to pecking at the other. As Herrnstein discussed at the end of the paper, even a change in the distance between the keys could have produced a different outcome.[91] Later experiments using rats that were given a choice between two levers that paid off at different rates found matching only when a particularly long COD was introduced.[92]

Any such doubts did not prevent an increasing number of Harvard graduate students in the 1960s from carrying out projects related to matching. Two who started together in 1962, William Baum and Howard Rachlin, were particularly active in the

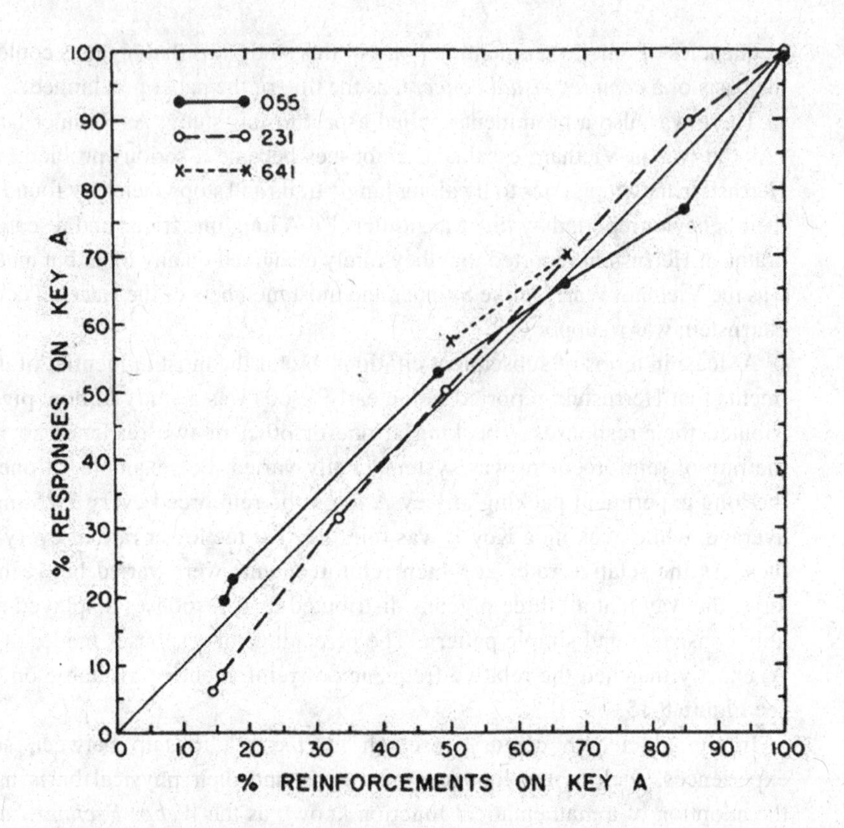

Figure 8.15 This graph is from Herrnstein's report of an experiment in which three pigeons were faced with a choice between two response keys, A and B. The rates of reinforcement for pecking at each of these keys were systematically varied. This was the first report of a 'matching function': Across a wide range, the percentage of responses on Key A equaled the percentage of reinforcements delivered for responding on A.
From Herrnstein (1961). Open access: Available from NIH National Library of Medicine.

study of pigeons' choices between various reinforcement contingencies. They were followed by several other Harvard Ph.D. students, including John Staddon[93] and Peter Killeen.[94] These experiments contributed to increasing knowledge on factors affecting the choice behavior of pigeons. However, Herrnstein wanted to broaden his theory beyond choice to encompass an animal's rate of responding when there was only one response key to peck and only one source of reinforcement.

Baum was present shortly after what appears to have been an eureka moment. "One day I met Dick as he came out of the bathroom, and he told me, 'Even if there's only one key, there's still choice. It's between pecking and everything else.' He raced back to his office, and that day the Herrnstein hyperbola was born."[95] The development of a new set of equations was later reported in a paper titled *The Law of Effect*. The final sentence of the abstract to the paper read: "All these results, plus several others, can be accounted for by a coherent system of equations, the most general of which states that the absolute rate of any response is proportional to its associated relative reinforcement."[96]

For Skinner these developments were anathema. As a graduate student, he had strongly criticized Hulls' attempt to base his 'principles of psychology' on a set of equations. In 1955 Skinner was asked to review a book by two of his Harvard colleagues, Robert Bush and Frederick Mosteller, called *Stochastic models for learning*, which had a strong influence on later learning theorists. It prompted in Skinner his deep disdain for the use of statistics in the study of behavior. He noted that "Bush and Mosteller had wasted vast quantities of impeccable mathematics on vast quantities of peccable data" and that he "was bothered by the unwarranted prestige of mathematics in a field which I did not think was ready for it."[97] And now in the 1960s many of the most promising students working in the lab that he had set up believed that the field was now ready for it.

In my first year as a graduate student in the Harvard department, 1963/64, its weekly Pigeon Lab meetings were still attended by many researchers working in various institutions scattered around the greater Boston area, although it was over a year since Skinner had attended these meetings. Some came from the Harvard Medical School and some from the Center for Programmed Instruction. One presentation still stands out: Ogden Lindsley showed us cumulative records produced by people with some form of psychosis pressing buttons on various schedules in order to obtain tokens that could later be exchanged for cigarettes or candy. I felt appalled by the approach and convinced that this kind research was never going to lead to better understanding of schizophrenia or better conditions for such people.

By my final year, 1965/66, the only people attending these weekly meetings were graduate students and other researchers working in the pigeon lab. Clearly, those from outside with a main concern for practical applications of operant conditioning could see nothing useful in the concern with quantification and equations. The rift between those in the Skinnerian tradition with a central concern for applications of operant conditioning and those who followed Herrnstein later became institutionalized, when the latter established the Society for the Quantification of Behavior, or SQUAB.[98]

In the long run, it turned out that the SQUAB tradition nonetheless produced some important real-world outcomes. A notable example stems from the experiments that Rachlin and others ran on choice between a response that could deliver an immediate small reinforcer and a response that could deliver a much larger reinforcement but after a delay.[99] These led to extensive use of such *delay discounting* schedules to measure self-control and impulsiveness in both animals and humans; this continues well into the 2020s. Rachlin's experiments also provided the foundations for what became known as *behavioral economics*.[100]

As for the matching law, this became progressively more complex in order to account for data that were not as neat as those reported by Herrnstein in 1961.[101] These developments are beyond the scope of this chapter. One extension of this line of research to the world outside the pigeon lab was to ethological studies of how in natural environments birds choose between different sources of food; studies that led to optimal foraging theory.[102]

The Expansion of the Experimental Analysis of Behavior Beyond the USA

In the 1950s operant conditioning laboratories started to spread overseas as well as throughout the USA. An early example in the UK was set up by Harry Hurwitz at Birkbeck College, a part of London University that still specializes in providing evening classes for students in full-time work. Hurwitz appears to have been inspired by a brief meeting with Skinner in Stockholm in 1951. Shortly after the meeting, he set up the first operant lab in Europe.

Hurwitz had to design his own Skinner box.[103] He was helped by a friend, Karl Weiss, who went on to manufacture operant conditioning equipment, eventually founding Campden Instruments in 1970; this became the major supplier to operant labs throughout Europe. Two other UK labs founded a little later at least had the benefit of being given Harvard-built equipment by Skinner. Both Lawrence Weiskrantz, as a Harvard Ph.D. student with a main interest in the primate brain, and Leslie Reid, as a visiting graduate student for a year, had attended Skinner's undergraduate course and later took up university appointments in the UK. Weiskrantz set up the first primate testing facility using Skinner boxes in the Cambridge Psychology Department and subsequently moved to Oxford University, where he became Professor and highly effective Head of the Psychology Department from 1967 to his retirement in 1993. During this time, he became one of the world's leading cognitive neuroscientists.[104] As for Reid, he set up operant labs first in his native Aberdeen and then at the University of Exeter, where in 1963 he became the foundation professor of psychology and until 1980 continued as the Head of a thriving department that always "had a tilt towards behaviorism."[105]

While Exeter was one of the existing English universities that first set up a Psychology Department in the 1960s, Sussex University was one of several entirely new universities to be established in that era. Perhaps as a reflection of the increase in the popularity of psychology, at Sussex, there were three separate mini-departments. One of them, the Laboratory of Experimental Psychology, was part of the School of Biological Sciences. Experimental Psychology's founding professor, Stuart Sutherland, came from Oxford; see Chapter 7. Perhaps as a retort to Skinner's scorn for researchers who still used jumping stands, Sutherland had set his Oxford undergraduates an essay: "The Skinner box is a device that essentially decorticates both the animal and the experimenter; discuss." Despite his antagonism towards Skinner's radical behaviorism, Sutherland made sure that his department contained a suite of Skinner boxes and that the Biology Animal House included colonies of pigeons as well as of rats. Furthermore, he sponsored researchers who would make good use of these facilities. For example, after arriving in Sussex Terry Bloomfield continued to use multiple reinforcement schedules to analyze behavioral contrast in pigeons, a series of experiments that he had begun when a Ph.D. student at Exeter;[106] see Chapter 7.

Returning to Hurwitz, one of his main research interests along with his student, Peter Harzem, was the *progressive ratio* (*PR*) schedule. This reinforcement schedule was first described in 1961 by Bill Hodos, who ran these experiments in Brady's lab at the Walter Reed Army Institute of Research. His report was titled: *Progressive Ratio as a Measure of Reward Strength.* He described the procedure he used as follows: "After a brief initial period of training to press the lever to receive 0.05 ml of sweetened condensed milk as a reward, the rats were placed on the progressive ratio schedule, which requires that the animal emit an increasing number of responses in order to obtain each reward. The ratios used in these experiments increased by an increment of two, so that the rats were required to emit two responses for the first reward, four for the second, six for the third, eight for the fourth, and so on. Each run of responses in this increasing schedule is called a ratio run. A timer in the circuit was set so that if at any time during the experiment, the animal failed to respond for a period of 15 minutes, the session was automatically terminated." In a series of neat experiments, Hodos found that the *breakpoint* a rat reached – the response requirement at which a rat stopped responding – increased in a very tidy function with both the concentration of the sweetened condensed milk used as reward and the hours for which it had been deprived of food.[107] The PR schedule has been used extensively ever since for just the purpose for which it was designed by Hodos, namely, to assess the reward value of some outcome.

On obtaining his Ph.D., Harzem set up an operant conditioning lab in the new Department of Psychology in Bangor, then part of the University of Wales. He was Head of the Department until 1978, when his student, Fergus Lowe, became Head and maintained the strong Skinnerian tradition there. Another of Hurwitz's students, David Dickins, set up an operant lab at Liverpool University. These were but two examples of the many universities throughout the British Isles that in the 1960s acquired Departments of Psychology containing operant conditioning labs. Others included Durham, Belfast, Leeds, Liverpool, Aberdeen, and Dublin.[108]

One important figure in this expansion was a very clever American who had obtained his Ph.D. at Columbia University with Schoenfeld as his advisor. Jock Millenson was an unusual academic whose private fortune allowed him to move from an honorary lecturing and research position at one university to another. Starting in 1959 he worked and taught with Hurvitz at Birkbeck College. Later he moved to Oxford. There he supervised Julian Lesley, who, after obtaining his Ph.D. in 1974, established a Skinnerian research tradition at the University of Ulster in Northern Ireland. From Oxford, Millenson moved to McGill University in Montreal. Despite these frequent moves, he managed to publish a large number of experimental papers, to publish a textbook in 1967 with the title, *Principles of behavior analysis*,[109] and develop a programming language, ACT, that was used in many of the early computers that took over control of operant experiments from racks full of relays, timers, and counters, of the kind illustrated earlier in Figure 8.11.

One of the undergraduates who worked in the operant lab that Millenson set up at McGill University was Stephen Pinker, who wrote: "I loved Jock's textbook for the elegance and formal rigor of the system of experimental analysis of behavior it laid out. Jock was one of the last of the radical behaviorists and advanced a particularly pure form of the scientific ideology. Though I was never a convert, and later became a diehard cognitive psychologist, I enjoyed Jock's lectures and conversations for their intellectual rigor and purity, and I got good at pretending to be a behaviorist."[110]

It seems that in the early 1960s Millenson may have been one of the founders of the UK-based *Experimental Analysis of Behaviour Group* – or *EABG*, pronounced 'ee-bag' by many members – that has continued to meet annually ever since. For many years it was the forum of choice for many UK researchers in which to give the first report of their experiments, whether the researchers were committed Skinnerians or – like myself and some Sussex colleagues – were grateful for operant technology but were less accepting of many of Skinner's claims.

Starting in the late 1960s Derek Blackman became coordinator of the EABG. He also played a major role in the subsequent expansion of operant labs and 'behavior analysis' within the UK. After completing an undergraduate degree at Essex University in 1966, he had moved to Queens University in Belfast, where he obtained his Ph.D. in 1968. This led to an appointment at Nottingham University, where he set up a highly productive lab and attracted several able graduate students. A few years later he repeated these achievements at Birmingham University, and then again at Cardiff University, where he became Head of the Department of Psychology. In 1974 he published a well-regarded book, *Operant conditioning: An experimental analysis of behavior.*[111] Unusually for someone highly involved in operant research, Blackman was also very active in the British Psychology Society. He became its president in 1981. By then, he had in effect acquired the position of Skinner's 'deputy' in the UK.[112]

As for continental Europe, what remained a lone outpost of operant conditioning for over a decade was set up in Belgium by Marc Richelle (1930–2021) at the University of Liége. As a graduate student, Richelle had spent two years at Harvard immediately prior to obtaining his Ph.D. from Liége in 1959. He was promptly appointed to a lectureship within the Department of Pharmacology where he set up an operant conditioning lab to study the effects of various drugs. In 1965 he was appointed Professor of Experimental Psychology. Thereafter, his dominant interests became the study of timing in a variety of species and the promotion within the French-speaking world of Skinner's radical behaviorism.[113] Belgium has been the one country in Western Europe to have maintained a strong tradition of research into learning and behavior ever since the 1960s, even though not following the direction that Richelle advocated.

The experimental analysis of behavior started to spread south from the USA in the 1960s. When nearing the time to retire from Columbia University, Keller received a letter from a former student who had returned to her native Brazil. He accepted the invitation to spend several months at the University of Sao Paulo, where he set up the first operant lab in South America in 1961. Keller returned to Brazil again in 1964 for a shorter stay that included the new capital, Brasilia, where he had been invited to set up a Department of Psychology in its university. The timing was unfortunate, as Keller and his wife found themselves in the middle of a coup d'état by an army that had no sympathy for universities, let alone the study of psychology.[114]

The following year Keller retired from Columbia and moved to Tempe, Arizona. A major element in this decision was that several former students of Keller and of Skinner had obtained appointments in the Psychology Department at Arizona State University.[115] The department had been named "by unkind critics as Fort Skinner in the Desert."[116] Meanwhile, in Brazil Keller's belief that an essential component of any psychology degree was a practical course involving operant conditioning spread to many other universities beyond Sao Paulo and Brasilia.[117]

The development of operant conditioning in Mexico had a very different history from that of Europe or Brazil. A key figure was Emilio Ribes. In 1960, at the unusually young age of 15, Ribes was in the first cohort of undergraduates to undertake a degree in psychology at UNAM – the Universidad Nacional Autónoma de México – in Mexico City. Within a curriculum dominated by philosophy, the single semester course in experimental psychology sparked Ribes' interest in learning theory. He became a member of a group of students, *Galileo Galilei*, dedicated to turning Mexican psychology into a scientific discipline. In 1964 Ribes moved to Xalapa on obtaining a position at the University of Vera Cruz where he continued to concentrate on learning theory. It was there a year later that the first Mexican rat was trained to press a lever in a Skinner Box, one that was donated by the University of Texas.[118]

This first-hand experience of operant conditioning did not, however, turn Ribes into a Skinnerian. Instead, he reported that he became a behaviorist by reading two books that Mowrer published in 1960; see Chapter 4. While traveling in Canada and the USA Ribes came to know many of the leading psychologists of that era. Sidney Bijou was particularly influential in that he managed to resolve Ribes' doubts about Skinner's radical behaviorism. Together they published a book on behavior modification in 1972.[119] Now back at UNAM Ribes was instrumental in both establishing psychology as an independent faculty and in setting up a full-scale animal lab in 1973. He also seems to have been very effective in obtaining financial support for a series of conferences that were attended by leading researchers from north of the border. In 1974 the first meeting of the Mexican Society for Behavior Analysis was held in Xalapa and the following year saw the launch of the Revista Mexicano por el Análysis de la Conducta.

The Cognitive Revolution within American Psychology; Conditioning without Awareness?

While the experimental analysis of behavior was expanding overseas during the 1960s, Skinner's influence within American psychology was receding. The turning point was the publication in 1957 of his *Verbal Behavior*. He completed the manuscript after working on it for over 13 years, following the challenge to develop a behaviorist explanation for novel sentences that had never been uttered before. Most other behaviorists had shied away from discussing language. Skinner regarded *Verbal Behavior* as his most important book. The more dedicated among his followers agreed. Others found it unreadable.

Over time the book might have been generally ignored, if it were not for a later review. Another publication in 1957, *Syntactic Structures*, was based on work carried out when Noam Chomsky was a member of the Harvard Society of Fellows, nearly thirty years after Skinner had been a member. It was the beginning of a revolution in theories of grammar, theories that had as one of their aims explaining how novel sentences are generated. The revolution does not seem to have attracted much attention at the time from psychologists. This occurred two years later with Chomsky's review of Skinner's *Verbal Behavior*. As Chomsky later wrote: "I had intended this review not specifically as a criticism of Skinner's speculations regarding language, but rather as a more general critique of behaviorist (I would now prefer to say 'empiricist') speculations as to the nature of higher mental processes. My reason for discussing Skinner's book in such detail was that it was the most careful and thoroughgoing presentation of such speculations."[120] His review was a long, very detailed, and sometimes amusing dissection of Skinner's claims about language. The core of his argument is summarized by the following quote: "The way in which these terms (*stimulus*, *control*, *response*, and *strength*) are brought to bear on the actual data indicates that we must interpret them as mere paraphrases for the popular vocabulary commonly used to describe behavior and as having no particular connection with the homonymous expressions used in the descriptions of laboratory experiments."[121]

It seems that for many psychologists with some previous sympathy for behaviorism the review made explicit their unformulated reservations about restricting explanations of psychological phenomena to the constraints of a vocabulary based on conditioning experiments in animals. One such person was a rising star within the Harvard department. In 1951 George Miller had published a book, *Language and Communication*, that contained a chapter on language acquisition that was sympathetic to Skinner's ideas on the topic. During a summer spent interacting with Chomsky, Miller became convinced of the importance of experimental studies of language within the framework provided by Chomsky's theories of grammar. Miller essentially founded the modern field of psycholinguistics. In 1960 he and Jerome Bruner established within the Harvard department the first Center for Cognitive Studies. This was a bitter blow for Skinner, the final end to his long-held ambition of converting the Harvard department to his radical behaviorism.[122]

In his review of *Verbal Behavior* Chomsky rejected Skinner's proposal that children acquire their native language through a process of operant conditioning. Instead, he indicated the need to understand the *Language Acquisition Device* within their brains that enabled humans, and only humans, to acquire a natural language within a relatively short time, even under impoverished conditions. The Harvard social psychologist, Roger Brown, then began experimental studies of the development of language in children, thus reviving a branch of experimental psychology that had been dormant for decades.

Another factor that was important for what became known as the *cognitive revolution* in American psychology was the development of artificial intelligence.[123] For example, computer scientists aiming to develop machine translation or speech recognition programs had more to learn from a psycholinguist than from a behaviorist. More generally, these developments showed that to refer to 'short-term memory,' 'storage of information' or to 'selective attention' was not a retreat into dualism that let "the ghost back into the machine."[124] The dismissal by Skinner and other behaviorists of all other approaches in psychology as examples of unscientific mentalism was no longer convincing.

Skinner was always convinced of the power of reinforcement to change people's behavior even when they are unaware of any relationship between their behavior and the occurrence of some reinforcing event. An example of which he was clearly proud took place when he was listening to a presentation by Erich Fromm, a German social psychologist and psychoanalyst. Presumably uninterested in what Fromm had to say, Skinner told a neighbor that he could shape Fromm's hand-raising behavior. Every time Fromm raised his hand, Skinner would look him in the eye with an interested expression; otherwise, Skinner looked away. Apparently, Fromm began to raise his hand with ever increasing frequency.[125]

We do not know whether Fromm had the habit of waving his hand with increasing frequency when giving a lecture. However, in the 1950s a number of experiments that included control conditions appeared to support Skinner's claims. The results from an experiment carried out by Joel Greenspoon when a Ph.D. student at Indiana University became famous enough to receive the title, the *Greenspoon effect*. Student volunteers were asked to produce an endless stream of words, while Greenspoon, sitting just behind them, occasionally made either an approving sound, 'Mmm-hmm,' or one normally signifying disapproval, 'Huh-huh.' For one group Greenspoon made the approving sound whenever the participant produced a plural word and this resulted in a steady increase in the frequency of plural nouns until Greenspoon stopped his 'Mmm-hmms'; at this point their frequency decreased, thus apparently demonstrating an extinction effect. No such pattern was seen in participants who had a completely silent Greenspoon sitting behind them or those for whom each plural noun was followed by a sound of disapproval.[126]

A quite different approach to the problem of conditioning without awareness was taken by Ralph Hefferline, one of Keller's many Ph.D. students at Columbia who remained a lifelong friend. The target response in this series of experiments was a participant's thumb twitch, one so small that it could be detected only through amplification available only to the experimenter. In an experiment reported in 1959 the contingency arranged by Hefferline and his coresearchers was a form of Sidman avoidance: The pleasant music to which a participant listened throughout a session lasting over an hour was regularly interrupted by a loud and unpleasant hum, unless the thumb-twitch response was made at a steady rate. Cumulative records from 10 of the 12 participants given this condition showed a steady increase in twitching that leveled off when extinction conditions were introduced. The researchers reported that, when debriefed, these participants "still believed that they had been passive victims with respect to the onset and duration of the noise, and all seemed astounded to learn that they themselves had been in control." No such effects were found in control groups.[127]

Many years later a highly critical review of such research was given the provocative title, *There is no convincing evidence for operant and classical conditioning in human beings*.[128] A key argument was that studies such as those of Greenspoon and Hefferline did not include systematic assessment of awareness and, as such assessments became more refined in later studies, evidence for conditioning without awareness became progressively harder to find.

One such example was an analysis of the Greenspoon effect. The original experiment was repeated but with more careful interrogation of participants as to what they believed was happening. Almost all of them had tried to make sense of the situation and, although no one guessed correctly that 'Mmm-Hmm' was contingent on any plural noun, most came up with 'correlated hypotheses' that produced the apparent conditioning effect. Thus, one participant concluded that he needed to produce an endless stream of vegetable names, in plural.[129]

Whether such criticisms or alternative explanations can undermine every claim for conditioning without awareness has stimulated a controversy that has continued for over fifty years.[130]

Skinner's Final Two Decades

To return to Skinner: 1971 saw him reach new heights of fame – or perhaps, notoriety – when he appeared on the front cover of an issue of Time Magazine. This reflected the mainly critical reception given to his *Beyond freedom and dignity*. The book extended his earlier arguments, going back to his novel of 1948, *Walden Two*, that a solution to the major problems facing the world – now including damage to the environment as well as the prospect of nuclear war – requires the "need to make vast changes in human behavior … What we need is a technology of behavior."[131]

The controversy over the political and philosophical issues raised by *Beyond Freedom and Dignity* have been treated in detail elsewhere[132] and are outside the scope of this chapter. Just two points are worth noting here. First, Skinner did not spell out in the book what a 'technology of behavior' would look like, possibly because such developments had been limited since his optimistic enthusiasm for the topic nearly two decades earlier in *Science and Human Behavior*. For example, despite investing a great deal of time over many years in advocating the adoption of teaching machines within the educational system, this effort had been unsuccessful.[133] Second, perhaps because of the partial failure to fulfill the promises made in 1953, *Beyond Freedom and Dignity* does not appear to have inspired many to become involved in operant conditioning research. In contrast to the trend in the 1960s, in the 1970s the number of psychology departments that included laboratories for purely behavioral research involving animals began to decline.

By his late 60s Skinner had adopted a daily routine that allowed him to produce a steady stream of articles and books until well into his 80s. He would set an alarm to get up at 5 AM and write for the next two hours. His days became fairly leisurely before he retired to bed at 9 PM and slept until the alarm woke him at midnight for another hour of writing, with once again a cumulative recorder hooked up to monitor the number of words he had produced. He maintained this routine of three hours of writing for seven days a week almost till he died.

An interruption to Skinner's daytime routine started with the arrival in 1977 of a 24-year-old graduate student, Robert Epstein, who started by helping Skinner sort out his papers. They then started running experiments together. These had the general aim of demonstrating that pigeons could be trained to produce similar results to those from studies suggesting that primates possessed special cognitive abilities. Now 73, Skinner still retained his handyman skills and built the equipment in his basement workshop. He was proud to show that one did not need a grant to do experiments.[134]

The first of several experiments in their 'Columban simulation' series was inspired by a report that two chimpanzees engaged in symbolic communication. One pigeon, Jill, was reinforced for providing information on the 'sample' that allowed a second pigeon, Jack, to choose the correct response key in a matching-to-sample problem. "All of the contingencies responsible for the behavior of both speaker and listener could be found in the early verbal histories of children," claimed Skinner.[135]

Another experiment in the series was a simulation of the classic box-and-banana problem that Koehler had first set his chimpanzees during the First World War. A banana is hung from the ceiling and can only be reached if the animal moves a box located in a corner of the room to a position immediately under the banana. A pigeon was trained to perform two separate behaviors: Pushing a box around the small arena in the absence of anything suspended from the ceiling; and mounting the box when it was placed immediately under a 'banana' containing grain. The subsequent test was whether the pigeon would integrate the two behaviors and, chimp-like, push the box from a corner of the arena to immediately below the grain-filled banana. The success of this and other experiments confirmed Skinner in the view he had held since a student, namely, that there was nothing special about primate behavior or evidence for problem-solving in other species that could not be explained in terms of conditioning.[136]

From the early 1980s, Skinner suffered from increasing ill-health but did not stop working. He never modified his youthful claim that understanding the lever-pressing behavior of a rat could somehow illuminate all the fascinating and important phenomena that psychologists study. At the age of 86, he gave a speech to a meeting of the American Psychological Association in August, 1990. Eight days later he died.[137]

9 How Animals Learn to Associate Events

The cognitive revolution within American psychology marked a decline in the general influence of behaviorism, both of Skinner's radical behaviorism and of the long-established tradition that Hull had founded. It seems no coincidence that this occurred at the same time as the cultural revolution within Western societies in the 60s. The top American journals for publishing reports on experiments in animal learning relaxed their rules against the use of 'mentalistic' terms; one top researcher was relieved to find that use of the word 'attention' was now permitted.[1] A new way of studying and theorizing about learning emerged. Four researchers from different intellectual traditions led this development. Two were trained in the research tradition founded by Mowrer that concentrated on fear and avoidance learning, as described in Chapter 4. One was from the Hullian tradition described in Chapter 2 and the fourth was from the British tradition that emphasized the role of selective attention in discrimination learning, described in Chapter 7.

A key feature of this new research tradition was that it did not aim for a general theory of behavior. Instead, the behavior of an animal in some experiment was of interest to the extent that it provided an index of what the animal had learned. If the procedure involved pairing, say, a tone with a shock, then the degree of fearful behavior, such as freezing, evoked by the tone was of interest only to the extent that it provided an index of the strength of an association formed between the tone and the shock.

Edward Tolman would have been pleased by this change of direction. It was consistent with his argument that animals in Pavlovian experiments form expectancies. Furthermore, unlike his opponents, the Hullians or the Skinnerians, it did not insist that there is only one important form of learning.

A change in emphasis from previous research was the move away from instrumental – or operant – conditioning to concentrate on Pavlovian conditioning procedures. Only after a decade or so was there considerable effort to apply the theoretical ideas based on Pavlovian experiments to instrumental conditioning.

One feature remained unchanged. Following Skinner's argument that nothing was to be gained by 'physiologizing,' little or no attempt was made to link the new associative theories to events within the brain. It took until the 1990s before such attempts became common.

Contiguity and Overshadowing

Since at least the time of Aristotle, many of the philosophers in the Western tradition who have speculated about the nature of the human mind placed great importance on associations. Aristotle proposed that three factors were important in whether people formed an association between two events. These were: *similarity*, *contiguity* and *contrast.* By 'contiguity' he meant the occurrence of the events close together in time or in space. The importance of associations based on contiguity was also emphasized by philosophers in the British empiricist tradition, from Locke to Hume to James Mill to John Stuart Mill. In the present context, a particularly interesting thinker in this tradition was David Hartley. In 1749 he proposed a neural basis for the formation of associations.[2]

A common assumption, often implied rather than explicit, was that, whenever one perceived event was followed closely in time by another such event, the two would become associated. In Hartley's theory, the assumption that temporal contiguity was a sufficient condition for an association to be formed was very explicit: If the pattern of nerve 'vibrations' set up by perception of one event was immediately followed by a different pattern set off by a second event, then the two patterns would become connected.[3]

What appears to be the first experimental evidence that contiguity between two events is not sufficient for them to become associated was reported in Pavlov's 1927 book. His students had been carrying out experiments in which a compound made up of two stimuli occurring at the same time preceded the US. Thus, Dr. Palladin carried out a conditioning experiment that combined two simultaneous events into a compound stimulus: one event was tactile – a touch to the dog's skin – and the second was thermal – applying something at 0 degrees C (ice perhaps?) to a different part of the dog's body. Once conditioning to this compound was well established, tests were made both of the compound stimulus and of the individual components applied singly. These tests revealed a strong response to the tactile stimulus but none at all to the thermal stimulus. Another example was an experiment carried out by Dr. Zeliony, in which "a conditioned alimentary reflex was established to the simultaneous application of the tone of a pneumatic tuning-fork … and a visual stimulus of three electric lamps placed in front of the dog." Subsequent tests revealed strong salivation to the compound, but none to the lamps when these were presented on their own.

In reporting the results of several such experiments, Pavlov wrote: "When the stimuli making up the compound act upon different analyzers, the effect of one of them when tested singly was found very commonly to overshadow the effect of the others almost completely." Separate experiment showed that the overshadowed stimuli could be effective conditioned stimuli when used on their own.[4] To summarize in terms that Pavlov did not use, whether an association is formed between a stimulus and an immediately subsequent event depends on what other stimuli are present at the same time.

Hull was sufficiently interested enough in Pavlov's reports of experiments involving compound stimuli to carry out an experiment on this topic. This used human participants that were given shock-based conditioning. The procedure involved separately conditioning two stimuli and then testing when they were combined. However, although concerned by possible implications of overshadowing and complaining that

Pavlov's claims were "on the basis of admittedly inadequate empirical data,"[5] Hull did not carry out a parallel experiment of conditioning the compound and then testing each component on its own. Instead, he decided that overshadowing was to be explained in terms of *afferent interaction*. "Concurrent afferent impulses (*s*) arising from the impact of distinct stimulus energies (*S*) on receptors are appreciably modified by each other before they reach that portion of the central nervous system where they initiate the efferent impulses (*r*) which ultimately evoke reactions (*R*)."[6] In other words, interactions between stimuli occur at a peripheral level and are not of great interest.

It was over 20 years before there was serious examination of the possibility that stimuli might compete with each other beyond the 'periphery' of the nervous system and that, just because A immediately preceded B, this did not necessarily mean that an A-B association would be formed. One of the first triggers to this examination was an unexpected result obtained from a control group in an experiment run at McMaster University in Hamilton, Ontario.

Leon Kamin, His Blocking Experiment and the Concept of 'Surprise'

As a Harvard graduate student, Kamin (1927–2017) was subpoenaed by the House Un-American Activities Committee; see Figure 9.1. He was convicted of contempt of the Senate for not revealing names of people who were members of the communist party.[7] As a result, he was blacklisted by US universities and instead, as noted in Chapter 4, obtained positions in a couple of Canadian universities that allowed him to continue his studies of Pavlovian conditioning. In 1957 he was appointed chair of the Psychology Department at McMaster University, where he made a series of appointments that contributed to it becoming one of the top Psychology Departments in Canada.

Kamin's experiments used the conditioned suppression procedure. Hungry rats were first trained to press a lever for food pellets on a variable-interval (VI) schedule that delivered a pellet every 2.5 minutes on average. Daily sessions lasted 2 hours. During the conditioning stage, four widely-spaced trials were introduced into each session. A trial would typically consist of a 3-minute presentation of some stimulus, such as a white noise or an overhead house-light, or a compound of two such stimuli. During conditioning, such a trial would be followed immediately by 0.5-second electric shock delivered through the steel bars of the floor. The number of lever-presses, B, made during a trial was a compared with the number, A, during the 3-minute period immediately prior to the trial, by calculating a suppression ratio, B/A+B. Initially, before a rat learned that a stimulus predicted the arrival of a shock, the ratio would be around 0.5; as the rat learned this relationship the ratio would drop to around 0, indicating full suppression of the lever-press response. Typically, only around 4–6 trials were needed under these conditions before responding was effectively suppressed.

The aim of a series of such experiments was to "examine the role of attention-like processes in conditioning."[8] A critical result was found in an experiment using the following design. In an initial conditioning stage one stimulus, say the noise, was paired with shock, and in a second stage, a simultaneous compound of both the noise and the

Figure 9.1 In 1955 Leon Kamin, as a graduate student at Harvard, was called to testify before the McCarthy Committee. He was convicted of "contempt of the Senate" for not naming former communist associates.
Image credit: Bettmann / Getty Images.

light was paired with shock. In a subsequent test, the rats showed no fear of the light; prior learning about the noise had *blocked* conditioning to the light, even though this stimulus (when in compound) had been followed immediately by the shock on eight trials. Such data made it "perfectly clear that, for a stimulus to be conditioned, mere temporal contiguity with the US is not sufficient."[9]

"Put very naively, our first notion was that, because of the prior training to Element A, that element might so 'engage the animal's attention' during presentation of the compound that it would not 'notice' the added Element B. The failure to notice the superimposed element might preclude any conditioning to it."[10] This notion was rejected following a series of experiments, of which the following was particularly important. In addition to a standard blocking condition, in which pairings of a noise with a 1-milliamp shock in the first stage preceded pairings of the noise-light compound with the same shock in the second stage, an additional group were given exactly the same training, except that in the second stage, the shock level was increased to 4 milliamps. On the subsequent test, complete suppression was found to the light in this group. Clearly, the rats had not failed to attend to the light when given the noise-light compound.

Kamin concluded from this and related experiments that conditioning took place to the extent that an animal was surprised by an outcome. When a rat that had already learned that a noise would be followed by a 1-milliamp shock was given the light-noise compound, then the occurrence of this shock was entirely predictable, and therefore no new learning took place. However, it would have been a complete surprise to a rat when the compound was followed by the 4-milliamp shock; as a result, the rat learned to associate the light with this new event.

In language that would have been used in the past by very few learning theorists, Kamin speculated that a surprising event might be needed to initiate the backward scanning in some short-term store that is needed for an association to be formed. "The predictability of the US might strip the US of a function is normally subserves in conditioning experiments – that of instigating some 'processing' of the memory store of recent stimulus input, which results in the formation of an association."[11]

That blocking is a general phenomenon was almost immediately shown in experiments carried out by a series of graduate students in the pigeon lab of Kamin's colleague, Herb Jenkins.[12] In these experiments pigeons were trained in go/no go discriminations between, say, green and red response keys. Initial training on such a discrimination was found to block subsequent control by auditory stimuli that were perfectly correlated with reinforcement. The auditory stimuli did control responding in control groups that were not initially trained on the color discrimination.[13]

After his conviction of contempt of the Senate was overturned on technical grounds, in 1968 Kamin returned to the USA as chair of the Psychology Department at Princeton University; see Figure 9.2.

Figure 9.2 Leon Kamin shortly after his appointment in 1968 as chair of Princeton University's Psychology Department.
Reproduced with permission from Marie-Claire Kamin.

Robert Rescorla, and Contingency vs. Contiguity

The person who most deserves to be known as Pavlov's heir was Robert Rescorla (1940–2020). A possibly apocryphal story is that for many years he kept a copy of Pavlov's 1927 book on *Conditioned reflexes* besides his bed and would often read a few pages before falling asleep.[14] Universally known as Bob, he was born in Pittsburgh, attended high school in Westfield, New Jersey and in 1958 began his undergraduate studies in Swarthmore College in Pennsylvania. The college's Psychology Department had a strong Gestalt tradition, stemming from the influence of Wolfgang Koehler, who after leaving Nazi Germany, had been a professor there for 20 years and left only three years before Rescorla arrived. Despite the absence of a graduate school, the Psychology Department at Swarthmore also had a strong research tradition. As an undergraduate, Rescorla was involved both in studies of human learning and experiments with monkeys.

The latter work was under the supervision of Henry Gleitman, whose Ph.D. advisor had been Tolman and who had been carrying out a series of experiments on place learning by rats. Four years before Rescorla's arrival, Gleitman had co-authored a highly critical review of Hull's account of extinction.[15] In 1964 Gleitman was appointed professor and chair of the Psychology Department at the University of Pennsylvania, where Rescorla had entered as a graduate student two years earlier. As noted in Chapter 4, Gleitman was actively interested in a variety of topics, was known as an outstanding teacher, and in 1981 published what is arguably the best single-authored textbook of psychology ever written.[16]

Just like Kamin, Rescorla's advisor was Richard Solomon. Like Kamin, Rescorla's early research as a graduate student was related to the development of the two-factor theory of avoidance learning; see Chapter 4. In this context, Rescorla, together with his fellow graduate student and lifelong friend, Vin LoLordo, carried out a series of experiments in which a behavioral baseline was first established and then some Pavlovian procedure involving visual and auditory stimuli and shocks was introduced.

Several of these studies involved rats subjected to a conditioned suppression procedure of the kind used in Kamin's blocking experiments, as described above. However, a particular important experiment differed in that, the behavioral baseline consisted of a dog jumping from one side to another in a shuttle box in which the two compartments were separated by a barrier nearly 40 centimeters high. This behavior was maintained by a Sidman avoidance schedule whereby a rate of jumping of at least two jumps per minute would prevent the delivery of a shock. Together with LoLordo, Rescorla had previously found that dogs trained on this schedule would jump even faster when a signal for an unavoidable shock was presented but slow down when a signal for a shock-free period was presented.[17]

The new experiment differed in a critical way from these earlier experiments. After three days of training on the Sidman avoidance schedule, the dogs were allocated to three groups. While confined to one compartment of the shuttle box, one group (R) was given 'random' training in which shocks were delivered independently of the sounding of 5-s tone. The dogs in Group P ('positive prediction') were trained in

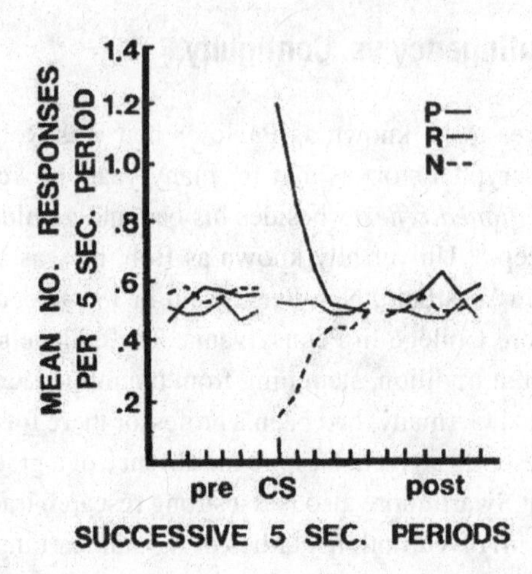

Figure 9.3 Test results showing responding by dogs on a Sidman shock avoidance schedule. In Group P presentation of a tone that predicted the occurrence of a shock increased response rates. In Group N a tone that predicted the absence of shock decreased responding. The tone had no effect in Group R for which there had been a random relationship between tone and shock. From Rescorla (1966). Reproduced with permission from Springer Nature.

the same way, except that they received only those shocks that were programmed to occur in the 30 s following each tone onset. The third group, N ('negative prediction'), received all the shocks that Group R were given, except for those programmed to occur 30 seconds after the end of the tone.

During a subsequent test when the dogs were again shuttling backwards and forwards to meet the demand of the avoidance schedule, the 5-second tones were presented every 2.5 minutes on average. The results from this test are shown in Figure 9.3: the dogs in Group P jumped faster when the tone arrived, whereas the dogs in Group N slowed down. The crucial new result was that the tone had no effect on dogs in Group R: "Despite the fact that Ss in Group R received at least as many pairings of the CS and US as Ss in Group P, only the Ss in Group P showed evidence of Pavlovian fear conditioning … These results suggest that we consider as a basic dimension of Pavlovian conditioning the degree to which the US is contingent upon prior CSs."[18]

This suggestion that Pavlovian excitatory conditioning depends on contingency rather than on pairings, that is, to the extent to which a conditioned stimulus predicts the unconditioned stimulus, was supported by further experiments that led to the following summary. "Two conceptions of Pavlovian conditioning have been distinguished by Rescorla (1967). The first, and more traditional, emphasizes the role of the number of pairings of CS and US in the formation of a CR. The second notion suggests that it is the contingency between CS and US which is important. The notion of contingency differs from that of pairing in that it includes not only what events are paired but also what events are not paired. As used here, contingency refers to the relative probability of occurrence of US in the presence of CS as contrasted with its probability in the absence of CS. The contingency notion suggests that, in fact, conditioning only occurs when these probabilities differ; when the probability of US is higher during CS than at other times, excitatory conditioning occurs; when the probability is lower, inhibitory conditioning results. Notice that the probability of a US can be the same in the absence and presence of CS and yet there can be a fair number of CS-US pairings. It is this that makes it possible to assess the relative importance of pairing and contingency in the development of a CR."[19]

In 1966 Rescorla received his Ph.D. and gave a job talk at Harvard that summarized evidence for the importance of contingency in Pavlovian conditioning. His appointment would have competed with Harvard's strong operant tradition. Whether Harvard made Rescorla an offer is not known. Maybe Yale University made a better one? In any case, that was where he took up his first academic position, literally alongside Allan Wagner, who had arrived at Yale seven years earlier.

Allan Wagner and the Relative Validity of Cues

Allan Wagner (1934–2018) was born in Springfield, Illinois. His parents were the children of German immigrants. Both encouraged the pursuit of learning and the study of science. After enrolling at the University of Iowa to study chemistry, in his junior year Wagner switched to a major in psychology. He was allowed to take Kenneth Spence's graduate course in learning theory and that decided his future. He remained at Iowa for graduate study, with Spence as his advisor. His first two research topics were on themes he continued for much of his career. One was on the partial reinforcement extinction effect and was related to the frustration theory of this effect developed by an earlier Spence student, Abe Amsel (see Chapter 2). The second topic was eye-blink conditioning, first in human participants and later in rabbit.[20]

Wagner obtained his Ph.D. in 1959, took up an assistant professorship at Yale, and in doing so left the orbit of one of the two top Hullians and entered into the orbit of the other. Neal Miller's lab at Yale was involved in several different projects to do with learning and motivation in animals. The research from this group that influenced Wagner most was on conditioned fear. His first research grant was to investigate the parallels between conditioned fear and conditioned frustration (see Chapter 2). This research used traditional maze running methods but at the same time, Wagner developed his skills in several very different experimental procedures. With his first graduate students, Shepard Siegel and Earl Thomas, he carried out Pavlovian conditioning experiments involving both eye-blink conditioning in rabbits and conditioned suppression in rats.[21]

A particularly influential colleague was Frank Logan, who had obtained his Ph.D. in 1951 with Spence as his advisor, whereupon – again like Wagner – he had obtained a position at Yale. The two taught a course on learning together, published a textbook together,[22] and then collaborated on one of the three key studies that led to the development of associative learning theory in the 1970s.

Wagner and his co-authors reported an identical pattern of results from three experiments that all followed the design shown in Figure 9.4. The experiments differed in

TABLE 1

Proportion of Exposures to Compound Stimuli A_1L and A_2L, and to the Absence of a Compound, $\overline{AL}$, which were Reinforced (+) and Nonreinforced (−) under the Two Experimental Treatments

	Correlated		Uncorrelated	
	+	−	+	−
A_1L	1.0	0.0	.5	.5
A_2L	0.0	1.0	.5	.5
$\overline{AL}$	0.0	1.0	0.0	1.0

Figure 9.4 The first experiments to reveal a *Relative Cue Validity* effect used the design shown in this table. A1 and A2 were auditory stimuli and L a light. Each animal was given two kinds of training in counterbalanced sequence. In the *Correlated* condition reinforcement always followed presentation of the compound, A_1L, and never followed presentation of A_2L. In the *Uncorrelated* condition there was a 50% chance of reinforcement following either of the compounds. Subsequent tests revealed a lack of responding to the light (L) in the Correlated condition, but considerable responding in the Uncorrelated condition, even though L had been followed by reinforcement equally often in the two conditions.
From Wagner, Logan, Haberlandt, and Price (1968). Reproduced with permission from the American Psychological Association.

the species – the first two used rats, while the third used rabbits – and in the methods that were used. The first used a discrete-trial operant procedure, in which rats had been trained to press a lever for food pellets. The second used a conditioned suppression procedure, similar to that used by Kamin, as already described. The third used conditioning of the eye-blink response.

Two kinds of stimuli were used in all three experiments. One was a bright overhead light, shown as L in Figure 9.4. The other – auditory (A) – stimuli were two tones, A_1 and A_2, one of 1kHz and the other of 2.5kHz. In both conditions, there were two types of occasional trials. During a trial, a compound stimulus was presented that consisted of one or other of the tones combined with the light (A_1L and A_2L). In the Correlated condition reinforcement (+) was available, or presented, only on A_1L trials and not (-) on A_2L trials. In the Uncorrelated condition, reinforcement occurred randomly on half of both kinds of trial. During the inter-trial intervals reinforcement was never available. The key feature of this design is that the light (L) is paired with reinforcement on half the trials in both conditions.

The important finding was that conditioning of the light occurred only in the Uncorrelated condition. Thus, the occurrence in the Correlated condition of a more valid cue, A_1, had prevented conditioning to the light. In terms that became more common a few years later, the general conclusion was that the degree to which a stimulus become associated with some outcome depends on its *relative validity*, compared to other stimuli, as a predictor of that outcome.[23]

These results, together with those demonstrating blocking and the importance of contingency, called out for a new theory of associative learning.

The Arrival of the Rescorla-Wagner Theory

Of my memories of excitement over a new theory or discovery, the strongest dates back to 1970. At the University of Sussex, major ingredients in Sutherland's recipe for raising the status of his new Laboratory of Experimental Psychology were on every suitable occasion to proclaim it as the best Department of Psychology in Europe and to extend invitations to spend time there to people he regarded as top researchers. Nineteen hundred and seventy was a bumper year for visitors. They included Herb Jenkins and Herb Terrace, together with Ulric Neisser who, long after co-authoring a critique of Hull's theory of extinction, had in 1967 published a seminal book, titled *Cognitive Psychology*,[24] that provided the first textbook in this new area of psychology.

Early in 1970, Sutherland let it be known that he had just been sent a pre-print of a paper by Rescorla and Wagner that set out their new theory of learning. He decided it was so important that he would hold a weekly seminar to discuss the theory, starting at 5 PM and ending at 6 PM when it was time to head up the hill to the village pub. His colleagues, including Sebastian Halliday and Euan MacPhail, postdoctoral researchers, Ph.D. students such as Westbrook and Dickinson, and the visitors were all invited to take part. Which we did, for several weeks arguing about the various assumptions and predictions of the theory but agreeing as to its importance.

As colleagues at Yale, Rescorla, and Wagner came from antagonistic intellectual traditions and maintained a highly competitive relationship. Interaction between them was minimal. Wagner chose not to attend the weekly research seminars, 'Learning Beer,' that Rescorla ran; see Figure 9.5. It seems that they discovered that their ideas were converging only when both were on their way to a symposium in Halifax, Nova Scotia. According to Mackintosh, because of a dense fog at the airport, he had to drive out to collect them. As they sat in the back of his car, they told each other about the papers they were due to present at the meeting; see Figure 9.6. It seems that they agreed to collaborate in developing a theory.[25]

The theory was eventually published in 1972 in a collection of chapters on Pavlovian conditioning in both humans and animals.[26] The key idea of the theory followed Kamin's suggestion that learning takes place only when an animal is surprised by an outcome. "The central notion suggested here can also be phrased in somewhat cognitive terms. One version might read: organisms only learn when events violate their expectations."[27] However, unlike Kamin, the authors went on to make this idea more precise by expressing it in mathematical terms. To do this they modified an equation

Figure 9.5 Bob Rescorla in 1970, the year he and Allan Wagner developed their theory. Reproduced with permission from Shirley Steele.

that Bush and Mosteller had proposed nearly 20 years earlier in an enterprise that Skinner had dismissed as 'premature mathematicising.'[28]

Bush and Mosteller's equation specified a change in the probability of a response as a result of a discrepancy on a given trial between the probability of a response and an asymptote of learning. Rescorla and Wagner modified this in two important ways. First, instead of referring to a response, they introduced a term represented by *V* that refers to the strength of an association between two events. Second, the discrepancy term now refers to the difference between the total associative strength of *all* current stimuli and an asymptote set by the strength of the unconditioned stimulus.

The key equation defines the increase or decrease in associative strength for a particular stimulus that occurs on a trial as follows:

$$\Delta V_{A} = \alpha_{A}.\beta(\lambda - \Sigma V)$$

In this equation, ΔV_A represents the increase in strength of an association between Stimulus A and a reinforcing event, such as a food pellet or a shock, that occurs as a result of a trial in some Pavlovian conditioning experiment. α_A represents the associability of Stimulus A; thus, this parameter is, for example, larger for a loud than for a soft noise. β represents the effectiveness of the reinforcing event, so that,

Figure 9.6 Allan Wagner and members of his lab in September, 1970. Gail Morse, his research assistant is on the far left; next to her is Lynn Eldridge, a doctoral student from another lab; and next to her Eva Popperova, a visiting researcher from Slovakia. On the far right is Maria Saavedra, who had just completed her doctoral dissertation on biconditional discriminations, then returned to Chile.
Reproduced with permission from Edgar Vogel.

for example, the rate of learning by a very hungry rat could be faster when a food pellet is used as the reinforcer than the rate of learning by a less hungry rat. λ represents the asymptotic strength of an association that can be supported by this reinforcing event, so that, for example, its value could be higher for a strong shock that a weak one. Finally, and crucially, ΣV represents the sum of the associative strength of all stimuli present on that trial.

How the theory explains a phenomenon such as Kamin's blocking effect, as described above, can be seen from the following example. In the first stage a noise, Stimulus X, is repeatedly paired with a shock until the strength of this association, V_X, has reached the asymptote, λ, that this particular shock can support. In the second stage a compound stimulus, AX, consisting of the noise combined with a light stimulus, A, is followed by the same shock. Because this compound will have a total strength, $\Sigma V = V_x + V_A$, equal to λ, there is no 'predictive error,' to use a term introduced later, and no learning will occur on such compound trials. Thus, V_A remains at zero; conditioning to the light has been blocked.

The theory also explains *unblocking*. Thus, following Kamin, if in the second stage, the shock intensity is increased, this leads to an increase in the asymptotic strength of an association that can be supported, from say λ_1 to λ_2. Now, when AX is presented and followed by the more intense shock, there is a prediction error, and learning about A will occur to the extent that λ_2 is larger than λ_1.

To account for Rescorla's contingency data, the theory needed to refer to the context in which conditioning occurs. "The important point to notice for this analysis is that the CS occurs against a background of uncontrolled stimuli. To speak of shocks occurring in the absence of the CS is to say that they occur in the presence of situational stimuli arising from the experimental environment. Although these stimuli are not explicitly manipulated by the experimenter, they can nevertheless be expected to influence the animal. Thus, one way to think about the occurrence of the CS is as an event transforming the background stimulus, A, into background-plus-CS, AX."[29] Following this idea, the finding that a CS that is reinforced at a rate no greater than the context alone (zero contingency) does not acquire any associate strength is to be understood in the same way as relative validity data: AX+ vs. A+ training fails to impart any associative strength to *X*, according to the R-W equations.

There have been many theories in psychology, some precise and some vague, but few have made clear predictions about interesting new phenomena to be found. The Rescorla-Wagner theory was exceptional in this respect. One example can be seen as the opposite of blocking. To follow the above example, if in Stage 1 a noise *X*, is trained as an inhibitor so that its associative strength, V_x, is negative, then at the start of Stage 2 when the compound, AX, of a light, A, with the noise, the compound will also have a negative value. As a result, the error term, $\lambda - V_{AX}$, is larger than it would have been if A on its own had been paired with the shock, $\lambda - 0$. Thus, a control group given the latter condition would show less conditioning than the compound group. Extra strong conditioning in the latter group is called *super conditioning*. Rescorla reported finding this predicted effect even before the theory was published.[30]

A second new phenomenon that the theory predicted was termed *over-expectancy* and is described later.

The Re-discovery of Konorski and Refining the Concept of Inhibition

As described in Chapter 1, Jerzy Konorski and Stefan Miller were the first to make a clear distinction between Pavlovian and instrumental conditioning. Following the death of Miller and destruction of their lab during World War II, Konorski was able to re-establish a small lab when the war ended. His 1948 book, published in the UK and in English, was critical of Pavlov, something that was never tolerated in Stalin's Soviet Union and barely tolerated in communist-run Poland.

With the death of Stalin in 1953, persecution of Konorski ended. He was able to expand the Nenski Institute after moving it to Warsaw. There were now opportunities for Western scientists to visit Warsaw and for eminent Eastern bloc scientists to visit laboratories in the West. In 1967 Konorski published his second book, again in English, *the Integrative activity of the Brain*. Around this time, he visited the Psychology Department at the University of Pennsylvania. There, just a couple of years earlier, the two graduate students, Rescorla and LoLordo had carried out experiments based on ideas of inhibition close to those of Konorski. However, it seems that they only became aware of his work much later.[31]

One reason why Western researchers had paid little attention to inhibitory conditioning is that, compared to excitatory conditioning, it can be difficult to measure inhibition in a direct way. Coming from a background of two-factor theories of avoidance, the solution found by Rescorla and LoLordo followed this reasoning: "If avoidance behavior is maintained in part by a conditioned fear reaction, then any stimulus that increases this fear should enhance the avoidance response and any stimulus that inhibits fear should weaken the avoidance response."[32] As described earlier, this idea led to the series of experiments demonstrating the importance of contingency rather than contiguity. In preparations other than those using an avoidance baseline, detecting inhibition may require a combination of a *retardation* test with a *summation* test, as Rescorla argued in 1969.[33]

Interest in inhibition increased quite separately in two other research areas during the 1960s. One area was what was then called physiological psychology: Lesions of certain sub-cortical brain areas were found to produce changes in behavior that seemed best described as a loss of the ability to inhibit responses. For example, rats with septal lesions failed to slow their response rate on a schedule – the *DRL* or *differential reinforcement of low response rates* – that reinforced such slowing[34] and also had difficulty learning when a discrimination was reversed.[35]

The other area was discrimination learning using operant conditioning schedules. Within this research tradition, experimenters had obtained inhibitory generalization gradients[36] and others had appealed to *response inhibition* to explain behavioral contrast;[37] see Chapter 7.

In 1970 Halliday and I felt that a great deal might be gained from a conference in which Western researchers from these different areas, as well as researchers on conditioning from Eastern Europe, could exchange views. The idea was both presumptuous, given our junior status, and naïve, given our almost complete lack of experience in even attending conferences. Those pre-internet days meant that we had to send out a large number of individual invitations in the mail. The response amazed us. Almost everyone we contacted agreed to take part, even Konorski from Warsaw and Asratian, the most senior Russian scientist following in Pavlov's footsteps. The response occurred even though we could not offer any financial assistance for travel and accommodation, except to Wagner who was the only one to benefit from the small grant we obtained from the Royal Society. The conference was held at the University of Sussex in 1971 and a selection of the papers were published a year later, under the title, *Inhibition and learning*.[38]

The paper that had the largest and longest impact was by Wagner and Rescorla.[39] They made it clear that the order of authors did not indicate that the first contributed any more than the second; rather both had contributed equally to both papers in which their theory and its implications were described. This second paper started with a description of the theory that was more clearly presented than in the first paper. This was followed by reports of experiments using a *conditioned inhibition* design, A+ vs. AX-, that yielded further support for the theory.

The final part of their paper considered three new issues that remained controversial and stimulated many subsequent studies. One was the so-called *latent inhibition* effect, the very general finding that simple pre-exposure to a stimulus (X-) retards subsequent conditioning (X+). A key finding was that such pre-exposure also retards conditioned inhibition learning (A+ vs. AX-). Therefore, they concluded that stimulus pre-exposure does not involve inhibitory learning, that is, 'latent inhibition' is a misnomer, and is better conceived as a manipulation that simply reduces the associability of a stimulus.

A second issue was whether simple discrimination training, A+ vs. B-, produces any inhibitory conditioning to B. Again, on the basis of retardation and summation tests, they included that such training does not involve any inhibitory learning. The third issue was a test of the theory's surprising prediction that, once a stimulus, X, is first established as an inhibitor by A+ vs. AX- training, it becomes resistant to extinction during subsequent repeated X- trials. Preliminary experiments failed to find any evidence that such repeated exposure of a conditioned inhibitor, X, had any effect on its ability to retard subsequent X+ training or to reduce responding to a second excitatory stimulus, B+, in a summation test, BX-. Subsequent experiments from Rescorla's lab provided further evidence for such resistance to extinction of a conditioned inhibitor.[40]

Nicholas Mackintosh's Attentional Theory of Associative Learning

For someone who was awarded his Ph.D. in 1963, followed by an appointment as a lecturer at Oxford University, Nick Mackintosh (1935–2015) spent a remarkably large amount of time in various American universities in the mid 1960s; see Chapter 7. Moving around the world, accompanied by his first wife and their baby son, somehow did not slow down Mackintosh's productivity. In 1967 he was appointed to a research-only position at Dalhousie University, in Halifax, Nova Scotia, as the Killam Professor of Psychology. The twenty or more major experimental and theoretical papers he had published since his Ph.D. must have been a key factor in this appointment. When Mackintosh was a schoolboy at Winchester College, a teacher reported that "Mackintosh has a fertile mind and a fluent pen and he uses both to avoid doing any work." Only this last observation could not be further from the truth about his adult life;[41] see Figure 9.7.

Almost all the research that Mackintosh carried out before he moved to Halifax was related to the attentional theory of discrimination learning that Sutherland had first

Figure 9.7 Nick Mackintosh in the early 1970s.
From Pearce (2018). Reproduced with permission from the Royal Society (UK).

proposed (see Chapter 7) and that had been major impetus for Wagner's experiments on relative validity. Mackintosh's experiments had employed the modified Lashley jumping stand, T-maze, or straight runway. At Dalhousie, he continued for a while to use such equipment but also added a set of Skinner boxes to his lab.

The symposium that Mackintosh and Honig had organized in 1968 included as invited speakers Kamin, Estes, and Jenkins, as well as Rescorla and Wagner.[42] This gave Mackintosh early familiarity with the ideas that developed into the Rescorla-Wagner theory. He was convinced of the theoretical importance of understanding Kamin's blocking effect: "we should be continuing with the work initiated by Kamin."[43] And this is the purpose for which his new Skinner boxes were intended.

One of Mackintosh's first experiments was to test the prediction from the Rescorla-Wagner theory and from Kamin's less formal account that overshadowing should not occur during a single conditioning trial. However, such one-trial overshadowing should be detectable according to Mackintosh's attentional analysis. Using a similar conditioned suppression procedure to that used by Kamin, Mackintosh obtained clear evidence of overshadowing by a light of conditioning to a tone, following a single trial in which a compound stimulus consisting of the light and the tone was followed by shock.

This first experiment was followed by one on blocking. The reasoning behind this was to test Kamin's claim that, following extensive pretraining on one stimulus – 8 Tone->shock pairings in this case – the shock is then perfectly predicted, so that, when subsequent Tone + Light compounds are followed by the same shock, no further learning should occur either to the Tone or to the Light. Mackintosh first gave all his rats training with Tone-shock pairings; then, one group was given a single Tone + Light->shock trial, and a second group was given sixteen such trials. The results from subsequent tests, in which the Tone was presented on its own, showed that the group given sixteen TL+ trials had learned to fear the tone even more strongly during this extra compound conditioning. According to Kamin's account, the lack of surprise when the TL compound was first followed by shock should have blocked both conditioning to the light and further conditioning to the tone. Mackintosh concluded: "Although the results of neither of these experiments are consistent with Rescorla and Wagner's theory, it is clear that the most they do is to show that other factors (presumably attentional) contribute to overshadowing and blocking."[44] The theory that these and other experiments led to was published in 1975.

In the Rescorla-Wagner theory the symbol, α, represents the associability of a stimulus; thus, α is at a maximum value, close to 1.0, for a very loud noise or very bright light, but closer to zero for quieter or less bright events. The theory proposes that α has a fixed value. A key feature of Mackintosh's attentional theory of 1975[45] is that α varies according to the extent that it serves as a predictor of events to follow. Thus, according to the theory blocking occurs because the animal learns that the added stimulus provides no extra information; consequently, its α-value is reduced – the animal has learned to ignore this stimulus – and this leads to a weaker association between the stimulus and the reinforcer.

Traditional attentional theories, such as the one that Sutherland and Mackintosh developed to account for features of discrimination learning (see Chapter 7), assume that an increase in attention to one stimulus or stimulus dimension is always accompanied by a decrease in attention to other stimuli or stimulus dimensions. Mackintosh dropped this 'inverse' assumption from his new theory; thus, the decrease in α-value of a blocked stimulus does not lead to an increase in the α-value of the blocking stimulus. Dropping the inverse assumption made sense of the failure to find mutual overshadowing by two stimuli, one salient and the other less salient, when they were compounded as a signal for reinforcement. According to the Rescorla-Wagner theory, not only should the presence of the strong stimulus decrease conditioning to the weak stimulus, but the weaker stimulus should somewhat decrease conditioning to the strong stimulus. The latter prediction proved very difficult to confirm.

As noted earlier, *latent inhibition* refers to the very general finding that animals learn more slowly about stimuli that are already familiar than they learn about novel stimuli. While the Rescorla-Wagner theory offered no explanation for this robust effect, Mackintosh's theory explains latent inhibition as a decrease in the α-value of a stimulus as it fails to predict any outcome.

The topic of attention is revisited in a later section.

Kamin Tests for Over-expectancy and Later Shuts Down His Lab

Kamin was not at first convinced that the Rescorla-Wagner theory fully captured his ideas that learning occurred only when some surprising event occurred. He had doubts about a remarkable new effect that their theory predicted, one termed *overexpectancy*. In an initial stage, both A and B are separately conditioned and then in the second stage, the compound AB is paired with the same reinforcer for a number of trials. The somewhat counterintuitive prediction made by the theory is that compound conditioning in the second stage should *reduce* the excitatory value of A and B. One of the triumphs announced by Rescorla and Wagner in 1972 was confirmation of just such a reduction.

Kamin spotted a further and even more surprising prediction from the theory. In the experiments on overexpectancy reported by Rescorla and Wagner, the two stimuli were of more or less equal salience. If, instead, one stimulus – for example, a bright light, *L*, is more salient than the second stimulus – for example, a moderate noise, *n*, the prediction for an overexpectancy experiment using a conditioned suppression procedure is that the animal should end up showing more fear to the weaker stimulus, *n*, that to *L*. This follows because, after both stimuli have been conditioned to asymptote, during subsequent conditioning of the *Ln* compound, *L* will lose strength more rapidly than *n*. Kamin persuaded a graduate student, Stephen Gaioni, to run such an experiment. The results came out exactly as the theory predicted.

At the end of the paper that reported their experiment, the authors decided to express the results in 'more cognitive' terms. "When the two elements are compounded for the first time, the animal expects a US much more intense than that to which it has previously been conditioned. The animal is now surprised by the 'weakness' of the US, and the decrements which accrue to the conditioned strength of each element are analogous to those which occur during normal extinction of a single conditioned element. The model further specifies an effect that was not at all intuitively obvious – that, in the compound case, the decrements should be distributed across elements in direct proportion to their relative saliences. Without the formal model, it is doubtful whether such a problem would have been considered and even more doubtful that the direction if such an effect would have been specified in advance."[46] The outcome of this experiment removed any reservation that Kamin had about the Rescorla-Wagner theory; as he commented later, "my ugly duckling of a theory was turned into a beautiful swan."[47]

Kamin had exceptional analytic skills when it came to detecting some problem with the data or a missing control group in an experimental report. In the Fall of 1971, I attended a research seminar in which Kamin demonstrated these skills by detecting a potential problem in a paper by Rescorla, a major challenge. A few weeks later Kamin applied these skills to a very different set of data.

Several months earlier Herrnstein had been invited to Princeton's Psychology Department to report on the research he had carried out on visual category learning by pigeons; see Chapter 8. Subsequently, in September 1971, he published an article in the *Atlantic Monthly* that endorsed theories that human intelligence was largely determined by a person's genes rather than by environmental factors. A small section of the article addressed the topic of racial differences in IQ. "Although there are scraps of evidence for a genetic component in the black-white difference, the overwhelming case is for believing that American blacks have been at an environmental disadvantage. To the extent that variations in the American social environment can promote or retard I.Q., blacks have probably been held back."[48]

Herrnstein's failure to reject the possibility of a genetic basis for racial differences in IQ led to student protests at a couple of campuses to which he had been invited as a speaker, including Princeton. The protesters demanded that on his visit Herrnstein should be challenged publicly on the arguments in his article. Half the psychology faculty agreed with the students; the other half argued that Herrnstein should give only the talk on pigeons that had been agreed. Kamin, as chair, needed to decide between the two camps. To do so, he decided, first, that he would take a careful look at data related to the heritability of IQ. The large desk in his office supported growing piles of books and papers on the topic. Meanwhile, Herrnstein decided to cancel his visit.

In the early 1970s, the most important body of evidence bearing on the heritability of IQ was that published some forty or more years earlier by the eminent and highly influential British educational psychologist, Cyril Burt. By measuring the IQs of identical twins who had been separated early in their lives, Burt concluded that the heritability of IQ was around 0.80, that is, very high. Kamin had an unusual ability with numbers. Apparently, he had been a child prodigy in arithmetic and as an adult, for example, could carry out a Mann-Whitney statistical test in his head.[49] The employment of these skills in Kamin's scrutiny of Burt's data revealed some suspicious features that persuaded Kamin that Burt's evidence was unreliable. He went on to write a book, *The science and politics of IQ*.[50] Its conclusion that there is no good evidence for the heritability of IQ, set off a vigorous controversy lasting many years and stimulated a large number of studies of human intelligence.

As a conscientious chair of a large department, Kamin had never been much involved in the day-to-day running of his lab. With his new, all-consuming interest in IQ, the paper on overexpectancy proved to be the last on animal learning on which Kamin was an author. Running the rat lab was left to his first and last postdoctoral researcher, Fred Westbrook. When the latter took up an appointment in Australia, the lab was closed.

SOP: Wagner's Extension and Process-based Version of the Rescorla-Wagner Theory

After collaborating on the two papers that explained their theory, Rescorla and Wagner parted company. Rescorla took the experimentalist path to analyze further aspects of associative learning. Wagner took a more theoretical path. A major respect in which he differed from his mentor, Spence, was that, while the latter regarded cognitive psychology as the enemy of the Hull-Spence theory, Wagner very much appreciated the research carried out by his cognitive contemporaries, especially studies of human memory.

The R-W theory was purely mathematical in that it made no appeal to underlying processes. Wagner's commitment over the following three decades was to specify such processes. His starting point was Kamin's suggestion that "the predictability of the US might strip the US of a function which it normally subserves in conditioning experiments, that of processing of the memory store of recent stimulus input, which results in the formation of an association."[51] Wagner proposed that an association is acquired only if post-trial rehearsal takes place in a limited capacity short-term memory store. It follows that, if such rehearsal is interrupted, for example, by a surprising event, either no association is formed or it is formed more slowly. Experiments on human memory had recently found that, if presentation of a to-be-remembered word was closely followed by some 'attention capturing' event like a photograph of a nude, then the participant was less likely to recall the target word.

Jerry Rudy arrived at Yale in 1970 with a fresh Ph.D. from the University of Virginia. Following a suggestion from Wagner that he come up with a way of studying rehearsal in animals, Rudy designed an experiment that he then commenced and that Bill Whitlow, a Ph.D. student, took over when Rudy left for a job at Princeton.[52]

Using the rabbit eyelid conditioning preparation, the design of the first of these experiments was as follows. In an initial stage, the animals were well trained on a discrimination in which stimulus A was always followed by a shock, and stimulus B was never reinforced. In a second stage a new stimulus, C, was followed by first the shock and then, 10 seconds later, by what was described as a posttrial episode (PTE). For the *Congruent* group of rabbits, the PTE was the non-reinforced presentation of B – an unsurprising event – whereas for the *Incongruent* group the PTE was the presentation of A followed by the surprising absence of any shock. For a control group, pairings of C with shock were not followed by any kind of PTE.

The question was whether in this second stage, the Incongruent group would acquire conditioned responses to C more slowly than the Congruent group. As seen in Figure 9.8, this was just the result the researchers obtained; indeed, the group given the Congruent condition learned just as rapidly as the *None* control group for which no kind of event followed the shock. Subsequent experiments in the series confirmed the general conclusion that "surprising episodes command rehearsal."[53]

As noted earlier, if a stimulus such as loud noise is repeatedly presented, when subsequently paired with some reinforcer conditioning is slower than if the loud noise had been novel; a *latent inhibition* effect. If, during repeated presentation of the noise, some aspect of the animal's reaction is measured, this response is seen to weaken; an effect known as *habituation*. Wagner proposed that both effects could be explained by the same processes. He distinguished between long-term and short-term habituation. He first described his ideas on habituation in 1974 in unpublished paper titled: *Priming in STM: An information processing mechanism for self-generated* (i.e. short-term) *and retrieval-generated* (i.e. long-term) *depression in performance.*[54]

Wagner's analysis of long-term habituation resembled the way the Rescorla-Wagner theory analyses contingency effects. As explained above, this proposes that an association is formed between a reinforcer and the context in which it occurs, so that if the context is a better predictor of the occurrence of the reinforcer than any potential

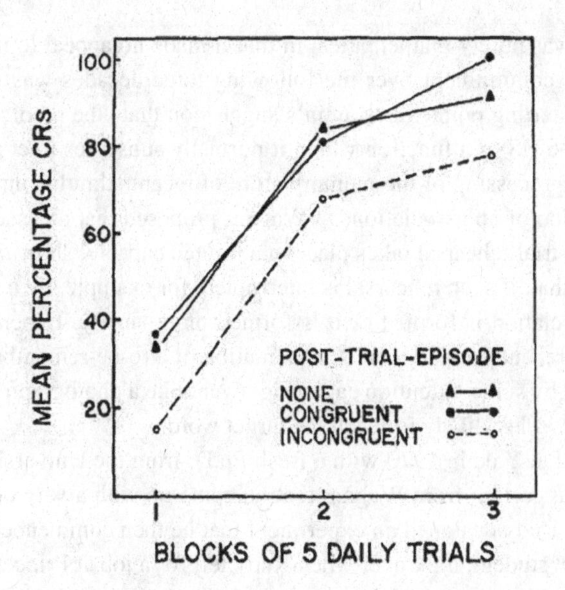

Figure 9.8 This graph shows the acquisition by rabbits of eyelid responses (CRs) to a light that was followed by a shock. For the *Congruent* group 10 seconds later a stimulus was presented that had previously been experienced in the absence of shock. As shown in this graph, this "unsurprising" event had no more effect on acquisition of the response to the light that in a group, *None*, that had never been given any post-trial event. In contrast, in the *Incongruent* group a post-trial ("surprising") event that had previously been followed by a shock, but not now, retarded conditioning.
From Wagner, Rudy, and Whitlow (1973). Reproduced with permission from the American Psychological Association.

conditioned stimuli, no conditioning occurs to the latter. Wagner extended this idea to the formation of associations between a context and all the stimuli that occur within it. As an animal learns that, say, a particular loud noise is likely to recur, it becomes increasingly predictable in that context; or in Wagner's terms, its representation is primed in short-term memory. As a result, it both reacts less vigorously and learns more slowly when the stimulus is subsequently followed by reinforcement.

Short-term habituation occurs when a stimulus is repeated after a short interval and the response to the second presentation is reduced. An experimental example of this effect, the acoustic startle reflex, had been studied by Michael Davis, a previous student of Wagner's who had obtained his Ph.D. in 1970.[55] Wagner now persuaded Whitlow to carry out a further series of experiments on habituation for his Ph.D. The basic method was to present pairs of loud tones ($S_1 - S_2$) to a rabbit and measure the resulting change of blood flow through a rabbit's ear. The tones could be of either high, 4,000 Hz, or low, 530 Hz, frequency. The time between the first and the second tone could be either 30, 60, or 150 seconds. The basic finding was that, when S_1 and S_2 were identical and separated by no more than 60 seconds, the response to S_2 was much smaller than when S_2 differed from S_1. As well as this demonstration of self-generated priming, a decline in the response to S1 when such trials were repeated demonstrated long-term habituation or *retrieval-generated priming*. Whitlow concluded his paper by noting that Wagner's theory "may offer a theoretical bridge between the phenomena of habituation and the more heavily researched phenomena of conditioning and memory."[56]

What was to be the first of several variants of Wagner's theory was described in 1981 in a chapter titled, *SOP: a model of automatic processing in animal behavior*, a somewhat odd title given that it described a theory of Pavlovian conditioning. As for 'SOP,' this initially stood for 'Sometimes Opponent Processes' and then, more commonly, for 'Standard Operating Procedures.' Following some suggestions by Konorski in 1967 and contemporary theories of human memory, the theory proposed that memories of events – or stimuli – are represented by elements that can either be in active states, *A1* and *A2*, where *A1* indicates that the memory is attended to and *A2* means the memory is at the periphery of attention, or could be in an inactive state, *I*. The detection of a previously encountered stimulus activates some of the elements representing the stimulus, with the result that these move from an inactive state to an *A1* state. Decay processes operating with fixed timing ensure that elements in *A1* decay relapse into *A2* and from there into an *I* state.

An association between two events, A and B, is formed and changes according to the extent that elements of A and elements of B are both in an *A1* state at the same time. An inhibitory relationship is formed, say between A and C, when elements of A are in an *A1* state at the same time as elements of C are in an *A2* state.

Further details of the theory are beyond the scope of the present chapter. It is sufficient to report that by allowing the learning of associations in a Pavlovian conditioning preparation to depend on both the surprisingness of the CS and of the US, SOP can explain both all of the phenomena covered by the Rescorla-Wagner theory and those that are additionally covered by attentional theory. In addition, while these theories make no reference to the timing of events, SOP is a time-based theory that can also predict the strength of both a conditioned and an unconditioned response.

Forty years later SOP remains one of the most powerful theories to explain a whole range of phenomena in associative learning. For example, after testing a number of simulations based on SOP, a report from Westbrook's lab concluded: "The present study has shown that Wagner's SOP model accounts for many findings that characterize experimental extinction, including its so-called signature characteristics (renewal) in all its various forms, reinstatement, and spontaneous recovery), the PREE, and variations in extinction with CS preexposure, the interval between extinction trials, and other stimuli present during those trials. Indeed, across the various classes of theories or models that have been applied to extinction (CS-processing, US-processing, configural theories, and retrieval interference theory), Wagner's model appears to be uniquely positioned to accommodate the full array of findings just described. These successes of the model follow from its unique form of stimulus representation and rules for generating associative change, which allow the model to explain conditioning/extinction phenomena by appeal to changes in processing of both CSs and USs; and its ability to incorporate potential changes in outcome certainty or uncertainty (or competition) in the processing capacity of A1."[57]

Rescorla's Experiments on Associative Learning in the 1970s

Throughout the 1970s Rescorla and his students carried out an impressive number of important experiments. Although now a well-established professor at Yale with many teaching and administrative duties, Rescorla remained a hands-on researcher. Initially, his lab was housed in crumbling rooms in the building dating back to the 1930s that had housed Hull's Institute for Human Relations; see Chapter 2. Then he was allocated a new lab in the heart of the campus. Rescorla liked to be the first one in the lab every day but was occasionally challenged by a graduate student. He would start his own experiments each day and sometimes those of his students.[58]

Some of the experiments extended ideas from the Rescorla-Wagner theory. Others examined types of associations that were beyond the scope of the theory. Only a few examples are discussed here.

An early concern was with the effects of manipulating the properties of a unconditioned stimulus after both first-order and second-order conditioning and then examining how this subsequently affected an animals response to the conditioned stimulus. Thus, one study used the standard procedure of conditioned suppression of lever-pressing for food pellets as a measure of conditioned fear to a light stimulus. What was unusual was the use of a loud noise instead of a shock as the unconditioned stimulus. In the following habituation stage, the noise was presented on its own again and again until it had little effect on the rats' lever pressing for food. In the test that followed the light was again presented to this group of rats with the result that it evoked far less fear than in a control group that had not been habituated to the loud noise.[59]

A conclusion that Rescorla drew from this result was that "first-order conditioning consists at least in part of learned associations between some internal representation of the CS and of the US."[60] Although important, this was far from being the first evidence to support such a view. What, however, was particularly interesting was that this effect of habituation on first-order conditioning was quite different from that found with second-order conditioning, as explained later. Meanwhile, we will turn to Rescorla's ideas on extinction, a topic for which the Rescorla-Wagner provided an inadequate analysis.

In his lecture on extinction, Pavlov wrote: "Left to themselves extinguished conditioned reflexes spontaneously recover their full strength after a longer or shorter interval of time, but this of course does not apply to conditioned reflexes which are only just in process of formation. Such reflexes, being weak and irregular, may require for their recovery after extinction, a fresh reinforcement by the underlying unconditioned reflex."[61] The generality of the first effect, *spontaneous recovery*, was soon well established by Western researchers. The latter generally rejected Pavlov's explanation in terms of the relative instability of the inhibitory processes that initially produced the extinction of a conditioned response, but without a clear alternative until the 1990s.[62]

The second effect referred to in the above quote from Pavlov has received far less attention. What became known as *reinstatement,* the recovery of a conditioned response following presentation of the original unconditioned stimulus, had been studied by Konorski.[63] Rescorla and his student, Donald Heth, revived interest in this effect in the context of Rescorla's ideas on conditioning and extinction. "Conditioning not only produces the formation of inter-event associations but also results in the building up of representations of those events."[64] As for extinction, "casually speaking, during extinction of a CS which has previously been paired with shock, the animal might reach either a conclusion about associations (that the US no longer follows that CS) or one about the US (that it has been dramatically reduced in strength). Thus, some of the responsibility for extinction-induced decrement might rest with a modified UCS representation."[65] Within this framework, reinstatement can be seen to result from restoring the strength of the US; a new shock reminding the animal of what shocks were like during conditioning. The results obtained from a series of experiments on reinstatement supported this claim.[66]

By the late 1950s, the debate between the Hullians and the Tolmanians over what is learned in conditioning experiments – crudely, habits or expectancies – had reached a stalemate as far as many theorists were concerned. In 1958 a student at the University of Chicago, Bill Rozeboom, wrote a conceptual thesis that argued that 'what is learned?' is a valid and important question and suggested a novel experimental procedure for finding an answer. In terms of Pavlovian conditioning, the critical new phase was to change the properties of the unconditioned stimulus following the initial conditioning phase. An example is provided by the experiment described earlier, in which fear conditioning was based on a loud noise to which the rats were subsequently habituated. If such a *revaluation* decreases responding to the CS, then one might conclude that the rat's response to the CS depends on an association between the CS and the loud noise; in other words, an expectancy. Rozeboom gave the confusing name, 'conditioned generalization,' to this procedure. The paper that summarized his arguments ended: "Because of … its crucial significance for the development of behavior theory, it is to be hoped that the conditioned generalization problem will soon be subjected to a concerted experimental attack."[67] Few, if any, researchers fulfilled Rozeboom's hope, until Rescorla ran with the idea.

An early test of the effects of devaluation in Rescorla's lab involved second-order conditioning. This refers to a procedure whereby in an initial first-order stage a stimulus, S1, is paired with an unconditioned stimulus such as food or a shock and this is followed by a second stage in which a second stimulus, S2, is paired with S1. Rescorla and his student, Ross Rizley, reported a series of experiments using a conditioned suppression measure in which the first-order stimulus (S1) was a light that was paired with shock. This was followed by a second-order stage in which a noise (S2) was followed by the light (S1) as a conditioned reinforcer. The effectiveness of second-order conditioning was revealed in subsequent tests in which the rats showed they were now frightened of the noise; various control groups did not show such fear. Surprising results were obtained after the light as S1 had been devalued by simply repeatedly presenting this stimulus in the absence of shock until the rats no longer displayed any fear in its presence. Despite such extinction of the first-order stimulus, a subsequent test revealed that the rats' fear of the noise as S2 was unaffected.

In discussing the conclusions to be drawn from these results, the authors considered two possibilities. "One possibility is that during the second-order phase the first-order CS re-arouses a representation of the US which then becomes associated with the second-order stimulus. Alternatively, the second-order conditioning procedure may result in a connection between the second-order stimulus and the CR, which is evoked in its presence by the first-order stimulus. The present experiments do not distinguish between these possibilities."[68]

Unusually for a Rescorla paper, it ended with discussion of possible real-world implications of these results. "It is common to regard certain phobic behaviors as examples of Pavlovian fear conditioning (e.g., Bandura, 1969). And it seems plausible that in many cases the phobia may be based upon second-order conditioning. If that is so, the present experiments suggest that elimination of fear of the original first-order stimulus upon which the phobia was based will not eliminate the phobia. The origin of the fear may have long ago lost its effect while secondary stimuli continue to produce anxiety. This observation may help to account for some of the apparent irrationality of such fears as well as to indicate that procedures aimed at their removal should not involve a search for their origin."[69]

This initial study of reinforcer devaluation using second-order conditioning was followed by one involving first-order conditioning. As already described briefly, this was achieved by using a loud noise as the reinforcer to condition fear to the 30-s presentation of a light as the S1 stimulus. In two experiments, subsequent repeated exposure to the noise resulted in habituation of responses to this sound. When the effect of such habituation on the light was tested, the rats no longer showed much fear. A different outcome was obtained in a third experiment that now involved second-order conditioning. Following first-order conditioning – light paired with a frighteningly loud noise – in a second-order phase a tone, as S2, was now paired with the light, as S1. Subsequent habituation of fear of the noise resulted in diminished fear of the light, the S1, but not of the tone, S2. Thus, these results replicated the earlier finding concerning the apparent autonomy of second-order conditioning under conditions in which first-order conditioning was found to remain sensitive to the current value of the original reinforcer.[70]

When Rescorla was invited to give a series of lectures at the University of Alberta in 1979, he chose to present an overview of research on second-order conditioning. A year later a minor edit of these lectures was published as Rescorla's only book.[71] The densely-packed information about dozens of complicated experiments makes one wonder how many people in the Alberta audience followed his lectures. However, at least one member of the audience was delighted: Bill Rozeboom.[72]

The point of studying Pavlovian conditioning, Rescorla argued, is that it provides understanding of "the learning of relations among events in the environment." He proposed that this involves finding answers to three main questions: "1. What are the conditions for learning to take place? 2. What is the content of learning? 3. How does such learning influence behavior?"[73]

Rescorla contrasted his approach to that of the 'reflex' concept of conditioning, dating back to Pavlov's concept of conditioning as a process by which a new stimulus comes to elicit the unconditioned response evoked by the unconditioned stimulus. This *principle of stimulus substitution* was still held by some theorists in the 1970s to indicate 'true' Pavlovian conditioning.[74]

Part of Rescorla's 1979 argument against the reflex concept came from the results from a second-order conditioning study that he and a graduate student, Peter Holland, had undertaken, one that used a relatively novel conditioning procedure developed by a Yale colleague, Fred Sheffield. This was inspired by the increase in activity level that hungry rats display when some event signals that they are about to be fed. Holland and Rescorla constructed a mechanical device for measuring a rat's activity, one very crude compared to later electronic versions. In one experiment they first arranged for a light to serve as S1, the signal that a food pellet was about to be delivered to a hungry rat. The result, as shown in the left-hand side of Figure 9.9, was that activity increased by a small amount in the presence of the light. Second-order conditioning was then carried out: a clicker, S2, was then paired with the light. This resulted in a steady increase in activity to the clicker, one much larger than that to the light,[75] as seen in the right-hand side of Figure 9.9.

The authors drew a number of conclusions from these results and that of related experiments. Some followed from the finding that, when the clicker was used as the first-order stimulus, S1, the response rate was much greater than when the light was used as S1. The first conclusion was that the degree of responding to the S1 can be a misleading measure of the strength of an association; as seen in Figure 9.9, the light must have been very strongly associated with food, since it served as a highly effective reinforcer in second-order conditioning. Second, the amount of responding seen in first-order conditioning depends on the type of stimulus involved and thus the principle of stimulus substitution is inadequate.

In his 1980 book, Rescorla provided several other examples of the use of second-order conditioning to detect learning. A study run with another student, Jim Nairne, used autoshaping in pigeons, a preparation that in Rescorla's lab came almost to replace fear conditioning in rats. When a tone, as S1, is paired with the delivery of food to a hungry pigeon, the bird does not start pecking or show any other sort of specific behavior in response to the auditory signal. However, when a response key is lit with a red light as an S2 that is followed by the tone in a second-order conditioning stage, the bird starts to peck at the key.[76]

Simultaneous conditioning provided a further and important example of how associations may be detected only when indirect measures are used. Presenting a conditioned stimulus at the same time as a reinforcer had always been regarded as a weak procedure at best, when compared to that of a delay condition in which the CS proceeded the US. A further student, Donald Heth, found evidence for strong associations between a stimulus and a shock when these were presented at the same time, as long as only 10 trials were given; when a greater number of conditioning trials was given the effect disappeared. In these experiments, the indirect test was the extent that such a stimulus could act as a punisher of lever-pressing.[77]

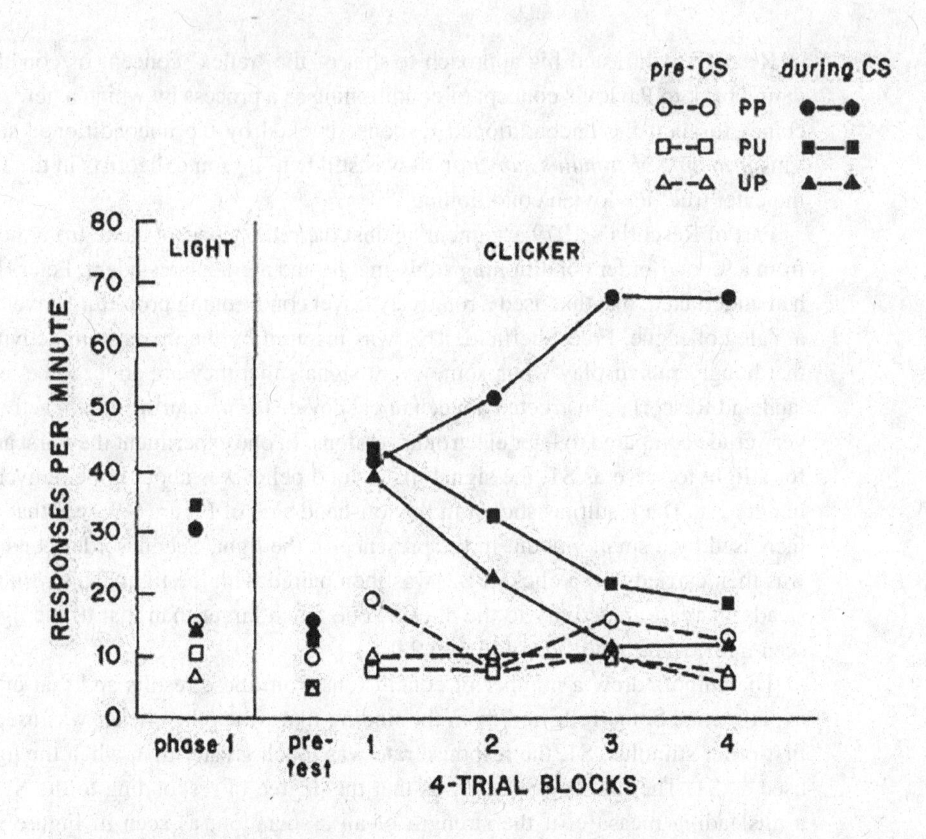

Figure 9.9 This graph shows data from an experiment on second-order conditioning in hungry rats. Increases in activity level were recorded to a light that had preceded delivery of food (Phase 1) or to a clicker (Phase 2). The important result was shown by the *PP* (= Paired-Paired) group. In Phase 1 the light paired with food evoked little change in activity, but in Phase 2 (second-order conditioning) it proved highly effective in reinforcing responding to a clicker. From Holland and Rescorla (1975a). Reproduced with permission from the American Psychological Association.

Another new conditioning procedure used in Rescorla's lab was that of flavor aversion learning; this occurs when exposure to a flavor is followed by a nausea-inducing event, such as injection of lithium chloride (see Chapter 6). One study of simultaneous associations – now named *within-compound associations* – was carried out by Rescorla and another of his students, Laura Freberg. These experiments used a *sensory preconditioning* design in which the key groups of rats were first given a solution containing both sucrose (S) and hydrochloric acid (H), thus producing a sweet, and sour taste. In a later session they were given a solution containing only H and five minutes later this was followed by an injection of lithium chloride. In a subsequent two-bottle preference test, with water in one of the bottles and a test solution in the other, rats showed both strong avoidance of a hydrochloric acid solution (H) and also of a sucrose solution (S). Further groups showed that exposing rats to either S or H prior to the taste aversion treatment had the effect of extinguishing the within-compound association between the two tastes.[78]

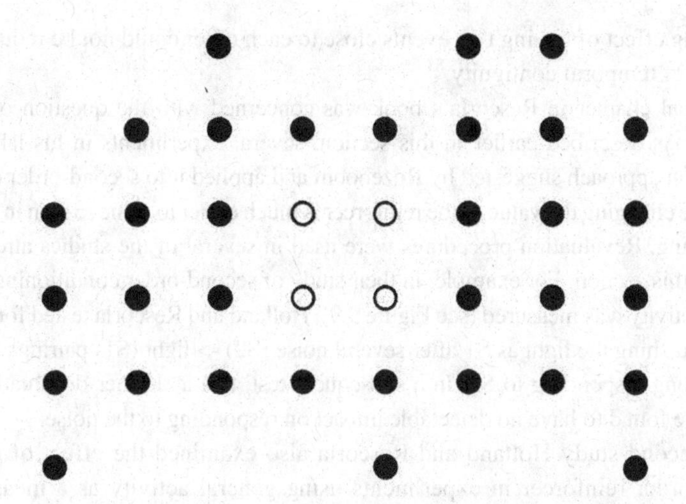

Figure 9.10 Why do we see a square formed by the four open circles and not squares formed by both open and filled circles? Wolfgang Koehler argued this was because of the importance of similarity in associating stimuli or events.
From Koehler (1941). Public domain.

The general conclusion from a variety of such experiments was that, whenever an animal is presented with a simultaneous compound consisting of two stimuli, it forms an association between them. As described earlier in this chapter, two of the key studies that led to the development of the Rescorla-Wagner theory involved the use of such compound stimuli, namely, in studies of blocking and cue validity. As Rescorla noted, the conclusions derived from these experiments had ignored the possibility that within-compound associations may have affected the results.

So far, the experiments outlined here were ones that demonstrated ways to detect or measure associations that were either hidden or appeared weak when only first-order Pavlovian conditioning was employed. In his lecture series, Rescorla moved on to describe studies using second-order conditioning to examine the conditions which influence associative learning. The influence of his Gestalt-flavored undergraduate introduction to psychology showed in some of these experiments, ones using highly ingenious designs. For example, in 1941 Wolfgang Koehler argued – partly on the basis of demonstrations like that shown in Figure 9.10 – that stimuli that are similar are more easily associated than are dissimilar stimuli.[79] A series of experiments from Rescorla's lab confirmed both that similar stimuli are more easily associated than are dissimilar ones and that part-whole relationships facilitate learning.[80]

Similarity was one of the principles suggested by the British Associationist philosophers cited by Rescorla. Another was that of spatial contiguity: events that occur close together in space were presumed to be more easily associated that spatially separated events. In the Hullian tradition, spatial contiguity was assumed to be a product of temporal contiguity; events occurring near each other tend to be experienced in more rapid succession than events distant from each other. The results of a further study carried out by Rescorla and yet another student, Chris Cunningham, showed that the facilitating effect of having two events close to each other could not be reduced to the principle of temporal contiguity.[81]

The final chapter in Rescorla's book was concerned with the question of 'what is learned?' As described earlier in this section, several experiments in his lab used the devaluation approach suggested by Rozeboom and applied it to second-order conditioning, where changing the value of the reinforcer is much easier to achieve than in first-order conditioning. Revaluation procedures were used in several of the studies already mentioned in this section. For example, in their study of second-order conditioning in which general activity was measured (see Figure 9.9), Holland and Rescorla tested for the effect by extinguishing the light as S1 after several noise (S2) -> light (S1) pairings had established strong responding to S2. In a subsequent test, this and other devaluation procedures were found to have no detectable impact on responding to the noise.

In a second study Holland and Rescorla also examined the effect of devaluing the first-order reinforcer in experiments using general activity as a measure. One form of devaluation was to establish an aversion to the food pellets used to produce first-order conditioning; this was achieved by the unusual procedure of spinning the rats on a wheel after consuming the pellets. The other was simpler and more conventional: rats that had been food-deprived throughout the experiments were given unrestricted access to the pellets. There was no difference between the effects of these two types of devaluation: both decreased responding to a first-order stimulus, but left intact responding to a second-order stimulus.[82]

Subsequent autoshaping experiments with pigeons used second-order conditioning to provide support for the idea that different aspects or components of a reinforcer can compete with each other, much as the components of a compound stimulus can compete, as in the overshadowing effect. Experiments that Rescorla described as "rather complicated, but elegant"[83] led to the following conclusion: "When one multifeatured event signals another, the resulting association includes a full representation of neither of these events. Instead, various factors act to favor one feature over another. This conclusion has been evident and important to theories for several years with regard to the signal; the contribution of the present experiments is to show that it also applies to the reinforcer.... The historical prejudice of describing these features in terms of 'stimulus' and 'response' may well prove not to be especially fruitful. Indeed, the issue of whether the organism learns about stimuli or responses seems best supplanted by the more general question of which reinforcer features are learned."[84]

A year after publishing his book on second-order conditioning Rescorla left Yale to return to the University of Pennsylvania. A major concern of the lab he set up there in 1981 was with instrumental conditioning. How this kind of learning became subject to analyses in terms of associative processes is the topic of the next section.

Do Rats Know that Pressing a Lever Produces a Food Pellet?

Another of my memories of the University of Sussex in the early 1970s is of the excitement generated by further preprints that Sutherland circulated among interested colleagues. This time the preprints were of chapters from the book that Mackintosh published in 1974, titled *The psychology of animal learning*.[85] This was within a tradition that began in 1940 with the publication by Hilgard and Marquis of their review of research on learning and conditioning. This was followed in 1961 by Kimble's updating of their book, a text that Herrnstein required my generation of graduate students in psychology at Harvard to read from cover to cover. Especially given the very large amount of research on conditioning and learning in non-human animals reported since 1961, just to review almost every paper of interest was a staggering achievement on Mackintosh's part. An American reviewer suggested that the book is "so impressive both in size and quality that it is likely to be the definitive secondary source in the field for many years…. The thoroughness of Mackintosh's coverage of his limited range of topics is thus unlikely ever to be surpassed…. From now on, anyone wanting to know about a topic in animal learning will probably begin by looking it up in Mackintosh."[86] No one has attempted such a book ever since.

To many of us, the chapter on theoretical analyses of instrumental learning was of particular interest. This included a detailed examination of the many studies that had addressed this issue since 1898, when Thorndike reported the first experiments on this topic. Mackintosh came to the following conclusion. "If instrumental responses are defined as those directly affected by their consequences, then it remains to be asked how these consequences come to exert control over a subject's behavior. The answer provided by Thorndike and Hull was that rewarding consequences strengthen the connection between preceding responses and the situation in which they occur. There is a considerable body of evidence inconsistent with the analysis provided by this law of effect. Studies of latent learning, irrelevant incentive learning, and contrast effects all suggest that animals form associations between their responses and the events contingent on those responses. Thus, instrumental, just like classical conditioning, may typically involve the establishment of associations between two events when a correlation is arranged between them."[87]

Most of his book was written while Mackintosh was at Dalhousie University. By the time it was published, he had returned to the UK. With the encouragement of Sutherland, Mackintosh obtained a Professorial Fellowship that enabled him to join his former Ph.D. supervisor at the University of Sussex in 1973. There, Tony Dickinson had just been awarded his Ph.D. He was promptly appointed to a post-doctoral research position on Mackintosh's new grant.

In his 1974 book, Mackintosh included a section on overshadowing and blocking in instrumental learning.[88] The only data to report were from a sketchy report by Konorski in 1948 of overshadowing of acquisition of a leg-flexion response in a dog and an unpublished paper by one of Mackintosh's Dalhousie students. Once Mackintosh's lab at Sussex was set up, the aim of several of the experiments run by Dickinson were to provide more substantial evidence for overshadowing and blocking in instrumental learning.

These experiments were reported at a further conference at the University of Sussex. Konorski had attended the previous one in 1971 and since then interest in his ideas had increased further. He had died in 1973, so the 1977 conference was organized

TABLE 6.1
Design of Experiment Replicating Konorski & Miller

Groups	Training Trials	Test Trials: Days 1–30	Test Trials: Days 26–30
T	[Tone / Forced Run] → Food	[Tone / Free Run] → Food	Free Run → Food
T+	[Tone / Forced Run] → Food; Tone → Food	[Tone / Free Run] → Food	Free Run → Food
T−	[Tone / Forced Run] → Food; Tone → No Food	[Tone / Free Run] → Food	Free Run → Food

Figure 9.11 The experiment using this design examined overshadowing of instrumental learning by hungry rats. It used a running wheel that could be either fixed in position, driven by a motor ("Forced run") or allowed the rat to run freely ("Free run"). The main question was whether a tone that was highly correlated with the delivery of a food pellet (in the T+ group) would overshadow learning that a forced run would be followed by the same reinforcer.
From Mackintosh and Dickinson (1979). Reproduced with permission from Taylor and Francis.

as a memorial event. The experiments reported by Mackintosh and Dickinson used an unusual procedure for studying instrumental learning. To reproduce the properties of the leg-flexion response used by Konorski and Miller, they wanted "to control the occurrence of the response and thus manipulate its relationship to reinforcement with the same freedom and precision as is granted the experimenter in studies of classical conditioning."[89]

The solution was provided by a running wheel that could be locked, driven by a motor, or free to rotate. Except during the 15-second trials signaled by a light, the wheel was locked. During a trial, the motor rotated the wheel at the slow speed of 6 revs per minute but the rat could override this and run faster. At the end of each trial, food pellets were delivered. One experiment using this procedure used a relative validity design; see Figure 9.11. For all rats a tone was sounded during every 'forced run' trial as just described; for different groups of rats the tone was either also presented during inter-trial periods and either followed by food (T+ group) or not (T- group) or only presented on "forced run" trials (T group).

The results were very clear. When the tone was the best predictor of food (T+ group), rats in this group did not run very much when the light was switched on for a 'free run' test. By contrast, the group for which the tone was a poor predictor

(T- group) ran over three times as much in this test. The authors suggested that this and a further experiment provide "considerable encouragement for the view first seriously suggested by Konorski and Miller that successful instrumental conditioning depends on the establishment of an association between representations of the instrumental response and of the reinforcer."[90]

At the end of their chapter the authors addressed the accusation that S-R theorists had traditionally made against such cognitive approaches, namely, that "they leave the animal buried in thought."[91] The animal may well learn that a certain response will produce a particular outcome but how is that knowledge translated into action? The proposed solution had two elements. First, that at least some associations are bidirectional: in addition to the idea of running prompting the idea of a food pellet, the idea of a food pellet may prompt the idea of running. Second, the concept of *propositional associations*. "If exposure to a contingency between wheel running and food establishes the propositional association, 'The response of wheel running produces food,' and, if a state of food deprivation engages an imperative premise of the form 'Perform any response that will produce food,' the rules of imperative inference will permit derivation of the instruction 'Perform the response of wheel running.'"[92] Many years later Dickinson proposed a somewhat different theory, his 'associative-cybernetic' model, for solving the 'thought into action' problem.[93]

In the meantime, a different approach to the analysis of instrumental learning had been adopted on the other side of the Atlantic. Experiments carried out at McMaster University[94] and at UCLA[95] both applied Rozeboom's revaluation logic, as described in the previous section, to instrumental learning, seemingly without being aware of his 1958 paper. In both studies, rats were first given several sessions of lever-press training on a VI schedule with either sucrose or saccharin solution as the reinforcer. Some of the rats were then given a reinforcer devaluation treatment. This consisted of pairing the sucrose or saccharin solution with an injection of lithium chloride, so as to produce an aversion to these sweet tastes. They then tested whether this treatment had affected the rats' lever-pressing rates in an extinction test. Neither study found any effect of devaluing the instrumental reinforcer.

In 1977 Tony Dickinson obtained a lectureship at the University of Cambridge. The project undertaken by his first Ph.D. student, Christopher Adams, was to further investigate the effect of devaluing a reinforcer that had been used to establish an instrumental response. By then, the studies in Rescorla's lab of revaluation in Pavlovian conditioning, as described in the previous section, had become well known, along with the argument that Rozeboom had made regarding the potential of revaluation procedures. By the late 1970s, it was also known that the effects of taste aversion learning could be context specific. Adams' experiments differed mainly from the previous studies of devaluation in that conditioning an aversion to the sucrose pellets used as the reinforcer for lever-pressing was carried out in the operant chambers used for lever-press training; in addition, subsequent tests were included to check that sucrose now failed to reinforce re-acquisition of lever-pressing.

Despite these improvements and confirmation that the devaluation procedure had been effective, two experiments failed to find any effect of the acquired aversion to the sucrose pellets on how much the rats pressed the lever in an extinction test. Adams noted that "these results suggest that the association controlling response emission is of the stimulus-response type and appears not to parallel the response-reinforcer relationship arranged by the conditioning procedure."[96]

The breakthrough came from two subsequent experiments in which the rats were trained on a variable-ratio reinforcement (VR) schedule, instead of the variable-interval (VI) schedules used in all previous experiments of this kind. In Experiment 1 rats were trained to press a lever on a VR schedule whereby on average every 9th response was followed by a pellet. Two types of pellet were used. One type – say, a sucrose pellet – functioned to reinforce lever-pressing, while the same number of another kind of pellet – a standard food pellet – was delivered independently of a rat's behavior. (Allocation of type of pellet to condition was counterbalanced across rats.) Three sessions in which lever-pressing was reinforced by one type of pellet alternated with three sessions of non-contingent pellet delivery. Then, an aversion was conditioned to one type of pellet. For Group P (Paired) consumption of response-dependent pellets was paired with injections of lithium chloride; for Group U (Unpaired) an aversion was conditioned to the response-independent pellets. In a subsequent extinction test, responding was lower in Group P than in Group U; see Figure 9.12. The final re-acquisition tests confirmed that aversions had been acquired to the appropriate type of pellet; as also seen in Figure 9.12.

A second experiment used improved control conditions and found the same result. A subsequent study confirmed that a crucial difference between this and earlier experiments of this kind was the use in lever-press training of a variable-ratio instead of a variable-interval schedule.[97]

Interestingly, in this and in the dozens of experiments that followed this initial finding of a reinforcer devaluation effect, the extinction test revealed some responding by rats in Group P despite their strong aversion to the reinforcer that had established the response; see Figure 9.12. Commenting on this aspect of the extinction data, Dickinson later suggested: "the instrumental training established lever-pressing partly as a goal-directed action, mediated by knowledge of the instrumental relation, and partly as an S-R habit impervious to outcome devaluation."[98]

The results opened the gate to a flood of research based on the distinction between habits and actions. One question addressed in the Cambridge lab was why reinforcer devaluation was more effective following training on a variable-ratio schedule than following VI training.[99] In 1978 Ruth Colwill graduated from the University of York after completing an undergraduate project with Hall and Pearce. She then moved to Cambridge, first as a research assistant to Dickinson and then as one of his students. On receiving her Ph.D. in 1982, she obtained a postdoctoral position in Rescorla's lab at the University of Pennsylvania, where she and Rescorla carried out a series of studies on associative structures in instrumental learning.[100] In Cambridge, a later student of Dickinson's, Bernard Balleine, analyzed motivational aspects of instrumental learning within an associative framework[101] and then began to explore the neural basis of actions and habits.[102]

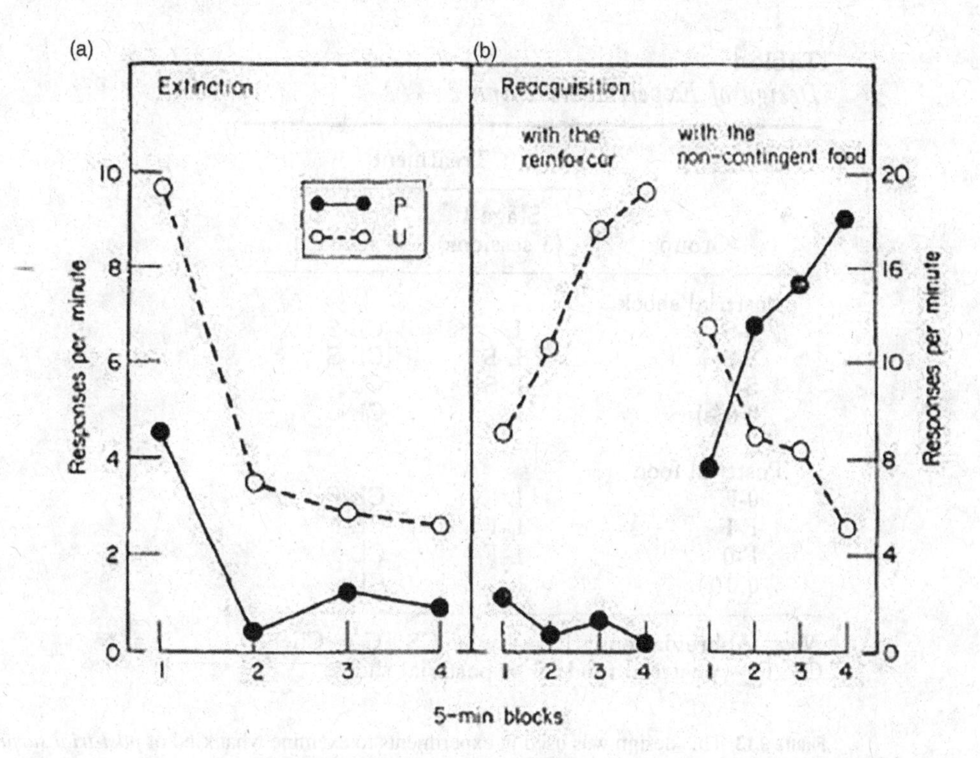

Figure 9.12 This figure shows key results from the first experiment to demonstrate that devaluing an instrumental reinforcer would reduce responding that had previously been acquired using this reinforcer. During an initial acquisition stage, the rats were given two types of food pellet. One (Type A, say) was used to reinforce lever-pressing. The other (Type B) was delivered independently of a rat's behavior. In the subsequent 'devaluation' stage for Group P ('paired') Type A pellets were made aversive by pairing their consumption with an injection of lithium chloride. In Group U this happened to the Type B pellets. As shown in Panel (a) of this figure, in a subsequent extinction test Group P made fewer responses than Group U. From Adams and Dickinson (1981). © SAGE Publications.

Paying Attention

When Mackintosh and Dickinson worked together in the mid 1970s, they were not just interested in instrumental learning. A major concern was the blocking effect and Kamin's claim that learning takes place only when a surprising event occurs. But what kind of events are surprising? One of Kamin's experiments had shown that the addition of an extra shock when the animal had learned to expect only a single shock was an effective 'surprise' that promoted learning about the added stimulus in a blocking experiment; in other words, surprise leads to *unblocking*. A study by Dickinson, Hall, and Mackintosh extended this result. In an initial phase, two groups of rats learned that a stimulus, A, was always followed by two shocks, in that a posttrial shock was delivered 8 seconds after the first one. In the next phase a second stimulus, B, was added to form a simultaneous compound, AB, and in one of the

Table 1
Design of Experiments 1 and 2

Group	Treatment: Stage 1 (3 sessions)	Treatment: Stage 2 (3 sessions)
Posttrial shock		
0-S	L	CL-S
S-S	L-S	CL-S
S-0	L-S	CL
0-0 (s)	L	CL
Posttrial food		
0-F	L	CL-F
F-F	L-F	CL-F
F-0	L-F	CL
0-0 (f)	L	CL

Note. Abbreviations: L = Light CS; C = Clicker CS; F = posttrial food; S = posttrial shock.

Figure 9.13 This design was used in experiments to examine what kind of *post-trial surprise* would produce *unblocking* of conditioning involving shock. In Experiment 1 (upper part of the table) *L* (light) and the compound *CL* (clicker plus light) were always followed immediately by shock; *S* indicates an additional *posttrial* shock eight seconds after the initial shock. In Experiment 2 (lower part of table) L and CL were always followed by food and F indicates posttrial food; that is, delivery of three pellets 8 seconds later. Whereas surprise involving shock produced unblocking of conditioning to the clicker, no effect of surprise involving food was detected. From Dickinson and Mackintosh (1979). Reproduced with permission from the American Psychological Association.

groups this was followed by a single shock, the posttrial shock was omitted. This surprising omission was enough to trigger excitatory learning about the added stimulus. A standard blocking effect of little excitatory learning about the added stimulus was found in the control groups where the outcome in the second phase was the same as that in the second phase.[103]

What if an expected shock was followed eight seconds later by an unexpected food pellet, or an expected food pellet followed by an unexpected shock? Would such kinds of posttrial surprise trigger learning about an added stimulus? As shown in Figure 9.13, one experiment using a conditioned suppression procedure included a group that in the second compound (AB) phase was given an unexpected delivery of three food pellets (*F* in Figure 9.13) 8 seconds after the expected shock and another group for which the delivery of three pellets they had come to expect following a shock in the first phase were omitted in the second, AB, phase. Neither the surprising arrival of pellets nor their surprising omission had any effect. A second experiment inverted this design by using a discriminative operant procedure with food pellets as the reinforcer. One group was given a shock for the first time following the first AB->food trial, while for another group an expected shock was omitted on this trial. Again, this kind of surprise had no effect.[104]

This last study also included a replication of the earlier finding that the omission of an expected shock produces unblocking. This result confirmed Kamin's original suspicion that the Rescorla-Wagner theory did not fully capture his concept of surprise, since the theory predicted that the omission of an expected shock should produce inhibitory learning to the added stimulus. In contrast, the result was consistent with Mackintosh's idea that a surprising event, even one that is – in this case – less painful than expected, can increase attention to the added stimulus and thus produce unblocking. But how to explain that the surprising addition or omission of an irrelevant reinforcer failed to produce unblocking? The authors appealed to an earlier discussion by Mackintosh that "entertained the idea that associated with a CS might be a variety of reinforcer-specific learning rate parameters, each determining the associability of the CS with a particular reinforcer or class of reinforcer."[105] They referred to the idea as *learned relevance*.

In 1980 John Pearce and Geoff Hall published a paper describing a different type of attentional theory, one apparently incompatible with the theory that Mackintosh had developed over the past decade. Hall had been a graduate student at Cambridge at a time when few others in its Department of Experimental Psychology were interested in animal learning. He carried out studies of discrimination learning in animals that were inspired by the attentional theory that Sutherland and Mackintosh had developed.[106] After being awarded his Ph.D. in 1971 he obtained a series of post-doctoral positions, first at the University of Sussex, then at Dalhousie University where he worked with Mackintosh, and then back to Sussex, where he worked with Mackintosh again, and also with Dickinson on the important shock-omission study described earlier. By 1975, when he obtained appointment as a lecturer at the University of York, he must have acquired an understanding of attentional analyses of animal learning as deep as anyone in the world.

Pearce had been a graduate student at the University of Sussex, where he worked closely with Tony Dickinson on interactions between appetitive and aversive stimuli. After being awarded his Ph.D. in 1976, he obtained a research fellowship that enabled him to work with Hall at the University of York for two years, before taking up a further postdoctoral appointment to work with Tony Dickinson in Cambridge until 1980. In that year he was appointed to a lectureship at Cardiff University.

In comparison to the large and well-equipped behavioral labs at Sussex and at Dalhousie, there were limited facilities for such research at York in the late 1970s. Despite these limitations, Hall and Pearce were able to complete a number of important studies. For example, they demonstrated that a stimulus that was a good predictor of an instrumental reinforcer could overshadow conditioning of lever-pressing.[107] This added to the finding by Dickinson and Mackintosh, as described in the previous section, by showing what these authors had found using a forced-response – running – could also be found when a free operant response was used. Incidentally, Pearce and

Hall saved some of the rats from this study for the two of the experiments that are described next, possibly the sign of a tight budget.

As Hall later described the situation: "This was a time of financial stringency and we struggled to set up a research lab, borrowing equipment from other universities and installing it in a leaky shed in a remote part of the campus used chiefly by the gardeners. In truly appalling circumstances, by modern (or any) standards, we set about using classical conditioning techniques to demonstrate that a stimulus that had reliably predicted its consequences would come to command attention, and thus be more readily learned about subsequently. Initially to our surprise, a series of experiments produced quite the opposite result – such a stimulus was learned about less readily."[108]

One of the first studies they carried out together was designed to follow on from the experiments on appetitive-aversive interactions that Pearce had been carrying out at Sussex with Dickinson. Thus, in the second of two similar experiments one group of rats, the *Correlated* group, was first trained that, for example, a light signaled a period in which shocks could occur and a tone signaled a safe period. The *Uncorrelated* group were given the same exposure to the two stimuli and to shocks but there was no relationship between these events. Next, the rats were trained in a two-lever appetitive discrimination in which, say, the light indicated that pressing the left lever would be followed by pellet delivery and the tone indicated that reinforcement would follow pressing the right lever.[109]

The expected outcome was faster discrimination learning by the *Correlated* group because the rats had previously learned to attend to the light and the tone, whereas the *Uncorrelated* group might have learned to ignore them. The results from both experiments yielded the opposite outcome to this prediction. Following discussion of various ways of making sense of the unexpected results, Hall and Pearce ended up with the following tentative suggestion: "Perhaps we should abandon attentional theory and adopt instead quite the opposite general hypothesis; that a cue, once it has entered into an association, will form new associations only with difficulty."[110]

To test this idea, they ran a study using rats in a conditioned suppression procedure. The same basic condition was used for the critical group in each of three experiments. During an initial stage a stimulus was paired with a weak shock and in the following stage it was paired with a strong shock. According to Mackintosh's theory, the first stage would have increased attention to the stimulus, represented by an increase in its α-value, so that in the second stage the stimulus should have become more rapidly conditioned than in various control groups. The opposite result was obtained. Pairing the stimulus with a weak shock had a similar effect to simply presenting the stimulus on its own, namely, a latent inhibition effect.

As seen in Figure 9.14, their second experiment contained a traditional latent inhibition group (Tone-alone) that was exposed in Stage 1 to a tone in the absence of any shock. After eleven sessions with six trials per session in Stage 1, all rats received conditioning in Stage 2 whereby the tone signaled a 0.8 milliamp shock. It may be seen that a typical latent inhibition effect was obtained in that conditioned suppression

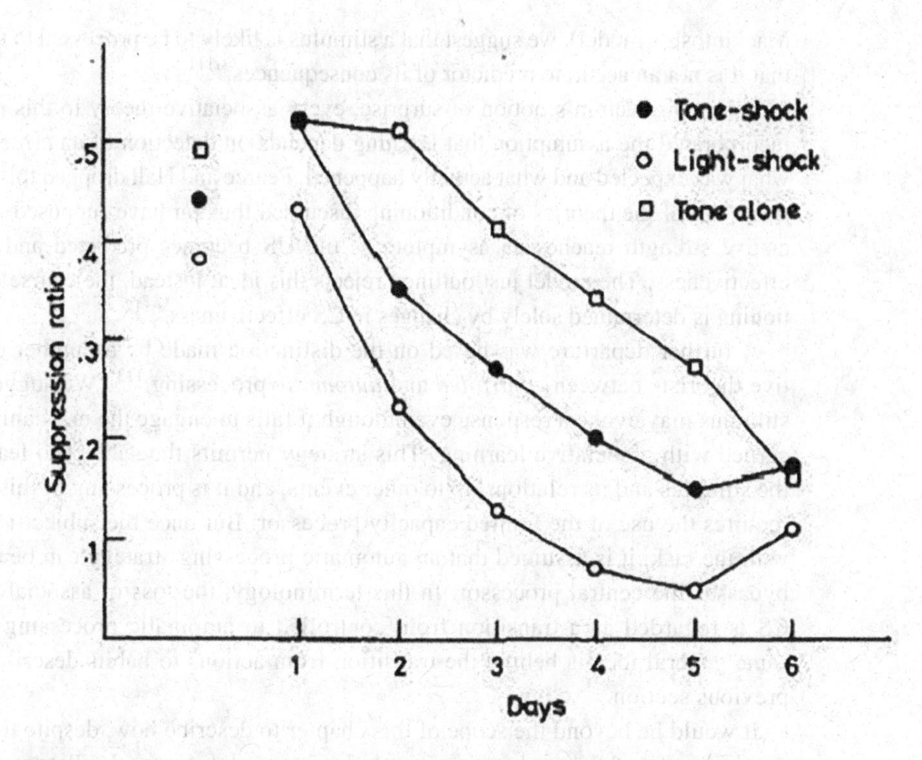

Figure 9.14 A conditioned suppression procedure was used to test the effect of giving rats a tone paired with a mild shock in Stage 1 on how rapidly they would learn to associate the same tone with a strong shock in Stage 2. Attention theory of the kind developed by Mackintosh predicted that this Stage 1 experience would produce faster learning than in control groups. Unexpectedly, as shown in the graph, this group (*Tone-shock*) learned more slowly than a control group (*Light-shock*) that had been given Light-mild shock pairings in the initial stage, although not as slowly as a group (*Tone-alone*) that had been exposed to the tone in the absence of any shock in Stage 1.
From Hall and Pearce (1979). Reproduced with permission from the American Psychological Association.

of the lever-press baseline developed more slowly than in a group for which the tone was novel at the start of Stage 2 (Light-shock group). The critical finding was that the Tone-shock group, given in Stage 1 66 trials in which the tone had been followed by a 0.4 milliamp shock, also showed slow development of suppression when given the Tone-0.8 milliamp shock pairing in Stage 2.

A third experiment confirmed that the same result could be found when the equivalent Tone-shock group showed more convincing evidence for conditioning in Stage 1. This was achieved by increasing the shock levels to 0.5 milliamps in Stage 1 and to a high 2 milliamps in Stage 2.

To explain these and related results Pearce and Hall developed a theory of Pavlovian conditioning that was radically different from those described earlier in this chapter. The core assumption was that animals pay attention to stimuli to the extent that they are uncertain as to what comes next. "In other words (and in direct contrast to

Mackintosh's model), we suggest that a stimulus is likely to be processed to the extent that it is not an accurate predictor of its consequences."[111]

Following Kamin's notion of surprise, every associative theory to this point had incorporated the assumption that learning depends on detection of an error between what was expected and what actually happened. Pearce and Hall dropped this assumption: "All of the theories of conditioning discussed thus far have supposed that associative strength reaches an asymptote as the US becomes predicted and loses its effectiveness. The model just outlined rejects this idea; instead, the course of conditioning is determined solely by changes in CS effectiveness."[112]

A further departure was based on the distinction made by a number of cognitive theorists between *controlled* and *automatic* processing.[113] "We suggest that a stimulus may evoke a response even though it fails to engage the mechanisms concerned with associative learning. This strategy permits the subject to learn about the stimulus and its relationship to other events, and it is processing of this sort that requires the use of the limited capacity processor. But once the subject is familiar with the task, it is assumed that an automatic processing strategy can be used that bypasses the central processor. In this terminology, the loss of associability of a CS is regarded as a transition from controlled to automatic processing."[114] The same general idea is behind the transition from actions to habits described in the previous section.

It would be beyond the scope of this chapter to describe how, despite the radical departures from the assumptions made by its predecessors, the Pearce-Hall theory accounted for most of the phenomena described earlier, such as conditioned inhibition or super-conditioning. How their theory fared in subsequent years and how it was later modified is described in textbooks such as those by Bouton[115] and Pearce.[116]

Two developments beyond 1980 are, however, worth mentioning. First, despite the rare agreement between the Mackintosh and the Pearce-Hall theory that no blocking should occur after a single compound trial, later clear evidence for such blocking was reported from a fear conditioning experiment. This first paired a light with a shock and then gave the animals a single compound of light plus tone, followed by the same shock. What may have been a critical feature of this experiment was the use of much shorter stimulus durations, just five seconds, than in previous studies of this kind. The rationale was to minimize the likelihood of second-order simultaneous conditioning taking place on the single compound trial, that is, to prevent rats acquiring fear of the tone because of a within-compound association with the light of the kind described earlier in this chapter.[117]

The second development was of hybrid models of attention that combined elements of both the Mackintosh and the Pearce-Hall theories.[118] A particularly interesting and persuasive resolution was proposed within the context of Pearce's later configural theory; "Attention is assumed to have two roles within this network. First, the salience of the stimuli at the input to the network can be increased if they are relevant to the occurrence of reinforcement and decreased if they are irrelevant. Second, the associability of configural units can increase on trials when the outcome is surprising and decrease when the outcome is not surprising."[119]

Epilogue

As its Chair, Stuart Sutherland (1927–1998) expanded the Laboratory of Experimental Psychology at the University of Sussex beyond its initial emphasis on animal learning to include outstanding research groups working on artificial intelligence, cognitive psychology, and psycholinguistics; see Figure 9.15. He had given up direct involvement in any kind of research when he suffered a mental breakdown in 1973. This experience and the various kinds of professional help he sought were described in his 1976 book, *Breakdown*, which deservedly sold very well and brought him some notoriety outside academia, to the extent that a play based on his book had a fair run in London and a second edition of the book was published in 1995.[120]

The treatment that worked best for his bipolar disorder was lithium carbonate, which required him to drink large quantities of water, of which a great deal was swigged in public from a 2-liter gin bottle. The breakdown seemed to have exaggerated his eccentricity, with many colleagues finding his manic periods harder to bear than his depressed periods. Much of his subsequent working life was devoted to writing book reviews, although in 1992 he published a book that introduced a general public to research on cognitive biases, *Irrationality: The enemy within*.[121]

Figure 9.15 Stuart Sutherland in his office at the University of Sussex in the late 1970s with his long-suffering secretary, Anne Doidge. Dick Stevenson remembered "a tutorial with Stuart, which began with him shouting 'Ann clean the f***ing board.' She did. But she did not look happy about it."
Reproduced with permission from the author (R.A. Boakes).

He was a chain smoker who often set fire to his paper bin and somehow persuaded the university to allow him, but no one else, to continuing to smoke cigarettes within an office with a specially fitted exhaust fan. His rigid habit of delaying any consumption of alcohol until after 6 PM – but copious amounts thereafter – may well have allowed him to just reach his 70s. Stuart Sutherland died in March 1998, aged 71.

As related earlier in this chapter, in 1973 Kamin published his book concluding that there was no good evidence for the heritability of IQ. It prompted Mackintosh to read up on the topic. Written in his characteristic style, Mackintosh's 14-page even-handed 'critical notice' of the book concluded that Kamin had not been even-handed. "Kamin has chosen not to provide an impartial, if critical, examination of the evidence said to demonstrate genetic influences on IQ, but was determined to demolish that evidence.... Kamin has shown that both in quantity and quality the evidence for the heritability of IQ is very much less than the consensus view has suggested.... (this) presents an extremely impressive achievement of scholarship. It would be absurd if all his arguments were dismissed as biased; but because he spends as much time advancing rather weak objections to reasonably sound data as he does advancing cogent objections to basically unsound data, the flaws in the former arguments may be thought to reflect on the validity of the latter."[122]

Having noted that Kamin had committed "the grave academic sin of trespassing outside his own speciality," Mackintosh went on to commit the same sin for almost the rest of his life. He continued to develop his knowledge of IQ research and also became actively involved in such research. He and a colleague analyzed data on differences between the IQ and school performance of West Indian and indigenous British children. The conclusion would have pleased Kamin. "Much of this difference in IQ scores between West Indian and indigenous children appears to be related to differences between them in such factors as parental occupation, income, size of family, degree of overcrowding, and neighborhood. All of these factors are related to IQ among whites, and when they are taken into account, the difference between West Indian and indigenous children is sharply reduced to somewhere between one and seven points. These findings tend to argue against those who would seek to provide a predominantly genetic explanation of ethnic differences in IQ."[123]

Mackintosh's expertise on IQ was recognized by a House of Commons committee that was concerned with the educational achievements of the children of recent immigrants. It awarded him a grant to study the relationship between ethnic and gender differences and achievement in UK schools. The conclusion he and his team reached was that IQ, but not ethnic background, was generally a good predictor of school achievement.[124] In 1998 Mackintosh published a book, *IQ and human intelligence*,[125] that showed the same impressive scholarship and ability to explain complex ideas as his 1974 book on animal learning.

In the meantime, Mackintosh had replaced Sutherland as Chair of Experimental Psychology at Sussex in 1978 and three years later had been appointed as Head of the Department of Experimental Psychology at Cambridge, a position he retained until retiring in 2002. Throughout this time he continued his active involvement in animal learning research, notably collaborating with his Ph.D. student, Ian McLaren, in

Figure 9.16 Andy Baker, Vin LoLordo, Lynn Nadel, and Nick Mackintosh at dinner at the Pearce's house prior to Cardiff University's Associative Learning Symposium in 2007. Reproduced with permission from the author (R.A. Boakes).

developing an elemental model of associative learning and testing its predictions.[126] During retirement, as well as tending a small flock of alpacas, he wrote a second edition of his book on intelligence and maintained his interest in animal learning research, continuing to attend meetings both in the UK and in Spain, where he had a strong influence on the development of research on learning; see Figure 9.16. Mackintosh was a towering intellect and – to misquote a close friend – was also a convivial person, with a great fondness for wine, whisky, and women. He died in February 2015 at the age of 79.[127]

In contrast, in the early 1970s Leon Kamin completely relinquished any involvement in animal learning research. Years later he explained that this was because he no longer believed that it made much of a contribution to understanding human behavior.[128] In its place, he pursued the path that started with his critique of research on IQ and, in particular, of claims of inherent racial differences. In 1984 he co-authored a book, *Not in our genes*,[129] that was highly critical of evolutionary psychology and argued for what many considered an implausibly extreme environmental position.

As Chair of the Psychology Department at Princeton from 1968 to 1974, Kamin was as successful in raising its standards and its reputation as he had been previously at McMaster University. He succeeded in increasing the number of Afro-American graduate students within the department, an achievement he repeated when in 1987

Figure 9.17 In 2018 Allan Wagner met with his collaborators, Susan Brandon (on the left) and Edgar Vogel (to the right). Shiva Revzan, on the far right, was a member of another research group in Yale's Department of Psychology.
Reproduced with permission from Edgar Vogel.

he left Princeton to become Chair of the Psychology Department at Northeastern University in his home city of Boston. On retiring in 1998, for many years he and his wife, Marie-Claire, spent every northern winter in South Africa, giving courses at the University of Cape Town. He died in December 2017, a week short of his 90th birthday.

Allan Wagner also served in several demanding administrative roles. He was Chair of Yale's Department of Psychology from 1983 to 1989, even Chair of its Philosophy Department from 1991 to 1993, and Director of its Division of the Social Sciences from 1993 to 1998. Nevertheless, he continued to maintain his lab, publish important experimental papers and, above all, further develop the theory that had begun as SOP in 1981, continuing well after he retired in 2012. His major collaborators from the late 1980s until close to his death were Susan Brandon and Ed Vogel; see Figure 9.17.

Figure 9.18 Bob Rescorla in his lab in 1992.
Reproduced with permission from Shirley Steele.

According to a close collaborator, "Allan was very active and happy until a few days before his death: He worked, played tennis, and enjoyed time with his family, specially, his daughters, grandchildren, and his lovely companion, Lois Meredith."[130] Wagner died of cancer in September, 2018, at the age of 84.

As for Bob Rescorla, he remained at the University of Pennsylvania until he retired in 2009. For four years he served as Chair of the Psychology Department, followed by another four years as Dean of Arts and Sciences. He also took very seriously a standard load of undergraduate teaching. Nevertheless, he maintained a highly productive lab until the day he retired.[131]

Unlike Mackintosh or Kamin, Rescorla did not become actively involved in other academic areas. He just continued to produce important experimental data using innovative designs relating to the nature of associative learning; see Figure 9.18. These studies included those on instrumental learning mentioned previously. Others used compound stimuli in classical conditioning to answer the question of whether the changes in associative strength of a given stimulus were determined solely by a common error term as the Rescorla-Wagner theory assumed.[132]

Rescorla completely ended his involvement with psychology as soon as he retired. He even turned down the invitation to attend a meeting to celebrate his achievements. His former student, Peter Holland, who also made a complete break with academia on retiring from his stellar career in behavioral neuroscience, similarly declined the invitation to take part in the meeting. Instead, he sent this message: “Bob taught me how to be a scientist. There wasn’t a minute in my learning theory and neuroscience careers when I wasn’t doing something inspired by things he and I talked about at Yale.”[133]

As soon as he retired, Rescorla and his third wife, Shirley Steele, an artist, moved to Austin, Texas, where he worked part-time in a library. His life-long friend, Vin LoLordo, remembers him as having “a strong moral sense, and despite his ferocious intellect he could be a very kind person.”[134] Rescorla died in March 2020, just short of his 80th birthday.

Notes

1 Ivan Pavlov, Conditioned Reflexes and Experimental Neuroses

1 Gray, 1979, pp. 24–30; Todes, 2014, pp. 148–150.
2 Todes, 2014, pp. 303–310.
3 Todes, 2014, pp. 252–285.
4 Todes, 2014, pp. 229–234.
5 Todes, 2014, pp. 233–234.
6 Todes, 2014, pp. 319–336.
7 Todes, 2014, pp. 240–242.
8 Todes, 2014, pp. 242–249.
9 Todes, 2014, pp. 250–251.
10 Pavlov, 1928, p. 241; Todes, 2014, pp. 112–121; 303–310.
11 Pavlov, 1928, p. 241.
12 Todes, 2014, pp. 227–229.
13 Todes, 2014, pp. 367–377.
14 Gantt, 1928, p. 28.
15 Gray, 1979, p. 122; Todes, 2014, pp. 411–439.
16 Todes, 2014, pp. 303–310.
17 Pavlov, 1927, pp. 20–21.
18 Pavlov, 1927, p. 59.
19 Pavlov, 1927, p. 113.
20 Pavlov, 1927, p. 320.
21 Gray, 1979, p. 103.
22 Todes, 2014, pp. 340–347.
23 Pavlov, 1928, Vol. 1, p. 348.
24 Todes, 2014, pp. 494–503.
25 Pavlov, 1941, p. 344; Todes, 2014, pp. 503–509.
26 Pavlov, 1941, p. 344.
27 Pavlov, 1932; reprinted in Pavlov, 1962.
28 Pavlov, 1932; reprinted in Pavlov, 1962, p. 52.
29 Gray, 1979, p. 121; for further details of Pavlov's involvement in psychiatry, see Todes, 2014, pp. 630–649.
30 Ruiz & Sánchez, 2016, p. 2.
31 Gantt, 1973; McGuigan, 1981; Ruiz & Sánchez, 2016.
32 Ruiz & Sánchez, 2016.
33 McGuigan, 1980.
34 Gantt, 1944.
35 Gantt, 1944, p. 23.
36 Gantt, 1944, p. 24.
37 Dykman & Gantt, 1960.

38 Mackintosh, 1974.
39 Brogden & Gantt, 1937.
40 Brogden, 1939.
41 Kimmel, 1977.
42 Pavlov, 1928.
43 For example, Gantt, 1937.
44 Gliedman, Gantt & Teitelbaum, 1957.
45 Gantt, 1936, 1937, 1970.
46 Dilger, Moore & Freeman, 1962.
47 Liddell, 1953, p. 52; see also Liddell, 1925.
48 Block, 1963b; Kirk & Ramsden, 2018.
49 Freeman, 1985; Kirk & Ramsden, 2018.
50 Liddell, 1953, p. 52.
51 Liddell, 1953, p. 54.
52 Liddell, 1953.
53 Freeman, 1985.
54 Block, 1963b, p. 172.
55 Thorndike, 1898.
56 Masserman, 1964, pp. 95–109.
57 Dimmick, Ludlow & Whiteman, 1939.
58 Winter, 2016.
59 Winter, 2016, p. 78.
60 Masserman, 1943; see also Masserman, 1950.
61 Winter, 2016.
62 Masserman, Wechkin & Terris, 1964.
63 Winter, 2016, p. 100.
64 Wolpe, 1952, p. 243.
65 Rachman, 2000.
66 Wolpe, 1952, p. 255.
67 Wolpe, 1952, p. 256.
68 Wolpe, 1952, p. 266.
69 Wolpe, 1952, p. 266.
70 Rachman, 2000.
71 Todes, 2014, p. 614.
72 Todes, 2014, pp. 575–613; 726–727.
73 Todes, 2014, p. 656.
74 Todes, 2014, p. 490.
75 Todes, 2014, p. 658.
76 Todes, 2014, pp. 662–664.
77 Todes, 2014, p. 689.
78 Konorski, 1948, p. 211.
79 Miller & Konorski, 1928; translated by B.F. Skinner, 1969; see also Konorski, 1948, pp. 211–235.
80 Windholz & Wyrwicka, 1996.
81 Todes, 2014, pp. 663, 668.
82 Konorski & Miller, 1937.
83 Konorski, 1948, p. 246.
84 McLaren & Dickinson, 1990.
85 Konorski, 1948, p. 128.
86 Wyrwicka, 1994. See also Todes, 2014, pp. 728–729.
87 Konorski, 1948, p. 40.
88 Halliday, 1979.

2 Developing Habits: Clark Hull and the Hullians

1 Thorndike, 1898, 1911.
2 Watson, 1907.
3 Watson, 1914.
4 Watson, 1924.
5 Beach, 1959; Hull, 1952a.
6 Beach, 1959; Leahey, 2004; Smith, 1986.
7 Hull, 1930, p. 242.
8 Hull, 1952a, p. 155.
9 Hull, 1929, p. 498.
10 Hull, 1929, p. 498.
11 Hull & Baernstein, 1929, p. 15.
12 Hull & Baernstein, 1929.
13 Hull, 1930, p. 242.
14 Hull, 1930.
15 Hull, 1931.
16 Hull, 1931, p. 512.
17 Hull, 1931, p. 514.
18 Hull, 1952a, p. 155.
19 May, 1971; Morawski,1986.
20 Hull, 1943, p. 27.
21 Hull, 1943, p. 393.
22 Hull, 1943, p. 79.
23 Hull, 1943, p. 386.
24 Grindley, 1932.
25 Konorski and Miller, 1937.
26 Skinner, 1938.
27 Koch, 1944.
28 Tinklepaugh, 1928.
29 Elliott, 1928.
30 Hull, 1931, p. 491.
31 Hull, 1931, p. 505.
32 Hull, 1951, p. v.
33 Miller, 1948.
34 Zangwill, 1977.
35 Zangwill, 1977.
36 Crespi, 1942.
37 Crespi, 1942.
38 Gantt, 1938.
39 Hull, 1951, p. 47.
40 Crespi,1942, 1944.
41 Crespi, 1942, p. 514.
42 Hull, 1943, p. 254.
43 Hull, 1952b, p. viii.
44 Brown, 1965.
45 Koehler, 1925.
46 Maier, 1929.
47 Birch, 1945.
48 Skinner, 1938.
49 Blodgett, 1929.

50 Blodgett, 1929; Tolman, 1932.
51 Spence & Lippit, 1940.
52 Deutsch, 1956.
53 Deutsch, 1960.
54 Deese, 1951.
55 Gleitman, Nachmias & Neisser, 1954.
56 Mackintosh, 1974, p. 440.
57 Papini, 2008; Rashotte, 2007.
58 Amsel, 1958, pp. 102, 108.
59 Amsel, 1958, p. 109.
60 Rashotte & Amsel, 1967.
61 Amsel & Rashotte, 1969.
62 Capaldi, 1964; Proctor et al., 2022.
63 Capaldi, 1966, p. 459.
64 Brown & Wagner, 1964, p. 507.
65 Rashotte, 2007, p. 2.
66 Mackintosh, 1974, pp. 434–467.
67 Amsel, 1994.
68 Campbell & Ellison,1997.
69 Sheffield, Wulff, & Backer,1951.
70 Sheffield & Roby, 1950.
71 Sheffield, 1965, p. 305.
72 Sheffield, 1965, p. 307.
73 Patten & Rudy, 1967.
74 Williams & Williams, 1969; see Chapter 6.
75 MacPhail, email October 18, 2021.
76 Miller, 1952, p. 380.
77 Egger & Miller, 1962, p. 97.
78 For example, Black, 1965.
79 Miller & DiCara, 1967, p. 12.
80 Black, 1965.
81 Trowill, 1967, p. 11.
82 Macphail, email October 18, 2021 to RAB; Booth, email October 20, 2021 to RAB.
83 Miller & DiCara, 1967.
84 Dworkin & Miller, 1986.
85 DiCara & Miller, 1968, p. 1486.
86 Dworkin & Miller, 1986, pp. 299, 312.
87 Clifton, email September 4, 2021; MacPhail, email October 18, 2021; Booth email, October 20, 2021. All to RAB.
88 Wikipedia, October, 2021.
89 Mackintosh, 1974, p. 231.
90 Brown, 1965.
91 Dollard et al., 1939.

3 Learning Where Things Are and Where Events Happen

1 Restle, 1957.
2 Burgess, 2014.
3 Watson, 1907.

4 Watson & Lashley, 1915.
5 Crutchfield, 1960.
6 Macfarlane, 1930.
7 Tolman, 1932, p. 80.
8 Blodgett, 1929.
9 Bruce, 1930; Tolman & Honzik, 1930.
10 Elliott, 1928.
11 Tinklepaugh, 1928.
12 Koehler, 1925.
13 Maier, 1929.
14 Tolman & Honzik, 1930.
15 Tolman, 1932, p. xvii.
16 Munn, 1950.
17 Spence & Lippitt, 1940; 1946.
18 Thistlethwaite, 1951.
19 Muenzinger, 1938; Tolman, 1948; p. 197.
20 Tolman, 1948, p. 200.
21 Tolman, Ritchie & Kalish, 1946a.
22 Tolman, Ritchie & Kalish, 1946b.
23 Restle, 1957.
24 Packard & McGaugh, 1996.
25 Tolman & Gleitman, 1949, p. 816.
26 Miller, 1935.
27 Gleitman, 1981.
28 Kendler, 1952.
29 Ritchie, 1953.
30 Restle, 1957.
31 Tolman, 1949.
32 Tolman, 1949.
33 Scoville & Milner, 1957.
34 Milner, 1965.
35 O'Keefe & Dostrovsky, 1971.
36 O'Keefe, 1976.
37 O'Keefe & Nadel, 1978.
38 Menzel, 1973.
39 Olton & Samuelson, 1976.
40 Olton, 1977.
41 Macfarlane, 1930.
42 Morris, 1981, p. 259.
43 Morris, Garrud, Rawlins & O'Keefe, 1982.
44 For example, Roberts, Cruz, & Tremblay, 2007.
45 Odling-Smee, 1975.
46 Kamin, 1968.
47 Tomie, 1976.
48 Baker, 1977.
49 Best, Best, & Mickley, 1973.
50 Rudy, Iwens, & Best, 1977.
51 For example, Rescorla & Cunningham, 1978.
52 Rescorla & Heth, 1975.
53 Bouton & Bolles, 1979a.
54 Bouton & Bolles, 1979b.

55 Balsam & Tomie, 1985.
56 Balsam, 1985; Rescorla, Durlach & Grau, 1985.

4 Fear, Avoidance, and Punishment

1 Todes, 2014, pp. 319–336.
2 Razran, 1956.
3 Warner, 1932.
4 Brogden & Culler, 1936.
5 Brogden, Lipman & Culler, 1938; Experiment 1.
6 Culler, 1938.
7 Mowrer, 1980, p. 26.
8 Mowrer, 1939, p. 564.
9 Mowrer, 1947, pp. 107–108.
10 Mowrer & Lamoreaux, 1942.
11 Mowrer, 1939, p. 553.
12 Mowrer & Lamoreaux, 1946.
13 Mowrer & Lamoreaux, 1946, p. 48.
14 Mowrer & Miller, 1942.
15 Miller, 1948, p. 97.
16 Miller, 1948, p. 99.
17 Mowrer, 1947.
18 Skinner, 1938.
19 Estes & Skinner, 1941.
20 Estes & Skinner, 1941, p. 391.
21 Thorndike, 1932.
22 Skinner, 1938, p. 151.
23 Skinner, 1938, p. 159.
24 Estes, 1945, p. 34.
25 Estes, 1945, p. 32.
26 see Mackintosh, 1974, pp. 272–277.
27 Estes, 1969.
28 Skinner, 1948b.
29 Skinner, 1948b.
30 Skinner, 1953.
31 For example, Skinner, 1954.
32 Skinner, 1954.
33 Mowrer, 1973.
34 Page, 2017, p. 2.
35 For example, Mowrer & L. N. Solomon, 1954.
36 Mowrer, 1960a, 1960b.
37 Page, 2017.
38 Page, 2017.
39 Kamin, 1956, p. 8.
40 Solomon & Wynne, 1950.
41 Rescorla, 1997, pp. 29–30.
42 Mowrer, 1948, p. 571.
43 Solomon & Wynne, 1953, p. 1.
44 Solomon & Wynne, 1953, p. 17.
45 Solomon, Kamin, & Wynne, 1953.

46 Glucksberg, 2017.
47 Mowrer & Lamoreaux, 1942.
48 Kamin, 1956.
49 Kamin, 1956.
50 Kamin, 1957.
51 Kamin, 1957.
52 Kamin, 1956, p. 423.
53 Estes & Skinner, 1941.
54 Estes, 1989, p. 114.
55 Walker, 1942.
56 Estes, 1943.
57 Estes, 1948.
58 Harlow & Stagner, 1933.
59 Smith, Brown, Toman & Goodman, 1947.
60 Black, 1957.
61 Solomon & Turner, 1962, p. 202.
62 Solomon & Turner, 1962, p. 218.
63 Annau & Kamin, 1961.
64 Kamin, Brimer & Black, 1963, p. 501.
65 Sidman, 1953.
66 Rescorla & LoLordo, 1965.
67 Konorski, 1967, pp. 380–383.
68 Rescorla & LoLordo, 1965, p. 412.
69 Rescorla, 1966.
70 Rescorla, 1967.
71 Rescorla & Solomon, 1967.
72 Overmier & Seligman, 1967; Table 1.
73 Overmier & Seligman, 1967, p. 33.
74 Mowrer & Viek, 1948, p. 193.
75 Seligman & Maier, 1967.
76 Maier, 1970.
77 Maier, Albin & Testa, 1973.
78 Seligman & Beagley, 1975, p. 534.
79 Seligman & Beagley, 1975.
80 Maier & Seligman, 2016, p. 349; see also LoLordo & Overmier, 2011.
81 Mowrer & Viek, 1948, p. 193.
82 Garcia, 1997; p. xi.
83 Petrinovich & Bolles, 1954.
84 Fanselow & Bouton, 1997
85 Bolles, 1967.
86 Bolles & Popp, 1964.
87 Bolles, 1967, pp. 406–407.
88 Hineline & Rachlin, 1969.
89 Bolles & Grossen, 1969, p. 90.
90 Bolles & Grossen, 1969, p. 99.
91 Rescorla, 1969b.
92 Weisman & Litner, 1969.
93 Ferster & Skinner, 1957.
94 Weisman & Litner, 1969.
95 Karpicke, Christoph, Peterson & Hearst, 1977.
96 Morris, 1979.
97 Zimmer-Hart & Rescorla, 1974.

5 Comparative Psychology: Species Differences in What Animals Can Learn?

1 Thorndike, 1898.
2 Boakes, 1984; Carmichael,1957.
3 Benjamin & Bruce, 1982.
4 Kellogg, 1931, p. 162.
5 Kellogg, 1931, p. 168.
6 Kellogg & Kellogg, 1933.
7 Kellogg & Kellogg, 1933, p. 142.
8 Kellogg & Kellogg, 1933, p. 282.
9 Kellogg, 1961; Kellogg & Rice, 1966.
10 Dewsbury, 2006, p. 100.
11 Benjamin & Bruce, 1982.
12 Benjamin & Bruce, 1982, pp. 464–465.
13 Hayes & Hayes, 1950, p. 108.
14 Hayes & Hayes, 1952, p. 451.
15 Hayes & Hayes, 1952, p. 451.
16 Yerkes,1925.
17 Gardner & Gardner, 1969, pp. 664–665.
18 Sullivan, 1995; New York Times.
19 Gardner & Gardner, 1969, p. 665.
20 Gardner & Gardner, 1969, p. 666.
21 Gardner & Gardner, 1969, p. 670.
22 Gardner & Gardner, 1969, p. 667.
23 Todes, 2014, pp. 644–645.
24 Email from Daniel Todes to RAB, March 27, 2022.
25 Chomsky, 1959.
26 Gardner & Gardner, 1969, p. 671.
27 Terrace, 1979, pp. 23–28.
28 Terrace, 1979, p. 31.
29 Terrace, 1979, p. 184.
30 Terrace, 1979, p. 221.
31 For example, Seidenberg & Petitto, 1979.
32 Gardner & Gardner, 1984.
33 For example, Fouts, 1973.
34 Pinker, 1994.
35 Premack, 1971.
36 Savage-Rumbaugh, Murphy, Sevcik et al., 1993, p. 24.
37 Savage-Rumbaugh et al., 1993.
38 Premack & Woodruff, 1978.
39 Wimmer & Perner, 1983.
40 Blum, 2002.
41 For example, Harlow, Harlow & Meyer, 1950.
42 For example, Butler, 1953.
43 Harlow, 1953, p. 29.
44 Blum, 2002, p. 111.
45 Harlow, 1953, p. 28.
46 Harlow, 1949.
47 Harlow, 1949, pp. 51–52.
48 Blum, 2002.

49 cited by Blum, 2002, p. 37.
50 Harlow, 1958, p. 673.
51 Warren, 1965a; see also Warren, 1965b.
52 Warren, 1965a, p. 275.
53 Kellogg & Rice, 1963, pp. 484–485.
54 Kellogg & Rice, 1963, p. 484.
55 Terrace,1963.
56 Kellogg & Rice, 1963, p. 493.
57 Kellogg & Rice, 1964, p. 143.
58 Breland & Breland, 1951.
59 Lilly, 1961, p. 72.
60 Lilly, 1961.
61 Lilly, 1961, p. 13.
62 Lilly, 1961, p. 126.
63 Wikipedia entry, John C. Lilly, July, 2019.
64 Evans & Bastian, 1969, p. 432–433; see account by Wood, 1973, pp. 113–118.
65 Boakes & Gaertner, 1977.
66 Obituary of Louis Hermann, NY Times, 2016.
67 Herman, 2010, p. 313.
68 Herman, Beach, Pepper, & Stalling, 1969.
69 Herman & Arbeit, 1973.
70 Herman & Arbeit, 1973, p. 391.
71 Herman, 1975; Thompson & Herman, 1977.
72 Herman, 2010, p. 314.
73 Herman, 2010, p. 314.
74 Herman, 2010, p. 314.
75 Herman, 2010.
76 Beach, 1950; cf. Shettleworth, 2009.
77 Warren, 1965a, p. 95.
78 Warren, 1965a, p. 110.
79 Bullock & Bitterman, 1962, p. 962.
80 Warren, 1965a, p. 108.
81 Warren, 1965a, p. 110.
82 Balsam, 2012.
83 Balsam, 2012, p. 72.
84 Bitterman, 1957, p. 145.
85 Wodinsky & Bitterman, 1957, p. 576.
86 Wodinsky & Bitterman, 1959.
87 For example, Mackintosh, 1974, p. 467.
88 Bitterman, 1965.
89 Mackintosh, 1969b, p. 138.
90 Mackintosh, 1969b.
91 Bitterman, 1969, p. 163.
92 Gonzalez, Berger, & Bitterman, 1966.
93 Bitterman, 1972, p. 164.
94 Bitterman, 1972, p. 173.
95 Moon & Harlow, 1955.
96 Ladygina-Kohts, 1926; see Boakes, 1984.
97 Skinner, 1950.
98 Cummings & Berryman, 1965.
99 For example, Zentall & Hogan, 1978.

100 Wilson, Mackintosh & Boakes, 1985a.
101 Katz & Wright, 2006.
102 Wilson, Mackintosh & Boakes, 1985b.
103 Mackintosh, Wilson, Boakes & Barlow, 1985, pp. 62–63.
104 Blough, 1959.
105 Blough, 1959, p. 159.
106 Berryman, Cummings, & Nevin, 1963.
107 Berryman, Cummings & Nevin, 1963, p. 107.
108 Herman & Gordon, 1974, p. 25.a
109 Roberts, 1998, pp. 64–120.
110 Macphail, 1987.
111 Macphail, 1982.
112 Macphail, 1987, p. 649.
113 Macphail, 1987, p. 649.
114 Macphail, 1987, p. 653.
115 Macphail, 1987, pp. 654–655.
116 Hulse, Fowler & Honig, 1978.
117 Roitblat, Bever, & Terrace, 1984.
118 Shettleworth, 1998; 2nd ed., 2009.
119 Roberts, 1998.

6 Imprinting and Constraints on Learning

1 Hinde, 1973.
2 Lorenz, 1935.
3 For example, Thorpe, 1958.
4 Lorenz, 1937, p. 262.
5 James, 1890.
6 Spalding, 1873/1954.
7 Sluckin, 1964.
8 Lorenz, 1937.
9 Lorenz, 1985.
10 Lorenz, 1985.
11 Sluckin, 1964.
12 Lorenz, 1937, pp. 264–265.
13 Lorenz, 1937, p. 266.
14 Lorenz, 1952.
15 Lorenz, 1966.
16 Bateson, 1978, p. 660.
17 Jaynes, 1956, p. 201.
18 Jaynes, 1956, 1957, 1958a, 1958b.
19 Hinde, Thorpe & Vince, 1956.
20 Hinde, 1987.
21 Bateson, Stevenson-Hinde & Clutton-Brock, 2018.
22 Hinde et al., 1956, p. 216.
23 Hinde et al., 1956, p. 240.
24 For example, Vince, 1956.
25 For example, Hinde & Spencer-Booth, 1967.
26 Hinde & Spencer-Booth, 1971.
27 Bateson et al., 2018. See Griffiths (2004) for the rejection of Lorenz's theory of instincts.

28 Hinde, 1987.
29 For example, Sluckin & Salzen, 1961.
30 For example, James 1959.
31 Sluckin & Salzen, 1961.
32 Rescorla, 1980b.
33 Hoffman & Hoffman, 1990.
34 Hoffman & Ison, 1980.
35 Hoffman & Ratner, 1973, pp. 530–531.
36 Hoffman & Ratner, 1973, p. 532.
37 Ratner & Hoffman, 1974; Experiment 2.
38 Gaioni, Hoffman, DePaulo & Stratton, 1978.
39 Boakes & Panter, 1985.
40 For example, Bateson, 1981.
41 Shettleworth, 1998, pp. 155–165.
42 Bateson, 1973.
43 Hogan, 1973.
44 Shettleworth, 1972.
45 Shettleworth, 1973, p. 243.
46 Thorndike, 1911.
47 Konorski, 1967.
48 Pearce, Hall, & Colwill, 1978, p. 270.
49 Brown & Jenkins, 1968.
50 Jenkins, 1973.
51 Williams & Williams, 1969.
52 Jenkins, 1973, p. 192.
53 Jenkins & Moore, 1973; Moore, 1973.
54 Zener, 1937.
55 Kimble, 1961, p. 54.
56 Hearst & Jenkins, 1974.
57 Domjan, Lyons, North & Bruell, 1986.
58 Wasserman, 1973.
59 Holland, 1977.
60 Timberlake & Grant, 1975, p. 691.
61 Jenkins, Barrera, Ireland, & Woodside, 1978.
62 Boakes, 1977.
63 Harris, Andrew & Kwok, 2013.
64 Breland & Breland, 1961.
65 Boakes, Poli, Lockwood & Goodall, 1978.
66 Bolles & Riley, 1973.
67 Revusky, 1977a.
68 Garcia, 1997.
69 Anonymous, 2013.
70 Freeman, & Riley, 2009.
71 Garcia, Kimmeldorf & Koelling, 1955.
72 Kimble, 1961.
73 Garcia & Koelling, 1966.
74 Revusky, 1977a.
75 Domjan & Wilson, 1972.
76 Garcia, Erwin & Koelling, 1966.
77 Rusiniak, Hankins, Garcia & Brett, 1979.
78 Clarke, Westbrook, & Irwin 1979.
79 Best & Gemberling, 1977.

80 Revusky, 1971.
81 Revusky, 1977b.
82 Durlach & Rescorla, 1980.

7 Discrimination Learning, Attention and Stimulus Generalization

1 Lashley,1930, p. 453.
2 Lashley, 1912.
3 Lashley, 1930.
4 Krech, 1974, p. 233.
5 Sutherland & Mackintosh, 1971, p. 25.
6 Lashley, 1929.
7 Krech, 1974, pp. 226–227.
8 Krechevsky, 1932.
9 Krechevsky, 1932, p. 532.
10 Krech, 1974, p. 227.
11 Lashley, 1942.
12 Spence, 1932.
13 Spence, 1936, p. 429.
14 Spence, 1936.
15 Spence, 1937a, pp. 437–440.
16 Spence, 1937a.
17 Hanson, 1959.
18 Spence, 1937b.
19 Kendler H, 1989.
20 Kendler, T, 2003.
21 Kendler, T, 2003, p. 256.
22 Kendler, T, 2003, p. 257.
23 Spence, 1942,
24 Kendler, T, 1950, p. 561.
25 Kendler, T, 1950.
26 Kendler & Kendler, 1962, p. 12.
27 Kendler & Kendler, 1962.
28 Kendler, T, 2003.
29 Spence, 1942, pp. 266–267.
30 Spence, 1942, p. 271.
31 Gonzalez, Gentry, & Bitterman, 1954, p. 385.
32 For example, Gonzalez, et al., 1954.
33 Spence, 1945, pp. 264.
34 Spence, 1945, p. 266.
35 Mackintosh, 1974, p. 572.
36 Spence, 1952.
37 Bitterman & Wodinsky, 1953.
38 Wagner, 2008, p. 172.
39 Wagner, 2008, p. 172.
40 Hilgard, 1967; Kimble, 1991.
41 Stanford University Memorial Archives: Douglas Lawrence (1918–1999).
42 Lawrence, 1949, p. 770.
43 Lawrence, 1949, p. 782.
44 Lawrence, 1950, p. 176.

45 Lawrence, 1950, p. 184.
46 Lawrence, 1950, p. 186.
47 James, 1890, pp. 513–515.
48 Pavlov,1927, pp. 121–125.
49 Lawrence, 1952.
50 Hilgard & Marquis, 1940.
51 Lawrence & DeRivera, 1954, p. 470.
52 Compare to, Baddeley, 2009.
53 Deutsch, 1953, 1960.
54 Deutsch & Clarkson, 1959a.
55 Deutsch, 1958.
56 Tolman & Honzik, 1930.
57 Deutsch & Clarkson, 1959b, p. 153.
58 Longuet-Higgins, 1998.
59 Mackintosh, 1998b.
60 Deutsch, 1955; Sutherland, 1959a.
61 Diana Deutsch, personal email, February, 2019.
62 Pearce, 2018, p. 301.
63 Pearce, 2018, p. 303.
64 Reid, 1953.
65 Mackintosh, 1969a.
66 Harlow, 1949.
67 Sutherland, 1959a.
68 Mackintosh, 1962.
69 Kendler & Kendler, 1962.
70 Mackintosh, 1962.
71 Mackintosh & Mackintosh, 1963.
72 Sutherland, Mackintosh & Mackintosh, 1963, p. 156.
73 For example, Lawrence, 1949.
74 Sutherland et al., 1963, p. 237.
75 Mackintosh, 1965, p. 124.
76 Broadbent, 1958.
77 Mackintosh, 1965, p. 124.
78 Mackintosh, 1965, p. 125.
79 Pearce, 2018.
80 Sutherland & Mackintosh, 1971.
81 Mackintosh, 1962.
82 Hall, 1974, p. 939.
83 Hall, 1974, p. 940.
84 Hall, 1974, p. 943.
85 Ferster, 1953.
86 Ferster, 1953, p. 274.
87 Lashley & Wade, 1946.
88 Skinner, 1965.
89 Guttman & Kalish, 1956, p. 80.
90 Hanson, 1959; p. 321.
91 Reynolds, 1961.
92 Jenkins & Harrison, 1960.
93 Jenkins & Harrison, 1962.
94 Shepard, 1965, p. 95.
95 Shepard, 1965, p. 96.
96 Reynolds, 1961, p. 70; italics as in original.

97 Skinner, 1938, pp. 203–206.
98 Terrace, 1963, p. 24.
99 Terrace, 1966, p. 340.
100 Rilling, 1977, p. 175.
101 Rilling, 1977, p. 475.
102 For example, Bloomfield, 1969.
103 For example, Halliday & Boakes, 1972.
104 Westbrook, 1973.
105 Williams & Williams, 1969.
106 Keller, 1974.
107 Schwartz, 1975.
108 Schwartz & Gamzu, 1977.
109 Schwartz & Gamzu, 1977, p. 91.

8 B.F. Skinner and the Experimental Analysis of Behavior

1 Skinner, 1971.
2 Bjork, 1993.
3 Skinner, 1976, p. 56.
4 Russell, 1927.
5 Skinner, 1979, p. 289.
6 cited by Bjork, 1993, p. 80.
7 Skinner, 1979, p. 38.
8 Keller, 1970.
9 Skinner, 1979, p. 289; see also Skinner, 1956, 1959.
10 Skinner, 1979, p. 98.
11 Skinner, 1979, p. 42.
12 but see, for example, Harris, Kwok & Gottlieb, 2019.
13 Skinner, 1979, p. 131.
14 Skinner, 1979, p. 70.
15 Skinner, 1938, p. 8.
16 Skinner, 1938, p. 22.
17 Skinner, 1938, p. 43.
18 Skinner, 1938, p. 428.
19 Skinner, 1938, p. 432.
20 Bjork, 1993, pp. 116–119.
21 Skinner, 1979, p. 212.
22 Skinner & Heron, 1937.
23 Skinner, 1938, p. 26.
24 Ellson, 1939, p. 566.
25 Skinner, 1979, p. 240.
26 Skinner & Estes, 1941.
27 Bjork, 1993, p. 117.
28 Skinner, 1979, p. 241.
29 Bjork, 1993, p. 124.
30 Skinner, 1979, p. 267.
31 Skinner, 1979, p. 283; see also Skinner, 1960.
32 Skinner, 1945.
33 Skinner, 1979, p. 286.
34 Bjork, 1993, p. 133.

35 Bjork, 1993, pp. 131–142.
36 Skinner, 1979, p. 283.
37 Skinner, 1948a.
38 Staddon, 1992.
39 Falk, 1961.
40 Falk, 1971.
41 Brown & Jenkins, 1968.
42 Rescorla, 1967.
43 Staddon, 2016, p. 158–160.
44 Staddon & Simmelhag, 1971.
45 Staddon, 1977.
46 see Boakes, Patterson, Kendig & Harris, 2015; Killeen & Pellón, 2013.
47 Skinner, 1979, pp. 231–235.
48 Skinner, 1979, p. 340.
49 Skinner, 1950.
50 Skinner, 1950, p. 215.
51 Skinner, 1951.
52 Estes, 1950.
53 Breland & Breland, 1951.
54 Kimble, 1986.
55 For example, Guttman & Kalish, 1956.
56 Keller, 1986, p. 141; see also Root, 2002.
57 Keller, 1941.
58 Keller & Schoenfeld, 1950.
59 Dinsmoor, 1996.
60 Keller, 1986.
61 Skinner, 1953, p. 5.
62 Skinner, 1953, p. 31.
63 Skinner, 1953, p. 104.
64 Skinner, 1983, p. 64–65.
65 Skinner, 1958, p. 979.
66 Bjork, 1993, pp. 167–190.
67 Rutherford, 2003; Skinner, 1983, pp. 52–55.
68 Ferster, 2002, p. 305.
69 Ferster, 2002.
70 Skinner, 1983, p. 36.
71 Morse, 2017.
72 Wikipedia, March 15, 2021.
73 Brady, 1958, p. 97.
74 Warren & Marshall, 1983.
75 Weiss, 1968.
76 Wikipedia, March 16, 2021.
77 Honig, 1966, p. 2.
78 Dinsmoor, 1987.
79 Catania, 2002.
80 Catania, 2002; Staddon, 2016, pp. 124–130.
81 Catania, 2002, pp. 341–342.
82 Baum, 1994.
83 Herrnstein & Boring, 1965.
84 Herrnstein, 1962, p. 677.
85 Herrnstein, 1962, p. 678.
86 Herrnstein & Loveland, 1964.

87 Skinner, 1983, p. 281.
88 Baum, personal email April, 2021.
89 Herrnstein, 1961.
90 Stevens, 1957.
91 Herrnstein, 1961, p. 272.
92 de Villiers, 1977.
93 For example, Staddon, 1968.
94 For example, Killeen, 1972.
95 Baum, 2002, p. 352.
96 Herrnstein, 1970, p. 243.
97 Skinner, 1983, p. 124.
98 Catania, 2012.
99 Rachlin, 1974.
100 For example, Bouton, 2016, pp. 266–276.
101 For example, Baum, 1979.
102 For example, Robinson & Woodward, 1989.
103 Hurwitz, 1953.
104 Goodale, 2020.
105 Lea, 2014.
106 For example, Bloomfield, 1966.
107 Hodos, 1961.
108 Dickins, 2023.
109 Millenson, 1967.
110 Steven Pinker, email to RAB May 3, 2022.
111 Blackman, 1974.
112 Blackman, in preparation.
113 Blackman, in preparation.
114 Wearden, 2021.
115 Keller, 2009, pp. 246–280.
116 Keller, 2009, p. 280.
117 Cirino, Miranda, & de Cruz, 2012.
118 Ribes, 2010, p. 38.
119 Bijou & Ribes, 1972.
120 Chomsky, 1967.
121 Chomsky, 1959.
122 Chomsky, 1959.
123 Skinner, 1983, p. 193.
124 Boden, 1977.
125 Ryle, 1949.
126 Skinner, 1983, p. 150.
127 Greenspoon, 1955.
128 Hefferline, Keenan & Harford, 1959.
129 Brewer, 1974.
130 Dulany, 1961
131 For example, Colagiuri & Livesey, 2016; Lovibond & Shanks, 2002.
132 Skinner, 1971, pp. 4–5.
133 For example, Bjork, 1993, pp. 192–213.
134 Bjork, 1993, pp. 167–190.
135 Skinner, 1983, pp. 381–383.
136 Skinner, 1953, p. 382; Epstein, Lanza & Skinner, 1980.
137 Skinner, 1983, p. 383; Epstein, 1987

9 How Animals Learn to Associate Events

1 Kamin, personal communication in 1990.
2 Leahey, 2004, pp. 190–191.
3 Oberg, 1976.
4 Pavlov, 1927, p. 141–142; 269–270.
5 Hull, 1943, p. 208.
6 Hull, 1943, p. 216.
7 Kamin, 2005.
8 Kamin, 1968.
9 Kamin, 1968.
10 Kamin, 1968.
11 Kamin, 1969b, p. 62; see also Kamin, 1969a.
12 Miles, 1965; vom Saal, 1967.
13 For example, vom Saal & Jenkins, 1970.
14 email from David Pittinger to RAB, 2021.
15 Gleitman, Nachmias & Neisser, 1954.
16 Gleitman, 1981.
17 Rescorla & LoLordo, 1965.
18 Rescorla, 1966, p. 384.
19 Rescorla, 1968, p. 1.
20 Anon, 1999.
21 Anon, 1999; Delamater & Whitlow, 2020.
22 Logan & Wagner, 1965.
23 Wagner, Logan, Haberlandt & Price, 1968.
24 Neisser, 1967.
25 Mackintosh, personal communication.
26 Black & Prokasy, 1972.
27 Rescorla & Wagner, 1972, p. 75.
28 Bush & Mosteller, 1955.
29 Rescorla & Wagner, 1972, p. 88.
30 Rescorla, 1971.
31 LoLordo, personal communication, July 14, 2021.
32 Rescorla & LoLordo, 1965, p. 406.
33 Rescorla, 1969a; see also Hearst, 1972.
34 Ellen, Wilson & Powell, 1964.
35 Schwartzbaum & Donovick, 1968.
36 Jenkins & Harrison, 1962.
37 For example, Halliday & Boakes, 1971.
38 Boakes & Halliday, 1972.
39 Wagner & Rescorla, 1972.
40 Zimmer-Hart & Rescorla, 1974.
41 Pearce, 2018.
42 Mackintosh & Honig, 1969.
43 Mackintosh, 1969a, p. 202.
44 Mackintosh, 1971, p. 124.
45 Mackintosh, 1975a.
46 Kamin & Gaioni, 1974, p. 597.
47 Westbrook, personal communication, 2021.
48 Herrnstein, 1971.

49 Westbrook, personal communication.
50 Kamin, 1974.
51 Kamin, 1969b, p. 295.
52 Rudy, email August 9, 2021.
53 Wagner, Rudy, & Whitlow, 1973, p. 408.
54 Whitlow, 1975, p. 205.
55 For example, Davis, 1970.
56 Whitlow, 1975; p. 205.
57 Holmes, Chan, & Westbrook, 2020, p. 229; see also Vogel et al., 2018.
58 Holland, personal communication, August, 2021
59 Rescorla, 1973.
60 Rescorla, 1973, p. 142.
61 Pavlov, 1927, p. 58.
62 For example, Bouton, 1993.
63 Konorski, 1948.
64 Rescorla & Heth, 1975, p, 95.
65 Rescorla & Heth, 1975, p. 88.
66 Rescorla & Heth, 1975.
67 Rozeboom, 1958, p. 32.
68 Rizley & Rescorla, 1972, p. 10.
69 Rizley & Rescorla, 1972, p. 11.
70 Rescorla, 1973.
71 Rescorla, 1980b.
72 D. Heth, email to RAB, September 8, 2021.
73 Rescorla, 1980b, p. 3.
74 For example, Gormezano & Kehoe, 1975.
75 Holland, & Rescorla, 1975a.
76 Nairne & Rescorla, 1981.
77 Heth, 1976.
78 Rescorla & Freberg, 1978.
79 Koehler, 1941.
80 Rescorla, 1980b, pp. 42–50.
81 Rescorla, 1980b, pp. 50–53.
82 Holland and Rescorla, 1975b.
83 Rescorla, 1980b, p. 97.
84 Rescorla, 1980b, pp. 100, 105.
85 Mackintosh, 1974.
86 Hill, 1966.
87 Mackintosh, 1974, p. 268.
88 Mackintosh, 1974, pp. 219–221.
89 Mackintosh & Dickinson, 1979, p. 155.
90 Mackintosh & Dickinson, 1979, p. 161.
91 Guthrie, 1952, p. 143.
92 Mackintosh & Dickinson, 1979, p. 166.
93 For example, Dickinson, 1994.
94 Morrison & Collyer, 1974.
95 Holman, 1975.
96 Adams,1980, p. 456.
97 Dickinson, Nicholas & Adams, 1983.
98 Dickinson, 1994, p. 52.
99 For example, Dickinson, 1985.

100 For example, Colwill & Rescorla, 1986.
101 For example, Balleine & Dickinson, 1994.
102 For example, Balleine & Dickinson, 1998.
103 Dickinson, Hall & Mackintosh, 1976.
104 Dickinson & Mackintosh, 1979.
105 Dickinson & Mackintosh, 1979, p. 175.
106 For example, Hall, 1973.
107 Pearce & Hall, 1978.
108 Hall, unpublished, p. 12.
109 Hall & Pearce, 1978.
110 Hall & Pearce, 1978, p. 548.
111 Pearce & Hall, 1980, p. 538.
112 Pearce & Hall, 1980, p. 539.
113 For example, Shiffrin & Schneider, 1977.
114 Pearce & Hall, 1980, p. 549.
115 Bouton, 2016.
116 Pearce, 2008.
117 Balaz, Kasprow, & Miller, 1982.
118 For example, Pearce & Mackintosh, 2010.
119 George & Pearce, 2012, p. 241.
120 Sutherland, 1976.
121 Sutherland, 1992.
122 Mackintosh, 1975b, p. 685.
123 Mackintosh & Mascie-Taylor, 1986, p. 122.
124 Mackintosh, Mascie-Taylor, & West, 1988.
125 Mackintosh, 1998a.
126 For example, McLaren & Mackintosh, 2000.
127 Pearce, 2018.
128 personal communication to RAB, 1990.
129 Lewontin, Rose, & Kamin 1984.
130 Ed Vogel, email to RAB, September 18, 2021.
131 LoLordo, 2021.
132 For example, Rescorla, 2000.
133 Holland, email, to RAB, September 2021.
134 LoLordo, 2021.

References

Adams, C. D. (1980). Post-conditioning devaluation of an instrumental reinforcer has no effect on extinction performance. *Quarterly Journal of Experimental Psychology, 32*, 447–458.

Adams, C. D. & Dickinson, A. (1981). Instrumental responding following reinforcer devaluation. *Quarterly Journal of Experimental Psychology, 33B*, 109–122.

Amsel, A. (1958). The role of frustrative nonreward in noncontinuous reward situations. *Psychological Bulletin, 55*(2), 102–119.

Amsel, A. (1994). Précis of frustration theory: An analysis of dispositional learning and memory. *Psychonomic Bulletin & Review, 1*, 280–296.

Amsel, A. (1995). Kenneth Wartenbee Spence. *Memoirs of the National Academy of Science,* 66, 334.

Amsel, A. & Rashotte, M. E. (1969). Transfer of experimenter-imposed slow response patterns to extinction of a continuously rewarded response. *Journal of Comparative and Physiological Psychology, 69*, 185–189.

Annau, Z. & Kamin, L. J. (1961). The conditioned emotional response as a function of the intensity of the US. *Journal of Comparative and Physiological Psychology, 54*, 428–432.

Anon (1999). Biographical sketch: Allan Wagner. *American Psychologist, 54*, 887–890.

Anonymous (2013). John Garcia (1917–2012) obituary. *Skagit Valley Herald.* March 7, 2013.

Baddeley, A. (2009). Psychology in the 1950s: A personal view. In P. Rabbitt (Ed.) *Inside psychology: A science of over 50 years* (pp. 27–35). Oxford: Oxford University Press.

Baker, A. (1977). Conditioned inhibition arising from a between-sessions negative correlation. *Journal of Experimental Psychology: Animal Behavior Processes, 3*, 144–155.

Balaz, M. A., Kasprow, W. J., & Miller, R. R. (1982). Blocking with a single compound trial. *Animal Learning & Behavior, 10*, 271–276.

Balleine, B. W. & Dickinson, A. (1994). Motivational control of goal-directed action. *Animal Learning & Behavior, 22*, 1–18.

Balleine, B. W. &. Dickinson, A. (1998). Goal-directed instrumental action: Contingency and incentive learning and their cortical substrates. *Neuropharmacology, 37*, 407–419.

Balsam, P. D. (1985). The functions of context in learning and performance. In P. D. Balsam & A. Tomie (Eds.) *Context and learning* (pp. 1–22). Hillsdale, NJ: Lawrence Erlbaum Associates.

Balsam, P. D. (2012). Obituary: M.E. Bitterman. *American Psychologist, 67*, 72.

Balsam, P. D. & Tomie, A. (1985). *Context and learning*. Hillsdale, NJ: Lawrence Erlbaum Associates.

Bateson, P. P. G. (1973). Internal influences on early learning in birds. In R. A. Hinde & J. Stevenson-Hinde (Eds.) *Constraints on learning: Limitations and predispositions* (pp. 101–116). New York: Academic Press.

Bateson, P. P. G. (1978). Sexual imprinting and optimal outbreeding. *Nature, 273*, 659–660.

Bateson, P. P. G. (1981). The control of sensitivity to the environment during development. In K. Immelman, G. Barlow, M. Main & L. Petrinovich (Eds.) *Behavioural development* (pp. 432–453). New York: Cambridge University Press.

Bateson, P. P. G., Stevenson-Hinde, J., & Clutton-Brock, T. (2018). Robert Aubrey Hinde CBE. *Biographical Memoirs of Fellows of the Royal Society, 65*, 151–177.

Baum, W. B. (1979). Matching, undermatching, and overmatching in studies of choice. *Journal of the Experimental Analysis of Behavior, 32*, 269–281.

Baum, W. B. (1994). Richard J. Herrnstein, a memoir. *The Behavior Analyst, 17*, 203–205.

Baum, W. B. (2002). The Harvard pigeon lab under Herrnstein. *Journal of the Experimental Analysis of Behavior, 77*, 347–355.

Beach, F. A. (1950). The Snark was a Boojum. *American Psychologist, 5*, 115–124.

Beach, F. A. (1959). *Clark Leonard Hull (1884–1952)*. Washington, DC: National Academy of Sciences.

Benjamin, L. T. & Bruce, D. (1982). From bottle-fed chimpanzee to bottlenose dolphin: A contemporary appraisal of Winthrop Kellogg. *Psychological Record, 32*, 46–482.

Berryman, R., Cumming, W. W., & Nevin, J. A. (1963). Acquisition of delayed matching in the pigeon. *Journal of the Experimental Analysis of Behavior, 6*, 101–107.

Best, M. E. & Gemberling, G. A. (1977). Role of short-term processes in the conditioned stimulus pre-exposure effect and the delay of reinforcement gradient in long-delay taste-aversion learning. *Journal of Experimental Psychology: Animal Behavior Processes, 3*, 253–263.

Best, P. J., Best, M. R. & Mickley, G. A. (1973). Conditioned aversion to distinct environmental stimuli resulting from gastrointestinal distress. *Journal of Comparative and Physiological Psychology, 86*, 250–257.

Bijou, S. W. & Ribes, E. (1972). *Behavior modification: Issues and extensions*. New York: Academic Press.

Birch, H. G. (1945). The relation of previous experience to insightful problem-solving. *Journal of Comparative Psychology, 38*, 367–383.

Bitterman, M. E. (1957). Review of *Spence's behavior theory and conditioning*. *American Journal of Psychology, 70*, 141–145.

Bitterman, M. E. (1965). Phyletic differences in learning. *American Psychologist, 20*, 396–410.

Bitterman, M. E. (1969). Habit reversal and probability learning: Rats, birds and fish. In R. M. Gilbert & N. S. Sutherland (Eds.) *Animal discrimination learning* (pp. 163–175). London: Academic Press.

Bitterman, M. E. (1972). Comparative studies of the role of inhibition in reversal learning. In R. A. Boakes & M. S. Halliday (Eds.) *Inhibition and learning* (pp. 153–176). London: Academic Press.

Bitterman, M. E. & Wodinsky, J. (1953). Simultaneous and successive discrimination. *Psychological Review, 60*, 371–376.

Bjork, D. J. (1993). *B.F. Skinner: A life*. New York: Basic Books.

Black, A. H. (1957). The extinction of avoidance responses under curare. *Journal of Comparative and Physiological Psychology, 51*, 519–524.

Black, A. H. (1965). Cardiac conditioning in curarized dogs: The relationship between heart rate and skeletal behaviour. In W. F. Prokasy (Ed.) *Classical conditioning: A symposium* (pp. 20–47). New York: Appleton-Century-Crofts.

Black, A. H. & Prokasy, W. F. (1972). *Classical conditioning II: Current research and theory*. New York: Appleton-Century-Crofts.

Blackman, D. E. (1974). *Operant conditioning: An experimental analysis of behavior*. London: Methuen.

Blackman, D. E. (in preparation). Beginnings of experimental analysis of behaviour in UK/Ireland and Europe. In Pellon, R., Blackman, D. E. & Arntzen, E. (Eds.) *Origins and evolution of behavior analysis in America and Europe.*

Block, J. D. (1963a). Howard S. Liddell, Ph.D.: Scientist and humanitarian. *Conditional Reflex, 1,* 171–180.

Block, J. D. (1963b). In memoriam: Howard S. Liddell (1895–1962). *Psychosomatic Medicine, 25,* 1–2.

Blodgett, (1929). The effect of the introduction of reward upon the maze performance of rats. *University of California Publications in Psychology, 4,* 113–134.

Bloomfield, T. M. (1966). Two types of behavioral contrast in discrimination learning. *Journal of the Experimental Analysis of Behavior, 9,* 155–161.

Bloomfield, T. M. (1969). Behavioral contrast and the peak shift. In R. M. Gilbert, & N. S. Sutherland (Eds.) *Animal discrimination learning* (pp. 215–241). New York: Academic Press.

Blough, D. S. (1959). Delayed matching in the pigeon. *Journal of the Experimental Analysis of Behavior, 2,* 151–160.

Blum, D. (2002). *Love at Goon Park: Harry Harlow and the science of affection.* New York: Berkley Books.

Boakes, R. A. (1977). Performance on learning to associate a stimulus with positive reinforcement. In H. Davis & H. M. B. Hurvitz (Eds.) *Operant-Pavlovian interactions* (pp. 67–97). Hillsdale, NJ: Erlbaum.

Boakes, R. A. (1979). Interactions between Type 1 and Type 2 processes involving positive reinforcement. In A. Dickinson & R. A. Boakes (Eds.) *Mechanisms of learning and motivation: A memorial volume to Jerzy Konorski* (pp. 233–268). Hillsdale, NJ: Lawrence Erlbaum Associates.

Boakes, R. A. (1984). *From Darwin to behaviourism.* Cambridge: Cambridge University Press.

Boakes, R. A. & Gaertner, I. (1977). The development of a simple form of communication. *Quarterly Journal of Experimental Psychology, 29,* 561–575.

Boakes, R. A. & Halliday, M. S. (1972). *Inhibition and learning.* London: Academic Press.

Boakes, R. A. & Panter, D. (1985). Secondary imprinting in the domestic chick blocked by previous exposure to a live hen. *Animal Behaviour, 33,* 353–365.

Boakes, R. A., Patterson, A. E., Kendig, M. D. & Harris, J. A. (2015). Temporal distributions of schedule-induced licks, magazine entries, and lever presses on fixed- and variable-time schedules. *Journal of Experimental Psychology: Animal Learning and Cognition, 41,* 52–68.

Boakes, R. A., Poli, M., Lockwood, M. J., & Goodall, G. (1978). A study of misbehavior: Token reinforcement in the rat. *Journal of the Experimental Analysis of Behavior, 29,* 115–134.

Boden, M. (1977). *Artificial intelligence and natural man.* New York: Basic Books.

Bolles, R. C. (1967). *Theory of motivation.* New York: Harper Row.

Bolles, R. C. (1970). Species-specific defense reactions and avoidance learning. *Psychological Review, 77,* 32–48.

Bolles, R. C. & Grossen, N. E. (1969). Effects of an informational stimulus on the acquisition of avoidance behavior in rats. *Journal of Comparative and Physiological Psychology, 68,* 90–99.

Bolles, R. C. & Popp, R. J. (1964). Parameters affecting the acquisition of Sidman avoidance. *Journal of the Experimental Analysis of Behavior, 7,* 315–321.

Bolles, R. C. & Riley, A. L. (1973). Freezing as an avoidance response: Another look at the operant-respondent distinction. *Learning & Motivation, 4,* 268–275.

Bouton, M. E. (1993). Context, time, and memory retrieval in the interference paradigms of Pavlovian learning. *Psychological Bulletin, 114,* 80–99.

Bouton, M. E. (2016). *Learning and behavior: A contemporary synthesis.* 2nd ed. Sunderland, MA: Sinauer Associates Inc.

Bouton, M. E. & Bolles, R. C. (1979a). Role of conditioned contextual stimuli in reinstatement of an extinguished fear. *Journal of Experimental Psychology: Animal Behavior Processes, 5*, 368–378.

Bouton, M. E. & Bolles, R. C. (1979b). Contextual control of the extinction of conditioned fear. *Learning and Motivation, 10*, 445–466.

Brady, J. V. (1958). Ulcers in "executive" monkeys. *Scientific American, 199*(4), 95–103.

Breland, K. & Breland, M. (1951). A field of applied animal psychology. *American Psychologist, 6*, 202–204.

Breland, K. & Breland, M. (1961). The misbehavior of organisms. *American Psychologist, 16*, 681–684.

Brewer, W. F. (1974). There is no convincing evidence for operant and classical conditioning in human beings. In W. B. Weimer & D. L. Palermo (Eds.) *Cognition and the symbolic processes* (pp. 1–42). Princeton, NJ: Lawrence Erlbaum Associates.

Broadbent, D. E. (1958). *Perception and communication.* London: Pergamon Press.

Brogden, W. J. (1939). Unconditioned stimulus-substitution in the conditioning process. *American Journal of Psychology, 52*, 46–55.

Brogden, W. J. & Culler, E. (1936). Device for motor conditioning of small animals. *Science, 83*, 269.

Brogden, W. J. & Gantt, W. H. (1937). Cerebellar conditioned reflexes. *American Journal of Physiology, 119*, 277–278.

Brogden, W. J., Lipman, E. A., & Culler, E. (1938). The role of incentive in conditioning and extinction. *American Journal of Psychology, 51*, 109–117.

Brown, P. L. & Jenkins, H. M. (1968). Auto-shaping of the pigeon's key-peck. *Journal of the Experimental Analysis of Behavior, 11*, 1–8.

Brown, R. W. (1965). On the combination of drive and incentive motivation. *Psychological Review, 72*, 310–317.

Brown, R. T. & Wagner, A. R. (1964). Resistance to punishment and extinction following training with shock or nonreinforcement. *Journal of Experimental Psychology, 68*, 503–507.

Bruce, R. H. (1930). The effect of removal of reward upon the maze performance of rats. *University of California Publications in Psychology, 4*, 203–214.

Bullock, D. H. & Bitterman, M. E. (1962). Habit reversal in the pigeon. *Journal of Comparative and Physiological Psychology, 55*, 958–962.

Burgess, N. (2014). The 2014 Nobel Prize in Physiology or Medicine: A spatial model for cognitive neuroscience. *Neuron, 84*, 1120–1125.

Bush, R. R. & Mosteller, F. (1955). *Stochastic models for learning.* New York: John Wiley & Sons, Inc.

Butler, R. A. (1953). Discrimination learning by rhesus monkeys to visual-exploration motivation. *Journal of Comparative and Physiological Psychology, 46*, 95–98.

Butler, R. A. (1954). Incentive conditions which influence visual exploration. *Journal of Experimental Psychology, 48*, 19–23.

Butler, R. A. & Harlow, H. F. (1954). Persistence of visual exploration in monkeys. *Journal of Comparative and Physiological Psychology, 47*, 260–265.

Campbell, B. A. & Ellison, G. D. (1997). Frederick Duane Sheffield (1914–1994). *American Psychologist, 57*, 67.

Capaldi, E. J. (1964). Effect of N-length, number of different N-lengths, and number of reinforcements on resistance to extinction. *Journal of Experimental Psychology, 68*, 230–239.

Capaldi, E. J. (1966). Partial reinforcement: A hypothesis of sequential effects. *Psychological Review, 73*, 459–477.

Carmichael, L. (1957). Robert Mearns Yerkes, 1876–1956. *Psychological Review, 64*, 1–7.

Carroll, D. (2017). *Purpose and cognition.* Cambridge, UK: Cambridge University Press.

Catania, A. C. (2002). The watershed years of 1958–1962 in the Harvard pigeon lab. *Journal of the Experimental Analysis of Behavior, 77*, 327–345.

Catania, A. C. (2012). The pursuit of experimental analysis. *European Journal of Behavior Analysis, 13*, 269–280.

Chomsky, N. (1959). A review of B.F. Skinner's *Verbal behavior. Language, 35*, 26–58.

Chomsky, N. (1967). Preface to the 1967 reprint of "A review of Skinner's Verbal Behavior." In L. A. Jacobovits & M. S. Miron (Eds.) *Readings in the psychology of language*. Prentice Hall.

Cirino, S. D., Miranda, R. L., & da Cruz, R. N. (2012). The beginnings of behavior analysis laboratories in Brazil: A pedagogical view. *History of Psychology, 15*, 263–272.

Clarke, J. C., Westbrook, R. F., & Irwin, J. (1979). Potentiation instead of overshadowing in the pigeon. *Behavioral and Neural Biology, 25*, 18–29.

Colagiuri, B. & Livesey, E. J. (2016). Contextual cueing as a form of nonconscious learning: Theoretical and empirical analysis in large and very large samples. *Psychonomic Bulletin & Review, 23*, 1996–2009.

Colwill, R. M. & Rescorla, R. A. (1986). Associative structures in instrumental learning. *Psychology of Learning and Motivation, 20*, 55–104.

Crespi, L. P. (1942). Quantitative variation in incentive and performance in the white rat. *American Journal of Psychology, 55*, 467–517.

Crespi, L. P. (1944). Amount of reinforcement and level of performance. *Psychological Review, 51*, 341–357.

Crutchfield, R. S. (1960). Edward Chace Tolman: 1886–1959. *American Journal of Psychology, 74,* 135–141.

Culler, E. (1938). Recent advances in some concepts of conditioning. *Psychological Review, 45*, 134–153.

Cummings, W. W. & Berryman, R. (1965). The complex discriminated operant: Studies of matching-to-sample and related problems. In D. Mostofsky (Ed.) *Stimulus generalization*, pp. 284–330. Stanford: Stanford University Press.

Davis, M. (1970). Effects of inter-stimulus length and variability on startle-response habituation in the rat. *Journal of Comparative and Physiological Psychology, 72*, 177–192.

Deese, J. (1951). The extinction of a discrimination without performance of the choice response. *Journal of Comparative and Physiological Psychology, 44*, 362–366.

Delamater, A. R. & Whitlow, J. W. (2020). Editorial: A special issue to commemorate the intellectual contributions of Allan R. Wagner. *Journal of Experimental Psychology: Animal Learning and Cognition, 46*, 165–169.

de Villiers, P. (1977). Choice in concurrent schedules and a quantitative formulation of the Law of Effect. In W. K. Honig & J. E. R. Staddon (Eds.) *Handbook of operant behavior* (pp. 233–287) Englewood Cliffs, NJ: Prentice Hall.

Deutsch, J. A. (1953). A new type of behaviour theory. *British Journal of Psychology, 44*, 305–317.

Deutsch, J. A. (1955). A theory of shape recognition. *British Journal of Psychology, 36*, 40–47.

Deutsch, J. A. (1956). The inadequacy of the Hullian derivations of reasoning and latent learning. *Psychological Review, 63*, 389–399.

Deutsch, J. A. (1958). Double drive learning in rats without previous selective reinforcement. *Quarterly Journal of Experimental Psychology, 10*, 207–210.

Deutsch, J. A. (1960). *The structural basis of behavior.* Chicago: Chicago University Press.

Deutsch, J. A. & Clarkson, J. K. (1959a). A test of the neo-behaviouristic theory of extinction. *Quarterly Journal of Experimental Psychology, 11*, 143–148.

Deutsch, J. A. & Clarkson, J. K. (1959b). Reasoning in the hooded rat. *Quarterly Journal of Experimental Psychology, 11*, 150–154.

Dewsbury, D. (2006). *Monkey farm: A history of the Yerkes Laboratory of Primate Biology, Orange Park, Florida, 1930–1965*. Lewisburg: Bucknell University Press.

DiCara, L. V. & Miller, N. E. (1968). Instrumental learning of vasomotor responses by rats: Learning to respond differentially in the two ears. *Science, 159*, 1485–1486.

Dickins, D. (2023). Bliss in that dawn: The beginnings of operant psychology in the UK. *History and Philosophy of Psychology, 23*, 34–49.

Dickinson, A. (1985). Actions and habits: The development of behavioural autonomy. *Philosophical Proceedings of the Royal Society of London, B, 308*, 67–78.

Dickinson, A. (1994). Instrumental learning. In N. J. Mackintosh (Ed) *Animal Learning and Cognition* (pp. 4–79). London: Academic Press.

Dickinson, A. & Mackintosh, N. J. (1979). Reinforcer specificity in the enhancement of conditioning by posttrial surprise. *Journal of Experimental Psychology: Animal Behavior Processes, 5*, 162–177.

Dickinson, A., Hall, G., & Mackintosh, N. J. (1976). Surprise and the attenuation of blocking. *Journal of Experimental Psychology: Animal Behavior Processes, 2*, 313–322.

Dickinson, A., Nicholas, D. J., & Adams, C. D. (1983). The effect of instrumental training contingency on susceptibility to reinforcer devaluation. *Quarterly Journal of Experimental Psychology, 35B*, 35–51.

Dilger, W. C., Moore, A. U., & Freeman, F. S. (1962). Obituary notice: Howard Scott Liddell. eCommons.Cornell.

Dimmick, F. L., Ludlow, N., & Whiteman, A. (1939). A study of "experimental neurosis" in cats. *Journal of Comparative Psychology, 28*, 39–43.

Dinsmoor, J. A. (1987). A visit to Bloomington: The first conference on the experimental analysis of behavior. *Journal of the Experimental Analysis of Behavior, 48*, 441–445.

Dinsmoor, J. A. (1996). Studies in the history of psychology: CVI. An appreciation of Fred S. Keller, 1899–1996. *Psychological Reports, 79*, 891–898.

Dollard, J., Miller, N. E., Doob, L. W., Mowrer, O. H., & Sears, R. R. (1939). *Frustration and aggression*. New Haven, CT: Institute of Human Relations.

Domjan, M. L. & Wilson, N. E. (1972). Specificity of cue to consequences in aversion learning in the rat. *Psychonomic Science, 26*, 143–145.

Domjan, M. L., Lyons, R., North, N. C., & Bruell, J. (1986). Sexual Pavlovian conditioned approach behavior in male Japanese quail (Coturnix coturnix japonica). *Journal of Comparative Psychology, 100*(4), 413–421.

Dulany, D. E. Jr. (1961). Hypotheses and habits in verbal 'operant conditioning'. *Journal of Abnormal and Social Psychology, 63*, 251–263.

Durlach, P. J. & Rescorla, R. A. (1980). Potentiation rather than overshadowing in flavor-aversion learning: An analysis in terms of within-compound associations. *Journal of Experimental Psychology: Animal Behavior Processes, 6*, 175–187.

Dworkin, B. R. & Miller, N. E. (1986). Failure to replicate visceral learning in the acute curarized rat preparation. *Behavioral Neuroscience, 100*, 299–314.

Dykman, R. A. & Gantt, W. H. (1960). A case of experimental neurosis and recovery in relation to the orienting response. *The Journal of Psychology, 50*, 105–110.

Egger, M. D. & Miller, N. E. (1962). Secondary reinforcement in rats as a function of information value and reliability of the stimulus. *Journal of Experimental Psychology, 64*, 97–104.

Ellen, P., Wilson, A. S., & Powell, E. W. (1964). Septal inhibition and timing behavior in the rat. *Experimental Neurology, 10*, 120–132.

Elliott, M. H. (1928). The effect of change of reward upon the maze performance of rats. *University of California Publications in Psychology, 4*, 19–30.

Ellson, D. G. (1939). The concept of reflex reserve. *Psychological Review, 46*, 566–575.

Epstein, R., Lanza, R. P., & Skinner, B. F. (1980). Symbolic communication between two pigeons. *Science, 207*, 543–545.

Epstein, R. (1987). The spontaneous interconnection of four repertoires of behavior in a pigeon. *Journal of Comparative Psychology, 101*, 197–201.

Estes, W. K. (1943). Discriminative conditioning: I. A discriminative property of conditioned anticipation. *Journal of Experimental Psychology, 32*, 150–155.

Estes, W. K. (1945). An experimental study of punishment. *Psychological Monographs, 57*, 1–40.

Estes, W. K. (1948). Discriminative conditioning: II. Effects of a Pavlovian conditioned stimulus upon a subsequently established operant response. *Journal of Experimental Psychology, 38*, 173–177.

Estes, W. K. (1950). Toward a statistical theory of learning. *Psychological Review, 57*, 94–107.

Estes, W. K. (1969). Outline of a theory of punishment. In B. A. Campbell & R. M. Church (Eds.) *Punishment and aversive behavior* (pp. 57–82). New York: Appleton-Century-Crofts.

Estes, W. K. (1989). Autobiography. In G. Lindzey (Ed.) *A history of psychology in autobiography* (Vol. 8, pp. 95–126). Stanford, Calif: Stanford University Press.

Estes, W. K. & Skinner, (1941). Some quantitative properties of anxiety. *Journal of Experimental Psychology, 29*, 390–400.

Evans, W. E. & Bastian, J. (1969). Marine mammal communication: Social and ecological factors. In H. T. Andersen (Ed.) *The biology of marine mammals* (pp. 425–475) New York: Academic Press.

Falk, J. L. (1961). Production of polydipsia in normal rats by an intermittent food schedule. *Science, 133*, 195–196.

Falk, J. L. (1971). The nature and determinants of adjunctive behaviour. *Learning & Behavior, 6*, 577–588.

Fanselow, M. S. & Bouton, M. E. (1997). The life and influence of Robert C. Bolles. In M. E. Bouton & M. S. Fanselow (Eds.) *Learning, motivation, and cognition: The functional behaviorism of Robert C. Bolles* (pp. 1–9). Washington, DC: American Psychological Association.

Ferster, C. B. (1953). The use of the free operant in the analysis of behavior. *Psychological Bulletin, 50*, 263–274.

Ferster, C. B. (2002). Schedules of reinforcement with Skinner. *Journal of the Experimental Analysis of Behavior, 77*, 303–311.

Ferster, C. B. & Skinner, B. F. (1957). *Schedules of reinforcement.* New York: Appleton-Century-Crofts.

Fouts, R. S. (1973). Acquisition and testing of gestural signs in four young chimpanzees. *Science, 180*, 978–980.

Freeman, F. S. (1985). A reflection: Howard Scott Liddell, 1895–1962. *Journal of the History of the Behavioral Sciences, 21*, 372–374.

Freeman, K., & Riley, A. (2009). The origins of conditioned taste aversion learning: An historical analysis. In S. Reilly & T. Schachtman (Eds.) *Conditioned taste aversion: Behavioral and neural processes* (New York: Academic Press.

Gaioni, S., Hoffman, H. S., DePaulo, P., & Stratton, V. N. (1978). Imprinting in older ducklings. *Animal Learning & Behavior, 6*, 19–26.

Gantt, W. H. (1928). Ivan Petrovitch Pavlov: A biographical sketch. In Pavlov, I. P. (Ed.) *Lectures on conditioned reflexes: Twenty-five years of objective study of the higher nervous activity (Behaviour) of animals* (pp. 11–31). London; Lawrence & Wishart, Ltd.

Gantt, W. H. (1936). An experimental approach to psychiatry. *American Journal of Psychiatry, 92*, 1007–1021.

Gantt, W. H. (1937). Contributions to the physiology of the conditioned reflex. *Archives of Neurology & Psychiatry, 87*, 848–858.

Gantt, W. H. (1938). A method of testing cortical function and sensitivity of the skin: An aid in differentiating organogenic and psychogenic disturbances. *Archives of Neurology and Psychiatry, 40*, 79–85.

Gantt, W. H. (1944). *Experimental basis for neurotic behavior.* New York: Hoebner Inc.

Gantt, W. H. (1970). The future of psychiatry. In W. H. Gantt, L. Pickenhain & Ch. Zwingmann (Eds.) *Pavlovian approach to psychopathology: History and perspectives.* Leipzig: Pergamon.

Gantt, W. H. (1973). Reminiscences of Pavlov. *Journal of the Experimental Analysis of Behavior, 20*, 131–136.

Garcia, J. (1997). Robert C. Bolles: From mathematics to motivation. In M. E. Bouton and M. S. Fanselow (Eds.) *Learning, motivation, and cognition: The functional behaviorism of Robert C. Bolles* (pp. xi–xiii). Washington, DC: American Psychological Association.

Garcia, J., & Koelling, R. A. (1966). Relation of cue to consequence in avoidance learning. *Psychonomic Science, 4*, 123–124.

Garcia, J., Clarke, J. C., & Hankins, W. G. (1973). Natural responses to scheduled rewards. In P. P. G. Bateson & P. H. Klopfer (Eds.) *Perspectives in ethology* (pp. 1–41). Springer.

Garcia, J., Erwin, F. R., & Koelling, R. A. (1966). Learning with prolonged delay of reinforcement. *Psychonomic Science, 5*, 121–122.

Garcia, J., Kimmeldorf, D. J., & Koelling, R. A. (1955). Conditioned aversion to saccharin resulting from exposure to gamma radiation. *Science, 122*, 157–158.

Gardner, B. T. & Gardner, R. A. (1971). Two-way communication with an infant chimpanzee. In A. M. Shrier & F. Stollnitz (Eds.) *Behavior of non-human primates: Modern research trends* (Vol 4, pp. 117–185). New York: Academic Press.

Gardner, R. A. & Gardner, B. T. (1969). Teaching sign language to a chimpanzee. *Science, 165*, 664–672.

Gardner, R. A. & Gardner, B. T. (1978). Comparative psychology and language acquisition. *Annals of the New York Academy of Sciences, 309*, 37–76.

Gardner, R. A. & Gardner, B. T. (1984). A vocabulary test for chimpanzees (Pan troglodytes). *Journal of Comparative Psychology, 98*, 381–404.

George, D. N. & Pearce, J. M. (2012). A configural theory of attention and associative learning. *Learning & Behavior, 40*, 241–254.

Gleitman, H. (1981). *Psychology*. New York: Norton.

Gleitman, H., Nachmias, J., & Neisser, U. (1954). The S-R reinforcement theory of extinction. *Psychological Review, 61*, 23–33.

Gliedman, L. H., Gantt, W. H., & Teitelbaum, H. A. (1957). Some implications of conditional reflex studies for placebo research. *American Journal of Psychiatry, 113*, 1103–1107.

Gluck, M. & Roediger, H. (2011). Remembering William K. Estes. *APS Observer*, Nov 2 edition.

Glucksberg, S. (2017). Obituary of Leon Kamin on the Department of Psychology, Princeton University website. https://psych.princeton.edu › news-events › news › me…

Gonzalez, R. C., Berger, B. D., & Bitterman, M. E. (1966). A further comparison of key-pecking with an ingestive technique for the study of discriminative learning in pigeons. *American Journal of Psychology, 79*, 217–225.

Gonzalez, R. C., Gentry, G. V., & Bitterman, M. E. (1954). Relational discrimination of intermediate size in the chimpanzee. *Journal of Comparative and Physiological Psychology, 47*, 385–388.

Gormezano, I. & Kehoe, E.J. (1975). Classical conditioning: Some methodological-conceptual issues. *Handbook of Learning and Cognitive Process, 2*, 143–179.

Goodale, M. A. (2020). Lawrence Weiskrantz, 1926–2018. *Biographical Memoirs of Fellows of the Royal Society, 69*, 539–559.

Gray, J. A. (1979). *Pavlov*. Brighton, Sussex: Harvester Press.

Greenspoon, J. (1955). The reinforcing effect of two spoken sounds on the frequency of two responses. *American Journal of Psychology, 68*, 409–416.

Griffiths, P. E. (2004). Instinct in the '50s: The British reception of Konrad Lorenz's theory of instinctive behavior. *Biology and Philosophy, 19*, 609–631.

Grindley, G. C. (1929). Experiments on the influence of amount of reward on learning in young chickens. *British Journal of Psychology, 20*, 173–180.

Grindley, G. C. (1932). The formation of a simple habit in guinea pigs. *British Journal of Psychology, 23*, 127.

Guthrie, E. R. (1952). *The psychology of learning*. New York: Harper.

Guttman, N. & Kalish, H. I. (1956). Discriminability and stimulus generalization. *Journal of Experimental Psychology, 51*, 79–88.

Hall, G. (1973). Response strategies after overtraining in the jumping stand. *Animal Learning & Behavior, 1*, 157–160.

Hall, G. (1974). Transfer effects produced by overtraining in the rat. *Journal of Comparative and Physiological Psychology, 87*, 938–944.

Hall, G. (unpublished). *Autobiography*.

Hall, G. & Pearce, J. M. (1978). Transfer of learning across reinforcers: Appetitive discrimination learning between stimuli previously associated with shock. *Quarterly Journal of Experimental Psychology, 30*, 539–549.

Hall, G. & Pearce, J. M. (1979). Latent inhibition of a CS during CS-US pairings. *Journal of Experimental Psychology: Animal Behavior Processes, 5*, 31–42.

Halliday, M. S. (1979). Jerzy Konorski and Western psychology. In A. Dickinson & R. A. Boakes (Eds.) *Mechanisms of learning and motivation: A memorial volume to Jerzy Konorski* (pp. 1–18). Hillsdale, NJ: Lawrence Erlbaum Associates.

Halliday, M. S. & Boakes, R. A. (1971). Behavioral contrast and response-independent reinforcement. *Journal of the Experimental Analysis of Behavior, 16*, 429–434.

Halliday, M. S. & Boakes, R. A. (1972). Discrimination learning involving response-independent reinforcement: Implications for behavioral contrast. In R. A. Boakes & M. S. Halliday (Eds.) *Inhibition and learning* (pp. 73–97). London: Academic Press.

Hanson, H. M. (1959). Effects of discrimination training on stimulus generalization. *Journal of Experimental Psychology, 58*, 321–334.

Harlow, H. F. (1949) The formation of learning sets. *Psychological Review, 56*, 51–65.

Harlow, H. F. (1953). Mice, monkeys, men and motives. *Psychological Review, 60*, 23–32.

Harlow, H. F. (1958). The nature of love. *American Psychologist, 13*, 673–685

Harlow, H. F. & Stagner, R. (1933). Effect of complete striate muscle paralysis upon the learning process. *Journal of Experimental Psychology, 16*, 283–294

Harlow, H. F., Harlow, M. K., & Meyer, D. F. (1950). Learning motivated by a manipulation drive. *Journal of Experimental Psychology, 40*, 228–234.

Harris, J., Andrew, B., & Kwok, D. (2013). Magazine approach during a signal for food depends on Pavlovian, not Instrumental, conditioning. *Journal of Experimental Psychology: Animal Behavior Processes, 39*, 107–116.

Harris, J. A., Kwok, D. W. S., & Gottlieb, D. A. (2019). The partial reinforcement extinction effect depends on learning about non-reinforced trials rather than reinforcement rate. *Journal of Experimental Psychology: Animal Learning and Cognition, 45*, 485–501.

Hayes, C. (1951). *The ape in our house*. Harper.

Hayes, K. J. & Hayes, C. (1950). The intellectual development of a home-raised chimpanzee. *Proceedings of the American Philosophical Society, 95*, 105–109.

Hayes, K. J. & Hayes, C. (1952). Imitation in a home-raised chimpanzee. *Journal of Comparative and Physiological Psychology, 45*, 450–459.

Hayes, K. J. & Hayes, C. (1954). The cultural capacity of chimpanzee. *Human Biology, 26*, 288–303.

Hearst, E. (1972). Some persistent problems in the analysis of conditioned inhibition. In Boakes, R. A. & Halliday, M. S. (Eds.) *Inhibition and learning* (pp. 5–39). London: Academic Press.

Hearst, E., & Jenkins, H. M. (1974). *Sign-tracking: The stimulus-reinforcer relation and directed action*. Psychonomic Society.

Hefferline, R. F., Keenan, B., & Harford, R. A. (1959). Escape and avoidance conditioning in human subjects without their observation of the response. *Science, 130*, 1338–1339.

Herman, L. M. (1975). Interference and auditory short-term memory in the bottlenose dolphin. *Animal Learning & Behavior, 3*, 43–48.

Herman, L. M. (2010). What laboratory research has told us about dolphin cognition. *International Journal of Comparative Psychology, 23*, 310–330.

Herman, L. M. & Arbeit, W. R. (1973). Stimulus control and auditory discrimination learning sets in the bottlenose dolphin. *Journal of the Experimental Analysis of Behavior, 19*, 379–394.

Herman, L. M. & Gordon, J. A. (1974). Auditory delayed matching in the bottlenose dolphin. *Journal of the Experimental Analysis of Behavior, 21*, 19–26.

Herman, L. M., Beach, F. A. III, Pepper, R. L., & Stalling, R. B. (1969). Learning-set formation in the bottlenose dolphin. *Psychonomic Science*, 14, 98–99.

Herrnstein, R. J. (1961). Relative and absolute strength of response as a function of frequency of reinforcement. *Journal of the Experimental Analysis of Behavior, 4*, 267–272.

Herrnstein, R. J. (1962). Placebo effect in the rat. *Science, 138*, 677–678.

Herrnstein, R. J. (1970). On the Law of Effect. *Journal of the Experimental Analysis of Behavior, 13*, 243–266.

Herrnstein, R. J. (1971). IQ. *Atlantic Monthly*, September issue.

Herrnstein, R. J. & Boring, E. B. (1965). *A source book in the history of psychology*. Cambridge, MA: Harvard University Press.

Herrnstein, R. J. & Brady, J. V. (1958). Interaction among components of a multiple schedule. *Journal of the Experimental Analysis of Behavior, 1*, 293–300.

Herrnstein, R. J. & Loveland, D. H. (1964). Complex visual concept in the pigeon. *Science, 146*, 549–551.

Heth, C. D. (1976). Simultaneous and backward fear conditioning as a function of number of CS-UCS pairings. *Journal of Experimental Psychology: Animal Behavior Processes, 2*, 117–129.

Hilgard, E. R. (1967). Kenneth Wartinbee Spence: 1907–1967. *American Journal of Psychology, 80*, 314–318.

Hilgard, E. R. & Marquis, D. G. (1940). *Conditioning and learning*. New York: D. Appleton-Century Company.

Hill, W. F. (1966). The psychology of animal learning by N.J. Mackintosh. *American Journal of Psychology, 89*, 190–192.

Hinde, R. A. (1973). Constraints on learning: An introduction to the problems. In R. A. Hinde & J. Stevenson-Hinde (Eds.) *Constraints on learning: Limitations and predispositions* (pp. 1–19). New York: Academic Press.

Hinde, R. A. (1987). William Homan Thorpe. *Biographical Memoirs of Fellows of the Royal Society, 33*, 621–639.

Hinde, R. A. & Spencer-Booth, Y. (1967). The behaviour of socially living rhesus monkeys in their first two and a half years. *Animal Behaviour, 15*, 169–196.

Hinde, R. A. & Spencer-Booth, Y. (1971). Effects of brief separation from mother on rhesus monkeys. *Science, 173*, 111–118.

Hinde, R. A., Thorpe, W. H., & Vince, M. A. (1956). The following response of young coots and moorhens. *Behaviour, 9*, 214–242.

Hineline, P. N. & Rachlin, H. (1969). Escape and avoidance of shock by pigeons pecking a key. *Journal of the Experimental Analysis of Behavior, 12*, 533–538.

Hodos, W. (1961). Progressive ratio as a measure of reward strength. *Science, 134*, 943–944.

Hoffman, A. M. & Hoffman, H. S. (1990). *Archives of memory: A soldier recalls World War II.* Lexington: University Press of Kentucky.

Hoffman, H. S. & Ison, J. R. (1980). Reflex modification in the domain of startle: 1. Some empirical findings and their implications for how the nervous system processes sensory input. *Psychological Review, 87*, 175–189.

Hoffman, H. S. & Ratner, A. M. (1973). A reinforcement mode of imprinting: Implications for socialization in monkeys and men. *Psychological Review, 80*, 527–544.

Hogan, J. A. (1973). How young chicks learn to recognize food. In R. A. Hinde & J. Stevenson-Hinde (Eds.) *Constraints on learning: Limitations and predispositions* (pp. 119–139). New York: Academic Press.

Holland, P. C. (1977). Conditioned stimulus as a determinant of the form of the Pavlovian conditioned response. *Journal of Experimental Psychology: Animal Behavior Processes, 3*, 77–104.

Holland, P. C. & Rescorla, R. A. (1975a). Second-order conditioning with food unconditioned stimulus. *Journal of Comparative and Physiological Psychology, 88*, 459–467.

Holland, P. C. & Rescorla, R. A. (1975b). The effect of two ways of devaluing the unconditioned stimulus after first- and second-order appetitive conditioning. *Journal of Experimental Psychology: Animal Behavior Processes, 1*, 355–363.

Holman, E. W. (1975). Some conditions for the dissociation of consummatory and instrumental behavior in rats. *Learning and Motivation, 6*, 358–366.

Holmes, N. M., Chan, Y. Y., & Westbrook, R. F. (2020). An application of Wagner's Standard Operating Procedures or Sometimes Opponent Processes (SOP) model to experimental extinction. *Journal of Experimental Psychology: Animal Learning and Cognition, 46*(3), 215–234.

Honig, W. K. (1966). Introductory remarks. In *Operant behavior: Areas of research and application*. (Edited by W. K. Honig). pp. 1–11. New York: Appleton-Century-Crofts.

Hull, C. L. (1929). A functional interpretation of the conditioned reflex. *Psychological Review, 36*, 498–511.

Hull, C. L. (1930). Simple trial-and-error learning: A study in psychological theory. *Psychological Review, 37*, 241–256.

Hull, C. L. (1931). Knowledge and purpose as habit mechanisms. *Psychological Review, 37*, 511–525.

Hull, C. L. (1943). *Principles of behavior.* New York: Appleton-Century-Crofts.

Hull, C. L. (1951). *Essentials of behavior*. New Haven: Yale University Press.

Hull, C. L. (1952a). Clark L. Hull. In E. G. Boring, H. S Langfeld, H. Werner & R. M. Yerkes (Eds.) *A history of psychology in autobiography* (Vol. 4). Worcester, MA: Clark University Press.

Hull, C. L. (1952b). *A behavior system.* New Haven: Yale University Press.

Hull, C. L. & Baernstein, H. D. (1929). A mechanical parallel to the conditioned reflex. *Science, 70,* 14–15.

Hulse, S. H., Fowler, H., & Honig, W. K. (1978). *Cognitive processes in animal behavior.* Hillsdale, NJ: Lawrence Erlbaum Associates.

Hurwitz, H. (1953). A new rat-pellet feeding machine. *Quarterly Journal of Experimental Psychology, 5,* 36.

James, H. (1959). Flicker: An unconditional stimulus for imprinting. *Canadian Journal of Psychology, 13,* 59–67.

James, W. (1890). *Principles of psychology.* New York: Holt.

Jaynes, J. (1956). Imprinting: The interaction of learned and innate behavior. I. Development and generalization. *Journal of Comparative and Physiological Psychology, 49,* 201–206.

Jaynes, J. (1957). Imprinting: The interaction of learned and innate behavior. II. The critical period. *Journal of Comparative and Physiological Psychology, 50,* 6–10.

Jaynes, J. (1958a). Imprinting: The interaction of learned and innate behavior. III. Practice effects on performance, retention and fear. *Journal of Comparative and Physiological Psychology, 51,* 234–237.

Jaynes, J. (1958b). Imprinting: The interaction of learned and innate behavior. IV. Generalization and emergent discrimination. *Journal of Comparative and Physiological Psychology, 51,* 238–242.

Jenkins, H. M. (1973). Effects of the stimulus-reinforcer relation on selected and unselected responses. In R. A. Hinde & J. Stevenson-Hinde (Eds.) *Constraints on learning: Limitations and predispositions* (pp. 189–203). New York: Academic Press.

Jenkins, H. M. & Harrison, R. H. (1960). Effect of discrimination training on auditory generalization. *Journal of Experimental Psychology, 59,* 246–253.

Jenkins, H. M. & Harrison, R. H. (1962). Generalization gradients of inhibition following auditory discrimination learning. *Journal of the Experimental Analysis of Behavior, 5,* 435–441.

Jenkins, H. M. & Moore, B. R. (1973). The form of the auto-shaped response with food or water reinforcers. *Journal of the Experimental analysis of Behavior, 20,* 163–181.

Jenkins, H. M., Barrera, F. J., Ireland, C., & Woodside, B. (1978). Signal-centered action patterns of dogs in appetitive classical conditioning. *Learning & Motivation, 9,* 272–296.

Kamin, L. J. (1956). The effects of termination of the CS and avoidance of the US on avoidance learning. *Journal of Comparative and Physiological Psychology, 49,* 420–424.

Kamin, L. J. (1957). The effects of termination of the CS and avoidance of the US on avoidance learning: An extension. *Canadian Journal of Psychology, 11,* 48–56.

Kamin, L. J. (1968). "Attention-like" processes in classical conditioning. In M. R. Jones (Ed.), *Miami Symposium on the prediction of behavior, 1967: Aversive stimulation* (pp. 9–31). Coral Gables, FL: University of Miami Press.

Kamin, L. J. (1969a). Predictability, surprise, attention, and conditioning. In R. Church & B. Campbell (Eds.) *Punishment and aversive behavior* (pp. 279–296). New York: Appleton-Century-Crofts.

Kamin, L. J. (1969b). Selective association and conditioning. In N. J. Mackintosh & W. K. Honig (Eds.) *Fundamental issues in associative learning* (pp. 42–64). Halifax, Nova Scotia: Dalhousie University Press.

Kamin, L. J. (1974). *The science and politics of IQ.* London: Routledge.

Kamin, L. J. (2005). McCarthyism at Harvard, cont'd. New York Review of Books, May 26th issue.

Kamin, L. J. & Gaioni, S. J. (1974). Compound conditioned emotional response conditioning with differentially salient elements in rats. *Journal of Comparative and Physiological Psychology, 87*, 591–597.

Kamin, L. J., Brimer, C. J., & Black, A. H. (1963). Conditioned suppression as a monitor of fear of the CS in the course of avoidance training. *Journal of Comparative and Physiological Psychology, 56*, 497–501.

Karpicke, J., Christoph, G., Peterson, G., & Hearst, E. (1977). Signal location and positive versus negative conditioned suppression in the rat. *Journal of Experimental Psychology: Animal Behavior Processes, 3*, 105–118.

Katz, J. S., & Wright, A. A. (2006). Same/different abstract-concept learning by pigeons. *Journal of Experimental Psychology: Animal Behavior Processes, 32*(1), 80–86.

Keller, F. S. (1941). Light-aversion in the white rat. *The Psychological Record, 4*, 235–250.

Keller, F. S. (1970). Psychology at Harvard (1926–1931): A reminiscence. In P. B. Dews (Ed.) *Festschrift for B.F. Skinner* (pp. 29–36). New York: Appleton-Century-Crofts.

Keller, F. S. (1986). A fire in Schermerhorn Extension. *The Behavior Analyst, 9*, 139–146.

Keller, F. S. (2009). *At my own pace: The autobiography of Fred S. Keller*. Cornwall-on-Hudson, NY: Sloan Publishing.

Keller, F. S. & Schoenfeld, W. N. (1950). *Principles of psychology: A systematic text in the science of behavior.* New York: Appleton-Century-Crofts.

Keller, K. (1974). The role of elicited responding in behavioral contrast. *Journal of the Experimental Analysis of Behavior, 21*, 249–257.

Kellogg, W. N. (1931). Humanizing the ape. *Psychological Review, 38*, 160–176.

Kellogg, W. N. (1961). *Porpoises and sonar.* Chicago: University of Chicago Press.

Kellogg, W. N. & Kellogg, L. A. (1933). *The ape and the child.* New York: Whittlesea House (McGraw-Hill).

Kellogg, W. N. & Rice, C. E. (1963). Visual discrimination in a bottlenose porpoise. *Psychological Record, 13*, 483–498.

Kellogg, W. F. & Rice, C. E. (1964). Visual problem-solving in a bottlenose dolphin. *Science, 143*, 1052–1055.

Kellogg, W. N. & Rice, C. E. (1966). Visual discrimination and problem-solving in a bottle-nose dolphin. In K. S. Norris (Ed.) *Whales, dolphins and porpoises.* Berkeley: University of California Press.

Kendler, H. H. (1952). "What is learned?" – A theoretical blind alley. *Psychological Review, 59*, 269–277.

Kendler, H. H. (1989). The Iowa tradition. *American Journal of Psychology, 44*, 1124–1132.

Kendler, H. H. & Kendler, T. S. (1962). Vertical and horizontal processes in problem solving. *Psychological Review, 69*, 1–16.

Kendler, T. S. (1950). An experimental investigation of transposition as a function of the difference between training and test stimuli. *Journal of Experimental Psychology, 40*, 552–562.

Kendler, T. S. (2003). A woman's struggle in academic psychology (1936–2001). *History of Psychology, 6*, 251–266.

Killeen, P. (1972). The matching law. *Journal of the Experimental Analysis of Behavior, 17*, 489–495.

Killeen, P. R. & Pellón, R. (2013). Adjunctive behaviors are operants. *Learning & Behavior, 41*, 1–24.

Kimble, G. A. (1961). *Hilgard and Marquis' Conditioning and Learning*. New York: Appleton-Century-Crofts.

Kimble, G. A. (1986). Norman Guttman (1920–1984). *American Psychologist, 41*, 579–580.

Kimble, G. A. (1991). Kenneth W. Spence: Theorist with an empiricist conscience. In G. A. Kimble & M. Wertheimer (Eds.) *Portraits of pioneers in psychology* (Vol. III, pp. 277–294). Washington, DC: American Psychological Association.

Kimmel, H. D. (1977). Notes from "Pavlov's Wednesdays": Sensory preconditioning. *American Journal of Psychology, 90*, 319–321.

Kirk, R. G. W. & Ramsden, E. (2018). Working across species down on the farm: Howard S. Liddell and the development of comparative psychopathology, c.1923–1962. *History and Philosophy of the Life Sciences,* 40, 24.

Koch, S. (1944). Hull's Principles of Behavior: A special review. *Psychological Bulletin, 41*, 269–286.

Koehler, W. (1925). *The mentality of apes*. New York: Harcourt Brace.

Koehler, W. (1941). On the nature of associations. *Proceedings of the American Philosophical Society, 84*, 489–502.

Konorski, J. (1948). *Conditioned reflexes and neuron organization*. Cambridge: Cambridge University Press.

Konorski, J. (1967). *Integrative activity of the brain*. Chicago: Chicago University Press.

Konorski, J. & Miller, S. (1937). On two types of conditioned reflex. *Journal of General Psychology, 16*, 264–272.

Krech, D. (1974). Autobiographical sketch. In G. Lindzey (Ed.) *A history of psychology in autobiography* (Vol. VI, pp. 221–250). Englewood Cliffs, NJ: Prentice-Hall.

Krechevsky, I. (1932). Antagonistic visual discrimination habits in the white rat. *Journal of Comparative Psychology, 14*, 263–277.

Ladygina-Kohts, N. (1926). The study of cognitive faculties of the chimpanzee. *Humana Mente, 1*, 391–393.

Lashley, K. S. (1912). Visual discrimination of size and form in the albino rat. *Journal of Animal Behavior, 2*, 310–331.

Lashley, K. S. (1929). *Brain mechanisms and intelligence*. Chicago: University of Chicago Press.

Lashley, K. S. (1930). The mechanism of vision: I. A method for rapid analysis of pattern vision in the rat. *Journal of Genetic Psychology, 37*, 453–460.

Lashley, K. S. (1942). An examination of the continuity theory as applied to discrimination learning. *Journal of Genetic Psychology, 26*, 241–265.

Lashley, K. S., & Wade, M. (1946). The Pavlovian theory of generalization. *Psychological Review, 53*, 72–87.

Lawrence, D. H. (1949). Acquired distinctiveness of cues, I: Transfer between discriminations on the basis of familiarity with the stimulus. *Journal of Experimental Psychology, 39*, 770–784.

Lawrence, D. H. (1950). Acquired distinctiveness of cues, II: Selective association in a constant stimulus situation. *Journal of Experimental Psychology, 40*, 175–188.

Lawrence, D. H. (1952). The transfer of a discrimination along a continuum. *Journal of Experimental Psychology, 45*, 511–516.

Lawrence, D. H. & DeRivera, J. (1954). Evidence for relational transposition. *Journal of Comparative and Physiological Psychology, 47*, 465–471.

Lea, S. (2014). Professor Leslie Reid (1924–2014). The psychologist. British Psychological Society.

Leahey, T. H. (2004). *A history of psychology*. 6th ed. Upper Saddle River, NJ: Pearson Education Inc.

Lewontin, R. C., Rose, S., & Kamin, L. J. (1984). *Not in our genes* (p. 7). New York: Pantheon Books.

Liddell, H. S. (1925). The behavior of sheep and goats in learning a simple maze. *American Journal of Psychology, 36*, 544–552.

Liddell, H. S. (1938). The experimental neurosis and the problem of mental disorder. *American Journal of Psychiatry, 94*, 1035–1043.

Liddell, H. S. (1953). *Emotional hazards in animals and man*. Springfield, IL: Thomas.

Lilly, J. C. (1961). *Man and dolphin*. New York: Doubleday.

Logan, F. A., & Wagner, A. R. (1965). *Reward and punishment*. Boston: Allyn & Bacon.

LoLordo, V. M. (2021). *Robert A. Rescorla: A biographical memoir*. Washington: National Academy of Sciences.

LoLordo, V. M. & Overmier, J. B. (2011). Trauma, learned helplessness, its neuroscience, and implications for posttraumatic stress disorder. In T. R. Schachtman & S. Reilly (Eds.) *Associative learning and conditioning theory: Human and non-human applications* (pp. 121–151). New York: Oxford University Press.

Longuet-Higgins, C. (1998). Obituary of Stuart Sutherland on the University of Sussex website.

Lorenz, K. (1935). Der Kumpan in der Umwelt des Vogels. *Journal fur Ornithology*, 83, pt. 2–3.

Lorenz, K. (1937). The companion in the bird's world. *Auk, 54*, 245–273.

Lorenz, K. (1952). *King Solomon's Ring*. London: Methuen.

Lorenz, K. (1966). *On Aggression*. London: Routledge.

Lorenz, K. (1985). My family and other animals. In D. A. Dewsbury (Ed.) *Leaders in the Study of Animal Behavior: Autobiographical Perspectives* (pp. 285–287). London and Toronto: Associated University Presses.

Lovibond, P. F., & Shanks, D. R. (2002). The role of awareness in Pavlovian conditioning: Empirical evidence and theoretical implications. *Journal of Experimental Psychology: Animal Behavior Processes, 28*, 3–26.

Macfarlane, D. A. (1930). The role of kinaesthesis in maze learning. *University of California Publications in Psychology, 4*, 277–305.

Mackintosh, N. J. (1962). The effect of overtraining on a reversal and a non-reversal shift. *Journal of Comparative and Physiological Psychology, 55*, 555–559.

Mackintosh, N. J. (1965). Selective attention in animal discrimination learning. *Psychological Bulletin, 64*, 124–150.

Mackintosh, N. J. (1969a). Further analysis of the overtraining reversal effect. *Journal of Comparative and Physiological Psychology Monograph*, 67(2), 1–18.

Mackintosh, N. J. (1969b). Comparative studies of reversal and probability learning: Rats, birds and fish. In R. M. Gilbert & N. S. Sutherland (Eds.) *Animal discrimination learning* (pp. 137–160). London: Academic Press.

Mackintosh, N. J. (1971). An analysis of blocking and overshadowing. *Quarterly Journal of Experimental Psychology, 23*, 118–125.

Mackintosh, N. J. (1974). *The psychology of animal learning*. London: Academic Press.

Mackintosh, N. J. (1975a). A theory of attention: Variations in the associability of stimuli with reinforcement. *Psychological Review, 82*, 276–298.

Mackintosh, N. J. (1975b). Critical notice: Kamin, L.J., *The science and politics of IQ*. *Quarterly Journal of Experimental Psychology, 27*, 672–686.

Mackintosh, N. J. (1990). B.F. Skinner (1904–1990). *Nature, 347*, 332.

Mackintosh, N. J. (1998a). *IQ and human intelligence*. Oxford: Oxford University Press.

Mackintosh, N. J. (1998b). Obituary: Professor Stuart Sutherland. *Independent Newspaper*, 18 Nov 1998.

Mackintosh, N. J. & Dickinson, A. (1979). Instrumental (Type II) conditioning. In A. Dickinson & R. A. Boakes (Eds.) *Mechanisms of learning and motivation: A memorial volume to Jerzy Konorski* (pp. 143–169). Hillsdale, NJ: Lawrence Erlbaum Associates.

Mackintosh, N. J. & Honig, W. K. (1969). *Fundamental issues in associative learning*. Halifax: Dalhousie University Press.

Mackintosh, N. J. & Mackintosh, J. (1963). Reversal learning in Octopus vulgaris Lamarck with and without irrelevant cues. *Quarterly Journal of Experimental Psychology, 15*, 236–242.

Mackintosh, N. J. & Turner, C. (1971). Blocking as a function of novelty of CS and predictability of UCS. *Quarterly Journal of Experimental Psychology, 23*, 359–366.

Mackintosh, N. J. & Mascie-Taylor, C. G. N. (1986). The IQ question. In C. Bagley & G. K. Verma (Eds.) *Personality, cognition and values* (pp. 77–131). London: Macmillan Press.

Mackintosh, N. J., Mascie-Taylor, C. G. N., & West, A. M. (1988). West Indian and Asian children's educational attainment. In G. Verma & P. Pumfrey (Eds.) *Educational attainments: Issues and outcomes in multicultural education.* London: Routledge.

Mackintosh, N. J., Wilson, B. J., Boakes, R. A. & Barlow, H. B. (1985). Differences in mechanism of intelligence among vertebrates. *Philosophical Transactions of the Royal Society of London. Series B. Biological Sciences, 308*, 53–65.

Macphail, E. M. (1982). *Brain and intelligence in vertebrates.* Oxford: Clarendon Press.

Macphail, E. M. (1987). The comparative psychology of intelligence. *Behavioral and Brain Sciences, 10*, 645–695.

Maier, N. R. F. (1929). Reasoning in white rats. *Comparative Psychology Monographs*, 6, 93.

Maier, S. F. (1970). Failure to escape traumatic electric shock: Incompatible skeletal-motor responses or Learned Helplessness? *Learning and Motivation, 1*, 157–169.

Maier, S. F. & Seligman, M. E. P. (2016). Learned helplessness at fifty: Insights from neuroscience. *Psychological Review, 123*, 349–367.

Maier, S. F., Albin, R. W., & Testa, T. J. (1973). Failure to learn to escape in rats previously exposed to inescapable shock depends on the nature of the escape response. *Journal of Comparative and Physiological Psychology, 85*, 581–592.

Masserman, J. H. (1943). *Behavior and neurosis.* Chicago: University of Chicago Press.

Masserman, J. H. (1950). Experimental neuroses. *Scientific American, 182*, 38–43.

Masserman, J. H. (1964). *Behavior and neurosis: An experimental psychoanalytic approach to psychobiologic principles.* New York: Hafner.

Masserman, J. H., Wechkin, H., & Terris, W. (1964). "Altruistic" behavior in rhesus monkeys. *American Journal of Psychiatry, 121*, 584–585.

May, M. A. (1971). A retrospective view of the Institute of Human Relations at Yale. *Behavior Science Notes, 3*, 141–172.

McGuigan, F. J. (1980). W. Horsley Gantt: In memoriam. *Pavlovian Journal of Biological Sciences, 15*, 1–4.

McGuigan, F. J. (1981). Obituary: W. Horsley Gantt (1892–1980). *American Psychologist, 36*, 417–419.

McLaren, I. P. L. & Dickinson, A. (1990). The conditioning connection. *Philosophical Transactions of the Royal Society, B, 329*, 179–185.

McLaren, I. P. L. & Mackintosh, N. J. (2000). An elemental model of associative learning: I. Latent inhibition and perceptual learning. *Animal Learning & Behavior, 26*, 211–246.

Menzel, E. W. (1973). Chimpanzee spatial memory organization. *Science, 182*, 943–945.

Miles, C. G. (1965). Acquisition of control by the features of a compound stimulus in discriminative operant conditioning. Unpublished doctoral dissertation, McMaster University,

Millenson, J. R. (1967). *Principles of behavior analysis.* New York: Macmillan.

Miller, N. E. (1935). A reply to "sign-gestalt or conditioned reflex?" *Psychological Review, 42*, 280–292.

Miller, N. E. (1948). Studies of fear as an acquirable drive. *Journal of Experimental Psychology, 38*, 89–101.

Miller, N. E. (1952). Comments on multiple-process conceptions of learning. *Psychological Review, 58*, 375–381.

Miller, N. E. & DiCara, L. (1967). Instrumental conditioning of heart rate changes in curarized rats: Shaping and specificity to discriminative stimulus. *Journal of Comparative and Physiological Psychology, 63*, 12–19.

Miller, S. & Konorski, J. (1928). Sur une forme particuliére des reflexes conditionels. *Les Comptes Rendues des Séances de Société de Biologie, 99*, 1155–1157. English translation by B.F. Skinner (1969), *Journal of the Experimental Analysis of Behavior, 12*, 187–189.

Milner, B. (1965). Visually-guided maze learning in man: Effects of bilateral hippocampal, bilateral frontal, and unilateral cerebral lesions. *Neuropsychologia, 3*, 317–338.

Moon, L. E., & Harlow, H. F. (1955). Analysis of oddity learning by rhesus monkeys. *Journal of Comparative and Physiological Psychology, 48*(3), 188–194.

Moore, B. R. (1973). The role of directed Pavlovian reactions in simple instrumental learning in the pigeon. In R. A. Hinde & J. Stevenson-Hinde (Eds.) *Constraints on learning: Limitations and predispositions* (pp. 159–186). New York: Academic Press.

Morawski, J. G. (1986). Organizing knowledge and behavior at Yale's Institute of Human Relations. *Isis, 77*, 219–242.

Morris, R. G. M. (1979). Preconditioning of reinforcing properties to an exteroceptive feedback stimulus. *Learning & Motivation, 6*, 289–298.

Morris, R. G. M. (1981). Spatial localization does not require the presence of local cues. *Learning and Motivation, 12*, 239–260.

Morris, R. G. M. (2008). Morris water maze. *Scholarpedia, 3*(8), 6315

Morris, R. G. M., Garrud, P., Rawlins, J. N. P., & O'Keefe, J. (1982). Place navigation impaired in rats with hippocampal lesions. *Nature, 297*, 681–683.

Morrison, G. R. & Collyer, R. (1974). Taste-mediated conditioned aversion to an exteroceptive stimulus following LiCl poisoning. *Journal of Comparative and Physiological Psychology, 86*, 51–55.

Morse, W. H. (2017). Contributions of Peter B. Dews (1922–2012) to the experimental analysis of behavir: A personal perspective and appreciation. *Journal of the Experimental Analysis of Behavior, 107*, 295–300.

Mowrer, O. H. (1939). A stimulus-response analysis of anxiety and its role as a reinforcing agent. *Psychological Review, 46*, 553–565.

Mowrer, O. H. (1947). On the dual nature of learning: A re-interpretation of 'conditioning' and 'problem-solving'. *Harvard Educational Review, 17*, 12–148.

Mowrer, O. H. (1948). Learning theory and the neurotic paradox. *American Journal of Orthopsychiatry, 18*, 571–610.

Mowrer, O. H. (1960a). *Learning theory and behavior.* New York: Wiley.

Mowrer, O. H. (1960b). *Learning theory and the symbolic processes*. New York: Wiley.

Mowrer, O. H. (1973). My philosophy of psychotherapy. *Journal of Contemporary Psychotherapy, 6*, 35–42.

Mowrer, O. H. (1974). Autobiography. In G. Lindzey (Ed.) *A history of psychology in autobiography* (pp. 327–364). Englewood Cliffs, NJ: Prentice-Hall.

Mowrer, O. H. (1980). Enuresis: The beginning work – What really happened. *Journal of the History of the Behavioral Sciences, 16*, 25–30.

Mowrer, O. H. & Lamoreaux, R. R. (1942). Avoidance conditioning and signal duration: A study of secondary motivation and reward. *Psychological Monographs*, 54(5), 34.

Mowrer, O. H. & Lamoreaux, R. R. (1946). Fear as an intervening variable in avoidance conditioning. *Journal of Comparative Psychology, 39*, 29–50.

Mowrer, O. H. & Miller, N. E. (1942). A multi-purpose learning-demonstration apparatus. *Journal of Experimental Psychology, 31*, 163–171.

Mowrer, O. H. & Solomon, L. N. (1954). Contiguity vs. drive reduction in conditioned fear: The proximity and abruptness of drive-reduction. *American Journal of Psychology, 67*, 15–25.

Mowrer, O. H. & Viek, P. (1948). An experimental analogue of fear from a sense of helplessness. *Journal of Abnormal and Social Psychology, 83*, 193–200.

Muenzinger, K. F. (1938). Vicarious trial-and-error at a choice point. I. A general survey of its relation to learning efficiency. *Journal of Genetic Psychology, 53*, 75–86.

Munn, N. L. (1950). *Handbook of psychological research on the rat.* New York: Houghton Mifflin.

Nairne, J. S. & Rescorla, R. A. (1981). Second-order conditioning with diffuse auditory reinforcers in the pigeon. *Learning & Motivation, 12*, 65–91.

Neisser, U. (1967). *Cognitive psychology*. New York: Appleton-Century-Crofts.

Oberg, B. B. (1976). David Hartley and the association of ideas. *Journal of the History of Ideas, 37*, 441–454.

Odling-Smee, F. J. (1975). The role of background stimuli and the inter-stimulus interval during Pavlovian conditioning. *Quarterly Journal of Experimental Psychology, 27*, 387–392.

O'Keefe, J. (1976). Place units in the hippocampus of the freely moving rat. *Experimental Neurology, 51*, 78–109.

O'Keefe, J. & Dostrovsky, J. (1971). The hippocampus as a spatial map: Preliminary evidence from unit activity in the freely moving rat. *Brain Research, 34*, 171–175.

O'Keefe, J. & Nadel, L. (1978). *The hippocampus as a cognitive map.* Oxford: Oxford University Press.

Olton, D. S. (1977). Spatial memory. *Scientific American, 236*, 82–99.

Olton, D. S. & Samuelson, R. J. (1976). Remembrance of places passed: Spatial memory in rats. *Journal of Experimental Psychology: Animal Behavior Processes, 2*, 97–116.

Overmier, J. B. & Seligman, M. E. P. (1967). Effects of inescapable shock upon subsequent escape and avoidance responding. *Journal of Comparative and Physiological Psychology, 63*, 28–33.

Packard, M. G. & McGaugh, J. L. (1996). Inactivation of hippocampus or caudate nucleus with lidocaine differentially affects expression of place and response learning. *Neurobiology of Learning and Memory, 65*, 65–72.

Page, C. (2017). Preserving guilt in the "age of psychology": The curious career of O. Hobart Mowrer. *History of Psychology, 20*, 1–27.

Papini, M. (2008). Integrating learning, emotion, behavior theory, development, and neurobiology: The enduring legacy of Abram Amsel (1922–2006). *American Journal of Psychology, 121*, 661–669.

Patten, R. L. & Rudy, J. W. (1967). The Sheffield omission training procedure applied to the conditioning of the licking response in rats. *Psychonomic Science, 8*, 463–464.

Pavlov, I. P. (1927). *Conditioned reflexes.* Oxford: Oxford University Press.

Pavlov, I. P. (1928). *Lectures on conditioned reflexes.* New York: International Publishers.

Pavlov, I. P. (1932). Neuroses in man and animals. *Journal of the American Medical Association, 99*, 1012–1013.

Pavlov, I. P. (1941). *Conditioned reflexes and psychiatry.* New York: International Publishers.

Pavlov, I. P. (1962). *Essays in psychology and psychiatry.* New York: Citadel Press.

Pearce, J. M. (2008). *Animal learning and cognition: An introduction*. 3rd ed. Hove: Psychology Press.

Pearce, J. M. (2018). Nicholas John Seymour Mackintosh. *Biographical Memoirs of Fellows of the Royal Society, 64*, 299–316. London: Royal Society Publishing.

Pearce, J. M. & Hall, G. (1978). Overshadowing instrumental conditioning of a lever-press response by a more valid predictor of the reinforcer. *Journal of Experimental Psychology: Animal Behavior Processes, 4*, 356–367.

Pearce, J. M. & Hall, G. (1980). A model for Pavlovian learning: Variations in the effectiveness of conditioned but not of unconditioned stimuli. *Psychological Review, 87*, 532–552.

Pearce, J. M. & Mackintosh, N. J. (2010). Two theories of attention: A review and possible integration. In C. J. Mitchell & M. E. LePelley (Eds.) *Attention and associative learning* (pp. 11–39). New York: Oxford University Press.

Pearce, J. M., Hall, G., & Colwill, R. M. (1978). Instrumental conditioning of scratching in the laboratory rat. *Learning & Motivation, 9*, 255–271.

Petrinovich, L. & Bolles, R. C. (1954). Deprivation states and behavioral attributes. *Journal of Comparative and Physiological Psychology, 47*, 450–453.

Pinker, S. (1994). *The language instinct.* New York: William Morrow and Company.

Premack, D. (1971). Language in chimpanzees? *Science, 172*, 808–822.

Premack, D. & Woodruff, G. (1978). Does the chimpanzees have a theory of mind? *Behavioural and Brain Sciences,* 1, 515–526.

Proctor, R. W., Martins, A. P. G., Altman, M. et al. (2022). Tribute to E. J. Capaldi: Celebration of a psychological scientist. *American Journal of Psychology, 135*, 199–124.

Rachlin, H. (1974). Self-control. *Behaviorism, 3*, 94–107.

Rachman, S. (2000). Joseph Wolpe (1915–1997). *American Psychologist, 55*, 431–432.

Rashotte, M. E. (2007). Abram Amsel (1922–2006): In memoriam. *Learning & Behavior, 35*, 1–10.

Rashotte, M. E. & Amsel, A. (1967). Acquisition and extinction, within subjects, of a continuously rewarded response and a response learned under discontinuous negatively correlated reward. *Psychonomic Science, 7*, 258.

Ratner, A. M. & Hoffman, H. S. (1974). Evidence for a critical period for imprinting in Khaki Campbell ducklings (Anas platyrhynchos domesticus). *Animal Behavior,* 22, 249–255.

Razran, G. (1956). Avoidant vs. unavoidant conditioning and partial reinforcement in Russian laboratories. *American Journal of Psychology, 69*, 127–129.

Reid, L. S. (1953). The development of non-continuity behavior through continuity learning. *Journal of Experimental Psychology*, 46, 107–112.

Rescorla, R. A. (1966). Predictability and number of pairings in Pavlovian fear conditioning. *Psychonomic Science, 4*, 383–384.

Rescorla, R. A. (1967). Pavlovian conditioning and its proper control procedures. *Psychological Review, 74*, 71–80.

Rescorla, R. A. (1968). Probability of shock in the presence and absence of CS in fear conditioning. *Journal of Comparative and Physiological Psychology, 66*, 1–5.

Rescorla, R. A. (1969a). Pavlovian conditioned inhibition. *Psychological Bulletin, 72*, 77–94.

Rescorla, R. A. (1969b). Establishment of a positive reinforcer through contrast with shock. *Journal of Comparative and Physiological Psychology, 67*, 260–263.

Rescorla, R. A. (1971). Variations in the effectiveness of reinforcement and non-reinforcement following prior inhibitory conditioning. *Learning and Motivation,* 2, 113–123.

Rescorla, R. A. (1973). Effect of US habituation following conditioning. *Journal of Comparative and Physiological Psychology, 82*, 137–143.

Rescorla, R. A. (1980a). *Pavlovian second-order conditioning: Studies in associative learning.* Hillsdale, NJ: Lawrence Erlbaum Associates.

Rescorla, R. A. (1980b). Simultaneous and successive associations in sensory preconditioning. *Journal of Experimental Psychology: Animal Behavior Processes, 6*, 207–216.

Rescorla, R. A. (1997). *Richard Lester Solomon: A biographical memoir.* Washington, DC: National Academies Press.

Rescorla, R. A. (2000). Associative changes in excitors and inhibitors differ when they are conditioned in compound. *Journal of Experimental Psychology: Animal Behavior Processes, 26*, 428–438.

Rescorla, R. A. & Cunningham, C. L. (1978). Recovery of the US representation over time during extinction. *Learning and Motivation, 9*, 373–391.

Rescorla, R. A. & Freberg, L. (1978). The extinction of within-compound flavor associations. *Learning & Motivation, 9*, 411–427.

Rescorla, R. A. & Heth, C. D. (1975). Reinstatement of fear to an extinguished conditioned stimulus. *Journal of Experimental Psychology: Animal Behavior Processes, 1*, 88–96.

Rescorla, R. A. & LoLordo, V. M. (1965). Inhibition of avoidance behavior. *Journal of Comparative and Physiological Psychology, 59*, 406–412.

Rescorla, R. A., & Solomon, R. L. (1967). Two-process learning theory: Relationships between Pavlovian conditioning and instrumental learning. *Psychological Review, 74*, 151–182.

Rescorla, R. A. & Wagner, A. R. (1972). A theory of Pavlovian conditioning: Variations in the effectiveness of reinforcement and non-reinforcement. In A. H. Black & W. F. Prokasy (Eds.) *Classical conditioning II: Current research and theory* (pp. 64–99). New York: Appleton-Century-Crofts.

Rescorla, R. A., Durlach, P., & Grau, J. W. (1985). Contextual learning in Pavlovian conditioning. In P. D. Balsam & A. Tomie (Eds.) *Context and learning* (pp. 23–56). Hillsdale, NJ: Lawrence Erlbaum Associates.

Restle, F. (1957). Discrimination of cues in mazes: A resolution of the "place-vs.-response" question. *Psychological Review, 64*, 217–228.

Revusky, S. (1971). The role of interference in association over a delay. In W. K. Honig & P. H. R. James (Eds.) *Animal memory* (pp. 155–213) New York: Academic Press.

Revusky, S. (1977a). Interference with progress by the scientific establishment: Example from flavor aversion learning. In N. W. Milgram, L. Krames. & T. M Alloway (Eds.) *Food aversion learning* (pp. 53–60). New York: Plenum Press.

Revusky, S. (1977b). Learning as a general process with an emphasis on data from feeding experiments. In N. W. Milgram, L. Krames. & T. M Alloway (Eds.) *Food aversion learning* (pp. 1–52). New York: Plenum Press.

Reynolds, G. S. (1961). Behavioral contrast. *Journal of the Experimental Analysis of Behavior, 4*, 57–71.

Ribes Inesta, E. (2010). Remembranzas y reflexiones autobiográphica. *Revista de Historia de la Psicología, 31*, 31–50.

Rilling, M. (1977). Stimulus control and inhibitory processes. In W. K. Honig & J. E. R. Staddon (Eds.) *Handbook of operant behavior* (pp. 432–480). Englewood Cliffs, NJ: Prentice Hall, Inc.

Ritchie, B. F. (1953). The circumnavigation of cognition. *Psychological Review, 60*, 216–221.

Rizley, R. C. & Rescorla, R. A. (1972). Associations in second-order conditioning and sensory pre-conditioning. *Journal of Comparative and Physiological Psychology, 81*, 1–11.

Roberts, W. A. (1998). *Principles of animal cognition.* New York: McGraw-Hill.

Roberts, W. A., Cruz, C., & Tremblay, J. (2007). Rats take correct novel routes and shortcuts in an enclosed maze. *Journal of Experimental Psychology: Animal Behavior Processes, 33*, 79–91.

Robinson, J. K. & Woodward, W. R. (1989). The convergence of behavioral biology and operant psychology: Toward an interlevel and interfield science. *The Behavior Analyst, 12*, 131–141.

Roitblat, H. L., Bever, T. G., & Terrace, H. S. (1984). *Animal cognition*. Hillsdale, NJ: Lawrence Erlbaum Associates.

Root, M. J. (2002). Keller, Fred S. (1899–1996), psychologist and educator. *American National Biography Online*. Oxford: Oxford University Press.

Rozeboom, W. W. (1958). 'What is learned?' – An empirical enigma. *Psychological Bulletin, 65*, 22–33.

Rudy, J. W., Iwens, J., & Best, P. J. (1977). Pairing novel exteroceptive cues and illness reduces illness-induced taste aversions. *Journal of Experimental Psychology, 3*, 14–25.

Ruiz, G. & Sánchez, N. (2016). W. Horsley Gantt, Nick, and the Pavlovian Science at Phipps Clinic. *Spanish Journal of Psychology, 19*, e71, 1–14.

Rusiniak, K. W., Hankins, W. G., Garcia, J., & Brett, L. P. (1979). Flavor-illness aversions: Potentiation of odor by taste in rats. *Behavioral and Neural Biology, 25*, 1–17.

Russell, B. (1927). *An outline of philosophy*. London: George Allen & Unwin.

Rutherford, A. (2003). Skinner boxes for psychotics: Operant conditioning at the Metropolitan State Hospital. *Behavior Analyst, 26*, 267–279.

Ryle, G. (1949). *The concept of mind*. London: Routledge.

Savage-Rumbaugh, E. S., Murphy, J., Sevcik, R. A. et al. (1993). Language comprehension in ape and child. *Monographs of the Society for Research in Child Development*, 58(3/4), 1–98.

Schindler, C. W. (1993). *Techniques in the behavioral and neural sciences* (Ch. 3). New York: Elsevier.

Schwartz, B. (1975). Discriminative stimulus location as a determinant of positive and negative behavioral contrast in the pigeon. *Journal of the Experimental Analysis of Behavior, 23*, 167–176.

Schwartz, B. & Gamzu, E. (1977). Pavlovian control of operant behavior: An analysis of autoshaping and implications for operant conditioning. In W. K. Honig & J. E. R. Staddon (Eds.) *Handbook of operant behavior* (pp. 53–97). Englewood Cliffs, NJ: Prentice Hall, Inc.

Schwartzbaum, J. S. & Donovick, P. J. (1968). Discrimination reversal and spatial alternation associated with septal and caudate dysfunction in the rat. *Journal of Comparative and Physiological Psychology, 65*, 83–92.

Scoville, W. B. & Milner, B. (1957). Loss of recent memory after bilateral hippocampal lesions. *Journal of Neurology, Neurosurgery & Psychiatry, 20*, 11–21.

Seidenberg, M. S. & Petitto, L. A. (1979). Signing behaviour in apes: A critical review. *Cognition, 7*, 177–215.

Seligman, M. E. P. & Beagley, G. (1975). Learned helplessness in the rat. *Journal of Comparative and Physiological Psychology, 88*, 534–541.

Seligman, M. E. P. & Maier, S. F. (1967). Failure to escape traumatic shock. *Journal of Experimental Psychology, 74*, 1–9.

Sheffield, F. D. (1965). Relation between classical conditioning and instrumental learning. In W. F. Prokasy (Ed.) *Classical conditioning: A symposium* (pp. 302–322). New York: Appleton-Century-Crofts.

Sheffield, F. D. & Roby, T. B. (1950). Reward value of a non-nutritive sweet taste. *Journal of Comparative and Physiological Psychology, 43*, 471–481.

Sheffield, F. D., Wulff, J. J., & Backer, R. (1951). Reward value of copulation without sex drive reduction. *Journal of Comparative and Physiological Psychology, 44*, 3–8.

Shepard, R. N. (1965). Approximation to uniform gradients of generalization by monotone transformations of scale. In D. I. Mostofsky (Ed.) *Stimulus generalization* (pp. 94–110). Stanford, CA: Stanford University Press.

Shettleworth, S. J. (1972). Constraints on learning. *Advances in the Study of Behavior, 4*, 1–68.

Shettleworth, S. J. (1973). Food reinforcement and the organization of behavior in golden hamsters. In R. A. Hinde & J. Stevenson-Hinde (Eds.) *Constraints on learning: Limitations and predispositions* (pp. 243–263). New York: Academic Press.

Shettleworth, S. J. (1975). Reinforcement and the organization of behavior in golden hamsters: Hunger, environment and food reinforcement. *Journal of Experimental Psychology: Animal Behavior Processes, 1*, 56–87.

Shettleworth, S. J. (1998). *Cognition, evolution and behavior.* New York: Oxford University Press.

Shettleworth, S. J. (2009). The evolution of comparative cognition: Is the snark still a boojum? *Behavioural Processes, 80*, 210–217.

Shiffrin, R. M. & Schneider, W. (1977). Controlled and automatic human information processing: Perceptual learning, automatic attending, and a general theory. *Psychological Review, 84*, 127–190.

Sidman, M. (1953). Avoidance conditioning with brief shock and no exteroceptive warning signal. *Science, 118*, 157–158.

Skinner, B. F. (1932). On the rate of formation of a conditioned reflex. *Journal of General Psychology, 6*, 22–37.

Skinner, B. F. (1935). Two types of conditioned reflex and a pseudo type. *Journal of Genetic Psychology, 12*, 66–77.

Skinner, B. F. (1938). *The behavior of organisms.* New York: D. Appleton-Century.

Skinner, B. F. (1945). Baby in a box. Ladies Home Journal, October issue.

Skinner, B. F. (1948a). *Walden Two*. New York: Macmillan.

Skinner, B. F. (1948b). 'Superstition' in the pigeon. *Journal of Experimental Psychology, 38*, 168–172.

Skinner, B. F. (1950). Are theories of learning necessary? *Psychological Review, 57*, 193–216.

Skinner, B. F. (1951). How to train an animal. Life Magazine.

Skinner, B. F. (1953). *Science and human behavior*. New York: Free Press.

Skinner, B. F. (1954). The science of learning and the art of teaching. *Harvard Educational Review, 24*, 86–97.

Skinner, B. F. (1956). A case history in scientific method. *American Psychologist, 11*, 221–233.

Skinner, B. F. (1958). Teaching machines. *Science, 128*, 969–977.

Skinner, B. F. (1959). *Cumulative record.* New York: Appleton-Century-Crofts.

Skinner, B. F. (1960). Pigeons in a pelican. *American Psychologist, 15*, 28–37.

Skinner, B. F. (1965). Stimulus generalization in an operant: A historical note. In D. I. Mostofsky (Ed.) *Stimulus generalization* (pp. 193–209). Stanford, California: Stanford University Press.

Skinner, B. F. (1971). *Beyond freedom and dignity*. Cambridge, MA: Hackett Publishing Company.

Skinner, B. F. (1976). *Particulars of my life*. London: Jonathan Cape.

Skinner, B. F. (1979). *The shaping of a behaviorist: Part Two of an Autobiography*. New York: Alfred A. Knopf.

Skinner, B. F. (1983). *A matter of consequences*. New York: Random House.

Skinner, B. F. & Estes, W. K. (1941). Some quantitative properties of anxiety. *Journal of Experimental Psychology, 29*, 390–400.

Skinner, B. F., & Heron, W. T. (1937). Effects of caffeine and benzedrine upon conditioning and extinction. *The Psychological Record, 1*, 340–346.

Sluckin, W. (1964). *Imprinting and early learning*. London: Methuen.

Sluckin, W. & Salzen, E. A. (1961). Imprinting and perceptual learning. *Quarterly Journal of Experimental Psychology, 13*, 65–77.

Smith, L. D. (1986). *Behaviorism and logical positivism: A re-assessment of the alliance*. Stanford, CA: Stanford University Press.

Smith, S. M., Brown, H. D., Toman, J. E. P, & Goodman, L. S. (1947). The lack of cerebral effects of d-tubocurarine. *Anesthesiology, 8*, 1–14.

Solomon, R. L., & Turner, L. H. (1962). Discriminative classical conditioning in dogs paralyzed by curare can later control discriminative avoidance responses in the normal state. *Psychological Review, 69*, 202–218.

Solomon, R. L. & Wynne, L. C. (1950). Avoidance conditioning in normal dogs and in dogs deprived of normal autonomic functioning. *American Psychologist, 5*, 264. (Abstract).

Solomon, R. L. & Wynne, L. C. (1953). Traumatic avoidance learning: Acquisition in normal dogs. *Psychological Monographs*, 67(4).

Solomon, R. L., Kamin, L. J., & Wynne, L. C. (1953). Traumatic avoidance learning: The outcomes of several extinction procedures with dogs. *Journal of Abnormal and Social Psychology, 48*, 291–302.

Spalding, D. A. (1954). Instinct, with original observations on young animals. *British Journal of Animal Behaviour, 2*, 2–11.

Spence, K. W. (1932). The order of eliminating blinds in maze learning by the rat. *Journal of Comparative Psychology, 14*, 9–27.

Spence, K. W. (1936). The nature of discrimination learning in animals. *Psychological Review, 43*, 427–449.

Spence, K. W. (1937a). The differential response in animals to stimuli varying within a single dimension. *Psychological Review, 44*, 430–444.

Spence, K. W. (1937b). Experimental studies of learning and the higher mental processes in infra-human primates. *Psychological Bulletin, 34*, 806–850.

Spence, K. W. (1942). The basis of solution by chimpanzees of the intermediate size problem. *Journal of Experimental Psychology, 31*, 257–271.

Spence, K. W. (1945). An experimental test of continuity and non-continuity theories of discrimination learning. *Journal of Experimental Psychology, 35*, 253–266.

Spence, K. W. (1952). The nature of the response in discrimination learning. *Psychological Review, 59*, 89–93.

Spence, K. W. & Lippitt, R. (1940). "Latent" learning of a simple maze problem with relevant needs satiated. *Psychological Bulletin, 37*, 429.

Spence, K. W. & Lippitt, R. (1946). An experimental test of the sign-gestalt theory of trial-and-error learning. *Journal of Experimental Psychology, 36*, 491–502.

Staddon, J. E. R. (1968). Spaced responding and choice: A preliminary analysis. *Journal of the Experimental Analysis of Behavior, 11*, 669–682.

Staddon, J. E. R. (1977). Schedule-induced behavior. In W. K. Honig & J. E. R. Staddon (Eds.) *Handbook of operant behavior* (pp. 125–152). Englewood Cliffs, NJ: Prentice Hall.

Staddon, J. E. R. (1992). The 'superstition' experiment: A reversible figure. *Journal of Experimental Psychology: General, 121*, 270–272.

Staddon, J. E. R. (2016). *The Englishman: Memoirs of a psychobiologist*. Buckingham, UK: University of Buckingham Press.

Staddon, J. E. R. & Simmelhag, V. L. (1971). The 'superstition' experiment: A re-examination of its implications for the principles of adaptive behaviour. *Psychological Review, 78*, 3–43.

Stevens, S. S. (1957). On the psychophysical law. *Psychological Review, 64*, 153–181.

Sullivan, Walter (1995). Beatrix T. Gardner Dies at 61; Taught Signs to a Chimpanzee. *New York Times, July 1, Section 1*, p. 8.

Sutherland, N. S. (1959a). A test of a theory of shape discrimination in *Octopus vulgaris* Lamarck. *Journal of Comparative and Physiological Psychology, 52*, 13–141.

Sutherland, N. S. (1959b). Stimulus analyzing mechanisms. In *Proceedings of a symposium on the mechanization of thought processes.* (Vol. 2, pp. 575–609). London: Her Majesty's Stationery Office.

Sutherland, N. S. (1976). *Breakdown: A personal crisis and a medical dilemma.* London: Weidenfeld and Nicholson.

Sutherland, N. S. (1992). *Irrationality: The enemy within.* London: Constable

Sutherland, N. S. & Mackintosh, N. J. (1971). *Mechanisms of animal discrimination learning.* New York: Academic Press.

Sutherland, N. S., Mackintosh, N. J., & Mackintosh, J. (1963). Simultaneous discrimination training of octopus and transfer of discrimination along a continuum. *Journal of Comparative and Physiological Psychology, 56*, 150–156.

Terrace, H. S. (1963). Discrimination learning with and without 'errors'. *Journal of the Experimental Analysis of Behavior, 6*, 1–27.

Terrace, H. S. (1966). Stimulus control. In W. K. Honig (Ed.) *Operant behavior: Areas of research and applications* (pp. 271–344). New York: Appleton-Century-Crofts.

Terrace, H. (1979). *Nim.* New York: Alfred Knopf.

Terrace, H. S., Petitto, L. A., Sanders, R. J. & Bever, T. G. (1979). Can an ape create a sentence? *Science, 206*, 891–902.

Thistlethwaite, D. (1951). A critical review of latent learning and related experiments. *Psychological Bulletin, 48*, 97–129.

Thompson, R. K. R. & Herman, L. M. (1977). Memory for lists of sounds by the bottle-nosed dolphin: Convergence of memory process with humans? *Science, 195*, 501–503.

Thorndike, E. L. (1898). Animal intelligence: An experimental study of the associative processes in animals. *Psychological Monographs*, 2.

Thorndike, E. L. (1911). *Animal intelligence.* New York: Macmillan.

Thorndike, E. L. (1932). Reward and punishment in animal learning. *Comparative Psychology Monographs*, 8(4), 65.

Thorpe, W. H. (1958). The learning of song patterns by birds, with especial reference to the song of the chaffinch fringilla coelebs. *Ibis, 100*, 535–570.

Timberlake, W. & Grant, D. L. (1975). Autoshaping in rats to the presentation of another rat predicting food. *Science, 190*, 690–692.

Tinklepaugh, O. L. (1928). An experimental study of representative factors in monkeys. *Journal of Comparative Psychology, 8*, 197–236.

Todes, D. F. (2014). *Ivan Pavlov: A Russian life in science.* New York, NY: Oxford University Press.

Tolman, E. C. (1932). *Purposive behavior in animals and men.* New York: Appleton-Century.

Tolman, E. C. (1948). Cognitive maps in rats and men. *Psychological Review, 55*, 189–208.

Tolman, E. C. (1949). There is more than one kind of learning. *Psychological Review, 56*, 357–369.

Tolman, E. C. & Gleitman, H. (1949). Studies in learning and motivation: I. Equal reinforcements in both end-boxes, followed by shock in one end-box. *Journal of Experimental Psychology, 39*, 810–819.

Tolman, E. C. & Honzik, C. H. (1930). "Insight" in rats. *University of California Publications in Psychology, 4*, 215–232.

Tolman, E. C., Ritchie, B. F., & Kalish, D. (1946a). Studies in spatial learning: I. Orientation and short-cut. *Journal of Experimental Psychology, 35*, 17.

Tolman, E. C., Ritchie, B. F., & Kalish, D. (1946b). Studies in spatial learning: II. Place learning versus response learning. *Journal of Experimental Psychology, 36*, 221–229.

Tomie, A. (1976). Interference with autoshaping by prior context conditioning. *Journal of Experimental Psychology: Animal Behavior Processes, 2*, 332–334.

Trowill, J. A. (1967). Instrumental conditioning of the heart rate in the curarized rat. *Journal of Comparative and Physiological Psychology, 63*, 7–11.

Vince, M. A. (1956). 'String-pulling' in birds: 1. Individual differences in wild adult great tits. *British Journal of Animal Behaviour, 4*, 111–116.

Vogel, E. H., Ponce, F. P., & Wagner, A. R. (2018). The development and present status of the SOP model of associative learning. *Quarterly Journal of Experimental Psychology*, *72*, 1–29.

vom Saal, W. (1967). Blocking the acquisition of stimulus control in operant discrimination learning. Unpublished Masters thesis, McMaster University.

vom Saal, W. & Jenkins, H. M. (1970). Blocking the development of stimulus control. *Learning and Motivation,* 1, 52–64.

Wagner, A. R. (2008). Some observations and remembrances of Kenneth W. Spence. *Learning & Behavior, 36*, 169–173.

Wagner, A. R. & Rescorla, R. A. (1972). Inhibition in Pavlovian conditioning: Application of a theory. In R. A. Boakes & M. S. Halliday (Eds.) *Inhibition and learning* (pp. 301–336). London: Academic Press.

Wagner, A. R., Logan, F. A., Haberlandt, K., & Price, T. (1968). Stimulus selection in animal discrimination learning. *Journal of Experimental Psychology, 76*, 171–180.

Wagner, A. R., Rudy, J. W. & Whitlow, J. W. (1973). Rehearsal in animal conditioning. *Journal of Experimental Psychology, 97*, 407–426.

Walker, K. C. (1942). Effects of a discriminative stimulus transferred to a previously unassociated response. *Journal of Experimental Psychology,* 31, 312.

Warner, L. H. (1932). The attention span of the white rat. *Journal of Genetic Psychology, 41*, 57–90.

Warren, J. M. (1965a). The comparative psychology of learning. *Annual Review of Psychology, 16*, 95–118.

Warren, J. M. (1965b). Primate learning in comparative perspective. In A. M. Schrier, H. F. Harlow, & F. Stollnitz (Eds.) *Behavior of non-human primates: Modern research trends* (Vol. 1, pp. 249–281). New York: Academic Press.

Warren, J. R. & Marshall, B. (1983). Unidentified curved bacilli on gastric epithelium in active chronic gastritis. *Lancet, 1*(8336), 1273–1275.

Wasserman, E. A. (1973). Pavlovian conditioning with heat reinforcement produces stimulus-directed pecking in chicks. *Science, 181*, 875–877.

Watson, J. B. (1907). Kinaesthetic and organic sensations: Their role in the reactions of the white rat to the maze. *Psychological Monographs*, 8(33).

Watson, J. B. (1914). *Behavior: An introduction to comparative psychology*. New York: Holt.

Watson, J. B. (1924). *Behaviorism*. New York: Norton.

Watson, J. B. & Lashley, K. S. (1915). *Homing and related activity of birds.* Washington, DC: Carnegie Institute.

Wearden, J. (2021). Marc Richelle (28th February 1930–6th January 2021). *Timing & Time Perception 9*, 123–126.

Weisman, R. G. & Litner, J. S. (1969). Positive conditioned reinforcement of Sidman avoidance behavior in rats. *Journal of Comparative and Physiological Psychology, 68*, 597–603.

Weiss, J. M. (1968). Effects of coping responses on stress. *Journal of Comparative and Physiological Psychology, 65*, 251–260.

Westbrook, R. F. (1973). Failure to obtain positive behavioral contrast when pigeons press a bar. *Journal of the Experimental Analysis of Behavior, 20*, 499–410.

Whitlow, J. W. (1975). Short-term memory in habituation and dishabituation. *Journal of Experimental Psychology, 104*, 189–206.

Williams, D. R. & Williams, H. (1969). Auto-maintenance in the pigeon: Sustained pecking despite contingent non-reinforcement. *Journal of the Experimental Analysis of Behavior, 12*, 511–520.

Wilson, B. J., Mackintosh, N. J., & Boakes, R. A. (1985a). Matching and oddity learning in the pigeon: Transfer effects and the absence of relational learning. *Quarterly Journal of Experimental Psychology, 37B*, 295–311.

Wilson, B. J., Mackintosh, N. J. & Boakes, R. A. (1985b). Transfer of relational rules in matching and oddity learning by pigeons and corvids. *Quarterly Journal of Experimental Psychology, 37B*, 313–332.

Wimmer, H. & Perner, J. (1983). Beliefs about beliefs: Representation and constraining function of wrong beliefs in young children's understanding of deception. *Cognition, 13*, 103–128.

Windholz, G. & Wyrwicka, W. (1996). Pavlov's position toward Konorski and Miller's distinction between Pavlovian and motor conditioning paradigms. *Integrative Physiological and Behavioral Science, 31*, 338–349.

Winter, A. (2016). Cats on the couch: The experimental production of animal neurosis. *Science in Context, 29*, 77–105.

Wodinsky, J. & Bitterman, M. E. (1957). Discrimination reversal in the fish. *American Journal of Psychology, 70*, 569–576.

Wodinsky, J. & Bitterman, M. E. (1959). Partial reinforcement in the fish. *American Journal of Psychology, 72*, 184–199.

Wolpe, J. (1952). Experimental neuroses as learned behavior. *British Journal of Psychology, 43*, 243–268.

Wood, F. G. (1973). *Marine mammals and man: The navy's porpoises and sea lions.* Washington and New York: Robert B. Luce, Inc.

Wyrwicka, W. (1994). Jerzy Konorski (1903–1973) on the 20th anniversary of his death. *Neuroscience and Biobehavioral Reviews, 18*, 449–453.

Yerkes, R. M. (1925). *Almost human.* New York: The Century Co.

Zangwill, O. (1977). Obituary: G.C. Grindley (1903–1976). *Quarterly Journal of Experimental Psychology, 29*, 1–5.

Zener, K. (1937). The significance of behavior accompanying conditioned salivary secretion for theories of the conditioned response. *American Journal of Psychology, 50*, 384–403.

Zentall, T. R. & Hogan, D. (1978). Same/different concept learning in the pigeon: The effect of negative instances and prior adaptation to transfer stimuli. *Journal of the Experimental Analysis of Behavior, 30*, 177–186.

Zimmer-Hart, C. L. & Rescorla, R. A. (1974). Extinction of Pavlovian conditioned inhibition. *Journal of Comparative and Physiological Psychology, 86.*

Index

Printed in the United States
by Baker & Taylor Publisher Services